T0271214

# Room Acoustics

Well established as a classic reference and specialized textbook since its first publication in 1973, *Room Acoustics* combines detailed coverage with a state-of-the-art presentation of the theory and practice of sound behaviour in enclosed spaces.

This seventh edition is developed to cover new measurement and simulation techniques, including sections on spatial and directional analysis and on recent psychophysical experimental approaches to determining auditory perception in concert halls. Other important topics include the various mechanisms of sound absorption and their practical application, as well as scattering through wall corrugations. The design and performance of sound reinforcement systems is also updated. As in previous editions, special emphasis is placed on the properties and calculation of reverberation.

The book particularly suits graduate students in the field, acoustical engineers, and architects.

**Heinrich Kuttruff** is Emeritus Professor and former Head of the Institute of Technical Acoustics at RWTH Aachen University, Germany. He has received the Rayleigh Medal from the British Institute of Acoustics, the Helmholtz Medal from the German Society for Acoustics, and the silver medal from the French Society of Acoustics. He is a member of the Göttingen Academy of Sciences, Honorary Fellow of the Institute of Acoustics, and an honorary member of the German Society for Acoustics.

**Michael Vorländer** has been Professor of Technical Acoustics at RWTH Aachen University, Germany, since 1996. He has received the Rayleigh Medal from the British Institute of Acoustics, the W.C. Sabine Medal from the Acoustical Society of America, and the EAA award for contribution to promotion of acoustics in Europe from the European Acoustics Association. His main research interests are auralization and acoustic virtual reality in their various applications in architectural acoustics, the automotive industry, and noise control.

# Room Acoustics

Seventh Edition

Heinrich Kuttruff and Michael Vorländer

CRC Press is an imprint of the
Taylor & Francis Group, an **informa** business

Cover image: Heinrich Kuttruff and Michael Vorländer

Seventh edition published 2024
by CRC Press
2385 NW Executive Center Drive, Suite 320, Boca Raton FL 33431

and by CRC Press
4 Park Square, Milton Park, Abingdon, Oxon, OX14 4RN

*CRC Press is an imprint of Taylor & Francis Group, LLC*



First edition published in 1973

Sixth edition published by CRC Press 2016



*Library of Congress Cataloging-in-Publication Data*
Names: Kuttruff, Heinrich, author. | Vorländer, Michael, author.
Title: Room acoustics / Heinrich Kuttruff and Michael Vorländer.
Description: Seventh edition. | Boca Raton, FL : CRC Press, 2024. |
Includes bibliographical references and index.
Identifiers: LCCN 2024002812 | ISBN 9781032478258 (hardback) |
ISBN 9781032486109 (paperback) | ISBN 9781003389873 (ebook)
Subjects: LCSH: Architectural acoustics.
Classification: LCC NA2800 .K87 2024 | DDC 729/.29—dc23/eng/20240507
LC record available at https://lccn.loc.gov/2024002812

ISBN: 978-1-032-47825-8 (hbk)
ISBN: 978-1-032-48610-9 (pbk)
ISBN: 978-1-003-38987-3 (ebk)

DOI: 10.1201/9781003389873

Typeset in Sabon
by Apex CoVantage, LLC

# Contents

# List of Symbols

## LATIN CAPITAL LETTERS

| | |
|---|---|
| $A$ | constant, equivalent absorption area, absorption cross section |
| $B$ | constant, frequency bandwidth, irradiation density |
| $C$ | constant, clarity index, circumference |
| $\mathbf{C}$ | correlation matrix |
| $D$ | diameter, thickness, definition, diffusion constant |
| $\mathbf{D}$ | diagonal matrix (in Section 7.8) |
| $E$ | energy |
| $F$ | force, arbitrary function |
| $G$ | strength factor |
| $G(\omega)$, $G(f)$ | transfer function, arbitrary function |
| $H$ | transfer function, distribution of damping constants |
| $\boldsymbol{I}$ | Sound intensity vector |
| $I$ | Sound intensity magnitude |
| IACC | interaural cross-correlation |
| $J_n(z)$ | Bessel function of order $n$ |
| $K_n$ | normalization constant |
| $K(\mathrm{r},\mathrm{r}')$ | kernel (of integral equation) |
| $\mathbf{K}$ | stiffness matrix (in Section 3.5) |
| $L$ | length, sound pressure level |
| LEF | lateral energy fraction |
| $M$ | mass |
| $\mathbf{M}$ | mass matrix (in Section 3.5) |
| $M'$ | specific mass (= mass per unit area) |
| $N$ | integer |
| $P$ | power, probability |
| PL | sound power level |
| $Q$ | volume velocity, quality factor |
| $Q$s | scattering cross section |
| $R$ | reflection factor, distance, radius of curvature, resistance |
| $R_\mathrm{r}$ | radiation resistance |
| $S$ | area |
| $S(\omega)$, $S(f)$ | spectral density or spectral function |
| SPL | sound pressure level |
| $T$ | period of an oscillation, transmission factor, reverberation or decay time |
| $V$ | volume |

| | |
|---|---|
| $W(\omega)$, $W(f)$ | power spectrum |
| $W(\mathbf{r},\mathbf{r}')$ | propagator (in Section 11.5) |
| $\mathbf{W}$ | transformation matrix (in Section 7.8) |
| $Y$ | admittance, wall admittance |
| $Y_n^m$ | Spherical harmonic of order $n$ and degree $m$ |
| $Z$ | impedance, wall impedance |
| $Z_r$ | radiation impedance |

## LATIN LOWER-CASE LETTERS

| | |
|---|---|
| $a$ | radius, constant |
| $a(t)$ | weighting function |
| $b$ | thickness, constant |
| $b(t)$ | weighting function |
| $c$ | sound velocity |
| $d$ | thickness, distance |
| $e(t)$ | envelope |
| $f$ | frequency |
| $f(\mathbf{r},\mathbf{v},t)$ | distribution function of particles |
| $g$ | directivity or gain |
| $g(t)$ | impulse response |
| $h$ | height, width |
| $h(t)$ | decaying sound pressure |
| $i$ | imaginary unit (= $\sqrt{-1}$) |
| $j(\mathbf{r},t,\mathbf{u})$ | energy flux density |
| $k$ | angular wave number |
| $l$ | integer, length |
| $m$ | integer, attenuation constant, modulation transfer function |
| $m(\Omega)$ | modulation transfer function |
| $n$ | integer, normal direction |
| $n_s$ | density of scatterers |
| $p$ | sound pressure |
| $q$ | amplifier gain, volume velocity per unit volume |
| $r$ | radius, distance, resistance |
| $r_c$ | critical distance or diffuse-field distance |
| $r_s$ | flow resistance |
| $r(t)$ | reflection response |
| $s$ | scattering coefficient |
| $s(t)$ | time function of a signal |
| $t$ | time or duration |
| $t_s$ | centre time (centre of gravity) |
| $\mathbf{u}$ | unit vector |
| $v$ | velocity, particle velocity |
| $v(r)$ | test function (in Section 3.5) |
| $w$ | energy density |
| $x, y, z$ | Cartesian coordinates |

## GREEK CAPITAL LETTERS

| | |
|---|---|
| $\Gamma(\vartheta, \varphi)$ | directional factor |
| $\Delta$ | difference, Laplacian operator |
| $\Lambda_n$ | shape function (in Section 3.5) |
| $\Xi$ | specific flow resistance |
| $\Psi$ | correlation coefficient |
| $\Omega$ | solid angle |

## GREEK LOWER-CASE LETTERS

| | |
|---|---|
| $\alpha$ | angle, absorption coefficient |
| $\beta$ | angle, specific admittance |
| $\beta'$ | phase constant |
| $\gamma$ | angle |
| $\gamma'$ | attenuation constant |
| $\gamma^2$ | relative variance of path length distribution |
| $\Delta$ | decay constant, difference |
| $\delta(t)$ | Dirac or delta function |
| $\varepsilon$ | angle |
| $\zeta$ | specific impedance |
| $\eta$ | imaginary part of the specific impedance |
| $\vartheta$ | angle, temperature in degrees centigrade |
| $\theta$ | angle |
| $\kappa$ | adiabatic or isentropic exponent |
| $\kappa(\mathbf{r},\mathbf{r}')$ | symmetric kernel |
| $\lambda$ | wavelength |
| $\xi$ | real part of the specific impedance |
| $\rho$ | density, reflection coefficient (= $1 - \alpha$) |
| $\rho_n$ | radius of *n*th Fresnel zone |
| $\sigma$ | porosity, standard deviation |
| $\tau$ | transit time or delay time |
| $\phi$ | angle, phase angle |
| $\varphi$ | angle, phase angle |
| $\chi$ | phase angle of reflection factor |
| $\Psi$ | phase angle |
| $\omega$ | angular frequency |

# Preface

This book is intended to present the fundamentals of room acoustics in a systematic and scientifically correct way and to give an overview of the present state-of-the-art techniques in room acoustics. We hope that it will contribute to a better understanding of the factors responsible for what is commonly called good or poor 'acoustics of a room'. One aspect of room acoustics concerns the physical laws of the generation, propagation, and absorption of sound in an enclosure. These laws are most effectively formulated in the language of mathematics. Therefore, to understand this book in its entirety, the reader should have a reasonable mathematical background and some elementary knowledge of wave propagation. (Certain derivations may be omitted without detriment by readers with more limited mathematical training.) However, the image conveyed by a purely physical description of sound propagation would be incomplete, if not useless, without regarding the physiological and psychological factors involved in the human perception of sound, since it is the person attending a concert or a lecture who is the ultimate consumer of acoustics.

More than five decades have passed since the publication of the first edition of this book, and in the meantime, many important insights and improvements in the techniques of room acoustical measurements and simulation have been introduced. The preceding editions of this book have tried to take regard of these progresses and, thus, in a way, reflect this development, which was mainly made possible by the rapid progress in digital techniques. The seventh edition contains some new findings on sound perception in rooms, as well as on the measurement and simulation of rooms. Another new section presents an integration of the finite element method into hybrid room acoustics simulation methods. In both cases, the theoretical fundamental has been laid long ago, but only the progresses in modern computer techniques have turned these methods into practical tools. An updated component is the section on 'virtual reality', the acoustical component of which is auralization. Furthermore, the fundamentals and applications of spherical harmonics for room acoustic measurements and directional analysis of sound fields were newly included.

To stay within the scope of this book, we have refrained from describing examples of completed rooms, apart from very few exceptions. As far as concert halls or opera theatres are concerned, the interested reader is referred to L. L. Beranek's famous collection *Concert and Opera Halls* (Beranek 1996), which presents technical data, drawings, and photos of as many as 67 halls along with many interesting observations made by musicians, critics, and experts in acoustics and other areas.

The literature on room acoustical subjects is so extensive that we have made no attempt to provide an exhaustive list of references. References have only been given in those cases where the work has been directly mentioned in the text, or in order to satisfy the possible demands for more detailed information.

**Heinrich Kuttruff and Michael Vorländer**
*Aachen, Germany, January 2024*

## REFERENCE

Beranek L.L. *Concert and Opera Halls: How They Sound*. Woodbury, NY: Acoustical Society of America, 1996.

# Acknowledgements

Despite many changes and amendments, this edition is still based on the competent and sensitive English translation of the original text provided by the late Professor Peter Lord of the University of Salford. Furthermore, we are indebted to several readers of this book for various helpful suggestions. We would also like to thank the Institute of Hearing Technology and Acoustics at RWTH Aachen University, in particular Marco Berzborn and Julian Hüther, for their support in the preparation of this edition. We also thank the publisher for the pleasant and successful cooperation.

# Introduction

We all know that a concert hall, theatre, lecture room, or church may have good or poor 'acoustics'. As far as speech in these rooms is concerned, it is relatively simple to make some sort of judgement on their quality by rating the ease with which the spoken word is understood. However, judging the acoustics of a concert hall or an opera house is generally more difficult, since it requires considerable experience, the opportunity for comparisons, and a critical ear. Even so, the inexperienced cannot fail to learn about the acoustical reputation of a certain concert hall should they so desire, for instance, by listening to the comments of others or by reading the critical reviews of concerts in the press.

An everyday experience (although most people are not consciously aware of it) is that living rooms, offices, restaurants, and all kinds of rooms for work can be acoustically satisfactory or unsatisfactory. Even rooms which are generally considered insignificant or spaces such as staircases, factories, passenger concourses in railway stations, and airports may exhibit different acoustical properties; they may be especially noisy or exceptionally quiet, or they may differ in the ease with which announcements over the public address system can be understood. That is to say, even these spaces have 'acoustics', which may be satisfactory or less than satisfactory.

Despite the fact that people are subconsciously aware of the acoustics to which they are daily subjected, there are only a few who can explain what they really mean by 'good or poor acoustics' and who understand factors which influence or give rise to certain acoustical properties. Even fewer people know that the acoustics of a room is governed by principles which are amenable to scientific treatment. It is frequently thought that the acoustical design of a room is a matter of chance, and that good acoustics cannot be designed in a room with the same precision as a nuclear reactor or space vehicle is designed. This idea is supported by the fact that opinions on the acoustics of a certain room or hall frequently differ as widely as the opinions on the literary qualities of a new book or on the architectural design of a new building. Furthermore, it is well known that sensational failures in this field do occur from time to time. These and similar anomalies add even more weight to the general belief that the acoustics of a room is beyond the scope of calculation or prediction, at least with any reliability, and hence, the study of room acoustics is an art rather than an exact science.

In order to shed more light on the nature of room acoustics, let us first compare it to a related field: the design and construction of musical instruments. This comparison is not as senseless as it may appear at first sight, since a concert hall, too, may be regarded as a large musical instrument, the shape and material of which determine, to a considerable extent, what the listener will hear. Musical instruments — string instruments, for instance — are, as is well known, not designed or built by scientifically trained acousticians but, fortunately, by people who have acquired the necessary experience through long and systematic practical training. Designing or building musical instruments is therefore not a technical or scientific discipline but a sort of craft, or an 'art', in the classical meaning of this word.

Nevertheless, there is no doubt that the way in which a musical instrument functions, that is, the mechanism of sound generation, the determining of the pitch of the tones generated and their timbre through certain resonances, as well as their radiation into the surrounding air, is all purely physical processes and can therefore be understood rationally, at least in principle. Similarly, there is no mystery in the choice of materials; their mechanical and acoustical properties can be defined by measurements to any required degree of accuracy. (How well these properties can be reproduced is another problem.) Thus, there is nothing intangible, nor is there any magic, in the construction of a musical instrument: many particular problems which are still unsolved will be understood in the near future. Then, one will doubtless be in a position to design a musical instrument according to scientific methods, that is, not only to predict its timbre but also to give, with scientific accuracy, details for its construction, all of which are necessary to obtain prescribed or desired acoustical qualities.

Room acoustics is different from musical instrument acoustics in that the end product is usually more costly by orders of magnitude. Furthermore, rooms are produced in much smaller numbers and have, by no means, geometrical shapes which remain unmodified through the centuries. On the contrary, every architect, by the very nature of his profession, strives to create something which is entirely new and original. The materials used are also subject to the rapid development of building technology. Therefore, it is impossible to collect in a purely empirical manner sufficient know-how from which reliable rules for the acoustical design of rooms or halls can be distilled. An acoustical consultant is confronted with quite a new situation with each task, each theatre, concert hall, or lecture room to be designed, and it is of little value simply to transfer the experience of former cases to the new project if nothing is known about the conditions under which the transfer may be safely made.

This is in contrast to the making of a musical instrument where the use of unconventional materials as well as the application of new shapes is either firmly rejected as an offence against sacred traditions or dismissed as a whim. As a consequence, time has been sufficient to develop well-established empirical rules. And if their application happens to fail in one case or another, the faulty product is abandoned or withdrawn from service — which is not true for large rooms in an analogous situation.

For the aforementioned reasons, the acoustician has been compelled to study sound propagation in closed spaces with increasing thoroughness and to develop the knowledge in this field much further than in the case with musical instruments, even though the acoustical behaviour of a large hall is considerably more complex and involved. Thus, room acoustics has become a science during the past century, and those who practise it on a purely empirical basis will fail sooner or later, like a bridge builder who waives calculations and relies on experience or empiricism.

On the other hand, the present level of reliable knowledge in room acoustics is not particularly advanced. Many important factors influencing the acoustical qualities of large rooms are understood only incompletely or even not at all. As will be explained later in more detail, this is due to the complexity of sound fields in closed spaces or, as may be said equally well, to the large number of 'degrees of freedom' which we have to deal with. Another difficulty is that the acoustical quality of a room ultimately has to be proved by subjective judgements.

In order to gain more understanding about the sort of questions which can be answered eventually by scientific room acoustics, let us look over the procedures for designing the acoustics of a large room. If this room is to be newly built, some ideas will exist as to its intended use. It will have been established, for example, whether it is to be used for the showing of cine films, for sports events, for concerts, or as an open-plan office. One of the first tasks of the consultant is to translate these ideas concerning the practical use into the language of objective sound field parameters and to fix values for them which he thinks will best meet the requirements. During this step, he has to keep in mind the limitations and peculiarities of our subjective listening abilities. (It does not make sense, for instance, to fix the duration of sound decay with an accuracy of 1% if no one can subjectively distinguish such small differences.) Ideally, the next step would be to determine the shape of the hall; to choose the materials to be used; to plan the arrangement of the audience, of the orchestra, and of other sound sources; and to do all this in such a way that the sound field configuration will develop, which has previously been found to be the optimum for the intended purpose. In practice, however, the architect will have worked out already a preliminary design, certain features of which he considers imperative. In this case, the acoustical consultant has to examine the objective acoustical properties of the design by calculation, by geometric ray considerations, by model investigations, or by computer simulation, and he will eventually have to submit proposals for suitable adjustments. As a general rule, there will have to be some compromise in order to obtain a reasonable result.

Frequently, the problem is refurbishment of an existing hall, either to remove architectural, acoustical, or other technical defects or to adapt it to a new purpose which was not intended when the hall was originally planned. In this case, an acoustical diagnosis has to be made first on the basis of appropriate measurement. A reliable measuring technique which yields objective quantities, which are subjectively meaningful at the same time, is an indispensable tool of the acoustician. The subsequent therapeutic step is essentially the same as described earlier: the acoustical consultant has to propose measures which would result in the intended objective changes in the sound field and, consequently, in the subjective impressions of the listeners.

In any case, the acoustician is faced with a twofold problem: on the one hand, he has to find and apply the relations between the structural features of a room — such as shape, materials, and so on — with the sound field which will occur in it, and on the other hand, he has to take into consideration as far as possible the interrelations between the objective and measurable sound field parameters and the specific subjective hearing impressions effected by them. Whereas the first problem lies completely in the realm of technical reasoning, it is the latter problem which makes room acoustics different from many other technical disciplines, in that the success or failure of an acoustical design has finally to be decided by the collective judgement of all 'consumers', that is, by some sort of average, taken over by the comments of individuals with widely varying intellectual, educational, and aesthetic backgrounds. The measurement of sound field parameters can replace, to a certain extent, systematic or sporadic questioning of listeners. But in the final analysis, it is the average opinion of listeners which decides whether the acoustics of a room is favourable or poor. If the majority of the audience (or that part which is vocal) cannot understand what a speaker is saying or thinks that the sound of an orchestra in a certain hall is too dry, too weak, or indistinct, then even though the measured reverberation time is appropriate or the local or directional distribution of sound is uniform, the listener is always right; the hall does have acoustical deficiencies.

Therefore, acoustical measuring techniques can only be a substitute for the investigation of public opinion on the acoustical qualities of a room, and it will serve its purpose better the closer the measured sound field parameters are related to subjective listening categories. Not only must the measuring techniques take into account the hearing response of the listeners, but the acoustical theory, too, will also only provide meaningful information if it takes regard of the consumer's particular listening abilities. It should be mentioned at this point that the sound field in a real room is so complicated that it is not open to exact mathematical treatment. The reason for this is the large number of components which make up the sound field in a closed space regardless of whether we describe it in terms of vibrational modes or, if we prefer, in terms of sound rays which have undergone one or more reflections from boundaries. Each of these components depends on the sound source, the shape of the room, and the materials from which it is made; accordingly, the exact computation of the sound field is usually quite involved. Supposing this procedure was possible with reasonable expenditure, the results would be so confusing that such a treatment would not provide a comprehensive survey and, hence, would not be of any practical use. For this reason, approximations and simplifications are inevitable; the totality of possible sound field data has to be reduced to averages or average functions which are more tractable and condensed to provide a clearer picture. Hence, we have to resort so frequently to statistical methods and models in room acoustics, whichever way we attempt to describe sound fields. The problem is to perform these reductions and simplifications once again in accordance with the properties of human hearing, that is, in such a way that the remaining average parameters correspond as closely as possible to particular subjective sensations.

From this it follows that essential progress in room acoustics depends, to a large extent, on the advances in psychological acoustics. As long as the physiological and psychological processes which are involved in hearing are not completely understood, the relevant relations between objective stimuli and subjective sensations must be investigated empirically — and should be taken into account when designing the acoustics of a room.

Many interesting relations of this kind have been detected and successfully investigated during the past few decades. But other questions which are no less important for room acoustics are unanswered so far, and much work remains to be carried out in this field.

It is, of course, the purpose of all efforts in room acoustics to avoid acoustical deficiencies and mistakes. It should be mentioned, on the other hand, that it is neither desirable nor possible to create the 'ideal acoustical environment' for concerts and theatres. It is a fact that the enjoyment when listening to music is a matter not only of the measurable sound waves hitting the ear but also of the listener's personal attitude and his individual taste, and these vary from one person to another, especially when non-acoustic factors, such as vision, are taken into account. For this reason, there will always be varying shades of opinion concerning the acoustics of even the most marvellous concert hall. For the same reason, one can easily imagine a wide variety of concert halls with excellent, but nevertheless different, acoustics. It is this 'lack of uniformity' which is characteristic of the subject of room acoustics, and which is responsible for many of its difficulties, but it also accounts for the continuous power of attraction it exerts on many acousticians.

**Heinrich Kuttruff and Michael Vorländer**

Chapter 1

# Some facts on sound waves, sources, and hearing

One of the purposes of this chapter is to introduce the physical quantities of sound and the laws by which they are related. The basic laws are illustrated by considering simple forms of sound wave, namely, the plane wave and the spherical wave. These waves are easy to comprehend and do not require complicated mathematics. Moreover, they are important and interesting in their own right and can also serve as the basis of the more complicated sound fields encountered in closed rooms. We can safely restrict our attention to sound propagation in gases, because in room acoustics, we are mostly concerned with air as a wave medium.

Another purpose of this chapter is to give a short description of the properties of human hearing and of the main sorts of sound and sound sources that we are concerned with in room acoustics.

First, we shall assume the sound propagation to be free of losses and ignore the effect of any obstacles such as walls, that is, we suppose the medium to be unbounded in all directions. Furthermore, we suppose that our medium is homogeneous and at rest. In this case, the velocity of sound is constant with reference to space and time. For air, its magnitude is

$$c = (331.4 + 0.6 \cdot \vartheta)\,\mathrm{m/s} \tag{1.1}$$

where $\vartheta$ is the temperature in degrees centigrade. We shall set $c$ = 343 m/s.

In large halls, variations of temperature and, hence, of the sound velocity with time and position cannot be entirely avoided. Likewise, because of temperature differences and air conditioning, the air is not completely at rest, and so our assumptions are not fully fulfilled. But the effects that are caused by these inhomogeneities are so small that they can be neglected.

## 1.1 BASIC RELATIONS — THE WAVE EQUATION

In any sound wave, the particles of the medium undergo vibrations about their mean positions. Therefore, a wave can be described completely by indicating the instantaneous displacements of those particles. It is more customary, however, to consider the velocity of particle displacement as a basic acoustical quantity rather than the displacement itself.

The vibrations in a sound wave do not take place at all points with the same phase. We can, in fact, find points in a sound field where the particles vibrate in opposite phase. This means that in certain regions, the particles are pushed together or compressed, and in other regions, they are pulled apart or rarefied. Therefore, under the influence of a sound wave, variations of gas density and pressure occur, both of which are functions of time and position. The difference between the instantaneous pressure and the static pressure is called the sound pressure.

DOI: 10.1201/9781003389873-1

The changes in gas pressure caused by a sound wave in general occur so rapidly that heat cannot be exchanged between adjacent volume elements. Consequently, a sound wave causes adiabatic variations of the temperature, and so the temperature, too, can be considered as a quantity characterising a sound wave.

The various acoustical quantities are connected by some fundamental laws which enable us to set up a general differential equation governing sound propagation. The conservation of momentum is expressed by the relation

$$\text{grad } p = -\rho_0 \frac{\partial v}{\partial t} \tag{1.2}$$

where $p$ denotes the sound pressure, $v$ a vector representing the particle velocity, $t$ the time, and $\rho_0$ the static value of the gas density (about 1.2 kg/m$^3$). The unit of sound pressure is Pascal: 1 Pascal (Pa) = 1 N/m$^2$ = 1 kg/ms$^2$. The one-dimensional version of this equation is

$$\frac{\partial p}{\partial x} = -\rho_0 \frac{\partial v_x}{\partial t} \tag{1.3}$$

Furthermore, the requirement of mass conservation leads to

$$\rho_0 \text{div } v = -\frac{\partial \rho}{\partial t} \tag{1.4}$$

$\rho$ being the variable part of the density. In these equations, it is assumed that the changes in the gas pressure and density, $p$ and $\rho$, are small compared with the static values $p_0$ and $\rho_0$, respectively; furthermore, the absolute value of the particle velocity $v$ is supposed to be much smaller than the sound velocity $c$.

Under the further assumption that we are dealing with an ideal gas, the following relations hold among the sound pressure $p$, the density variation $\rho$, and the temperature variation $\delta\vartheta$:

$$\frac{p}{p_0} = \kappa \frac{\rho}{\rho_0} = \frac{\kappa}{\kappa - 1} \cdot \frac{\delta\vartheta}{\vartheta + 273} \tag{1.5}$$

Here, $\kappa$ is the adiabatic or isentropic exponent ($\kappa$ = 1.4 for air), and $\vartheta$ is the temperature in degrees centigrade.

The particle velocity $v$ and the variable part $\rho$ of the density can be eliminated from Equations 1.2 through 1.5. This yields the differential equation

$$c^2 \Delta p = \frac{\partial^2 p}{\partial t^2} \tag{1.6}$$

where

$$c^2 = \kappa \frac{p_0}{\rho_0} \tag{1.7}$$

With this relation, the first part of Equation 1.5 becomes

$$p = c^2 \rho \tag{1.8}$$

$\Delta$ (= $\nabla^2$ = div grad) is the Laplacian operator. In Cartesian coordinates $x, y, z$, this operator reads

$$\Delta p = \frac{\partial^2 p}{\partial x^2} + \frac{\partial^2 p}{\partial y^2} + \frac{\partial^2 p}{\partial z^2}$$

The differential equation (Equation 1.6) governs the propagation of sound waves in any lossless fluid and is therefore of central importance for almost all acoustical phenomena. We shall refer to it as the 'wave equation'. It holds not only for sound pressure but also for density and temperature variations.

## 1.2 PLANE WAVES AND SPHERICAL WAVES

Now we assume that the acoustical quantities depend only on the time and on one single direction, which may be chosen as the x-direction of a Cartesian coordinate system. Then, Equation 1.6 reads

$$c^2 \frac{\partial^2 p}{\partial x^2} = \frac{\partial^2 p}{\partial t^2} \tag{1.9}$$

The general solution of this differential equation is

$$p(x,t) = F(ct - x) + G(ct + x) \tag{1.10}$$

where $F$ and $G$ are arbitrary functions, the second derivatives of which exist. The first term on the right represents a pressure wave travelling in the positive $x$-direction with a velocity $c$, because the value of $F$ remains unaltered if a time increase $\delta t$ is associated with an increase in the coordinate $\delta x = c\delta t$. For the same reason, the second term describes a pressure wave propagated in the negative $x$-direction. It follows that the constant $c$ is the sound velocity.

Each term of Equation 1.10 represents a progressive 'plane wave'. As shown in Figure 1.1a, the sound pressure $p$ is constant in any plane perpendicular to the $x$-axis. These planes of constant sound pressure are called 'wave fronts', and any line perpendicular to them is a 'wave normal'.

According to Equation 1.2, the particle velocity has only one non-vanishing component, which is parallel to the gradient of the sound pressure, that is, to the x-axis. This means sound

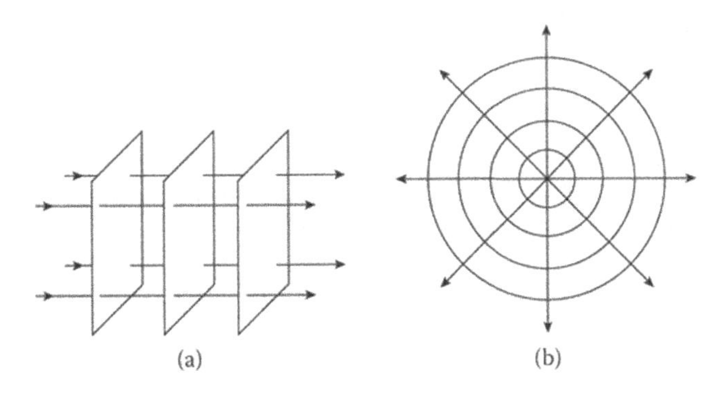


*Figure 1.1* Simple types of waves: (a) plane wave and (b) spherical wave.

waves in fluids are longitudinal waves. The particle velocity may be obtained by applying Equation 1.2 in Equation 1.10:

$$v(x,t) = \frac{1}{\rho_0 c}\left[F(ct - x) - G(ct + x)\right] \tag{1.11}$$

As may be seen from Equations 1.10 and 1.11, the ratio of sound pressure to particle velocity in a plane wave propagated in the positive direction ($G = 0$) is frequency-independent:

$$\frac{p}{v} = \rho_0 c \tag{1.12}$$

This ratio is called the 'characteristic impedance' of the medium. For air at 20°C, its value is

$$\rho_0 c = 414 \frac{\text{Pa}}{\text{m/s}} = 414 \text{kg/m}^2\text{s} \tag{1.13}$$

If the wave is travelling in the negative x-direction, the ratio $p/v$ is negative.

Of particular importance are harmonic waves, in which the time and space dependence of the acoustical quantities, for instance, of the sound pressure follows a sine or cosine function. If we set $G = 0$ and specify $F$ as a cosine function, we obtain an expression for a plane, progressive harmonic wave

$$p(x,t) = \hat{p}\cos\left[k(ct - x)\right] = \hat{p}\cos(\omega t - kx) \tag{1.14}$$

with the arbitrary constants $\hat{p}$ and $k$. Here, the angular frequency

$$\omega = kc \tag{1.15}$$

was introduced, which is related to the temporal period

$$T = \frac{2\pi}{\omega} \tag{1.16}$$

of the harmonic vibration represented by Equation 1.14. At the same time, this equation describes a spatial harmonic vibration with the period

$$\lambda = \frac{2\pi}{k} \tag{1.17}$$

This is the 'wavelength' of the harmonic wave. It denotes the distance in the $x$-direction where equal values of the sound pressure (or any other field quantity) occur. According to Equation 1.15, it is related to the angular frequency by

$$\lambda = \frac{2\pi c}{\omega} = \frac{c}{f} \tag{1.18}$$

where $f = \omega/2\pi = 1/T$ is the frequency of the vibration. It has the dimension second$^{-1}$; its units are hertz (Hz), kilohertz (1 kHz = $10^3$ Hz), megahertz (1 MHz = $10^6$ Hz), and so on.

The quantity $k = \omega/c$ is the propagation constant or the (angular) wave number of the wave, and $\hat{p}$ is the pressure amplitude.

A very powerful representation of harmonic oscillations and waves is obtained by applying the relation (with $i = \sqrt{-1}$ denoting the imaginary unit):

$$\exp(ix) = \cos x + i \sin x$$

$$\text{or} \tag{1.19}$$

$$\cos x = \frac{1}{2}\left[\exp(ix) + \exp(-ix)\right] \text{ and } \sin x = \frac{1}{2i}\left[\exp(ix) - \exp(-ix)\right]$$

This is the complex or symbolic notation of harmonic vibrations and will be employed quite frequently in this book. Accordingly, $\cos x$ can be considered as the real part $\mathrm{Re}\{\exp(ix)\}$ and $\sin x$ as the imaginary part $\mathrm{Im}\{\exp(ix)\}$ of the function $\exp(ix)$. Using this relation, Equation 1.14 can be written in the form

$$p(x,t) = \mathrm{Re}\left\{\hat{p} \cdot \exp\left[i(\omega t - kx)\right]\right\}$$

or omitting the sign Re:

$$p(x,t) = \hat{p} \cdot \exp\left[i(\omega t - kx)\right] \tag{1.20}$$

The complex notation has several advantages over the real representation of Equation 1.14. Differentiation or integration with respect to time is equivalent to multiplication or division by $i\omega$, respectively. Furthermore, only the complex notation allows a clear-cut definition of impedances and admittances (see Section 2.1). It fails, however, in all cases where vibrational quantities are to be multiplied or squared. If doubts arise concerning the physical meaning of an expression, it is advisable to recall the origin of this notation, that is, to take the real part of the expression.

As with any complex quantity, the complex sound pressure in a plane wave may be represented in a rectangular coordinate system with the horizontal and the vertical axis corresponding to the real and the imaginary part of the pressure, respectively. The quantity is depicted as an arrow, often called 'phasor', pointing from the origin to the point that corresponds to the value of the pressure (see Figure 1.2). The length of this arrow corresponds to the magnitude of the complex quantity, while the angle it includes with the real axis is its phase angle or 'argument' (abbreviated as arg $p$). In the present case, the magnitude of the

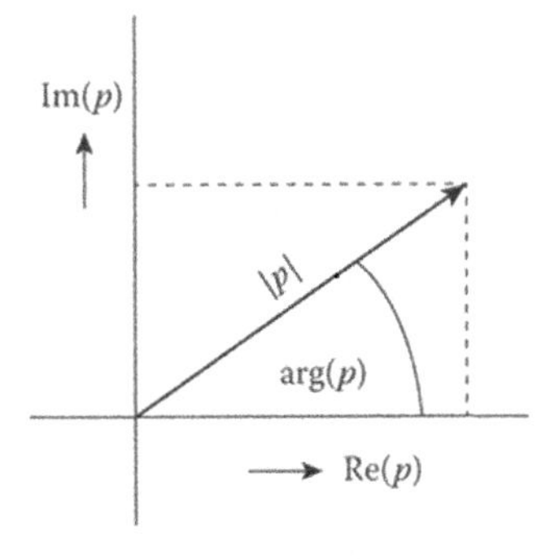


*Figure 1.2* Phasor representation of a complex quantity $p$.

phasor, represented by its length, equals the amplitude $\hat{p}$ of the oscillation; the phase angle depends on time $t$ and position $x$:

$$\arg p = \omega t - kx$$

This means that, for a fixed position, the arrow or phasor rotates around the origin with an angular velocity ω, which explains the expression 'angular frequency'.

So far, it has been assumed that the wave medium is free of losses. If this is not the case, the pressure amplitude does not remain constant in the course of wave propagation but decreases according to an exponential law. Then, Equation 1.20 is modified in the following way:

$$p(x,t) = \hat{p}\exp(-mx/2)\cdot\exp\left[i(\omega t - kx)\right] \tag{1.21}$$

We can even use the representation of Equation 1.20 if we conceive the wave number $k$ as a complex quantity, with half the attenuation constant $m$ as its imaginary part:

$$k = \frac{\omega}{c} - i\frac{m}{2} \tag{1.22}$$

Now, we need to find a representation of a plane wave propagating into an arbitrary direction, indicated by a unit vector **n** with the components cos α, cos β, and cos γ, where α, β, and γ are the angles which **n** makes with the three coordinate axes (see Figure 1.3); they have to fulfil the condition $\cos^2\alpha + \cos^2\beta + \cos^2\gamma = 1$. On the contrary,

$$x\cdot\cos\alpha + y\cdot\cos\beta + z\cdot\cos\gamma = \text{const}$$

is the equation of a plane with the normal **n**; the constant is the distance of the plane from the origin of the coordinate system. Hence, the plane wave is represented by

$$p(x,y,z,t) = \hat{p}\exp\left[i(\omega t - kx\cos\alpha - ky\cos\beta - kz\cos\gamma)\right] = \hat{p}\exp\left[i(\omega t - k\mathbf{nr})\right] \tag{1.23}$$

where **nr** is the scalar product of the vectors **n** and **r** = $(x, y, z)$.

Another simple wave type is the spherical wave, in which the surfaces of constant pressure, that is, the wave fronts, are concentric spheres (see Figure 1.1b). In their common centre, we have to imagine some very small source which introduces or withdraws fluid. Such a source is called a 'point source'. The appropriate coordinates for this geometry are polar coordinates (see Figure 1.3). Because of the spherical symmetry, the sound pressure is independent of the angles $\theta$ and $\phi$; the relevant space coordinate is the distance $r$ from the centre. Accordingly, the differential equation does not contain any angular derivatives of the pressure. Thus, transformed into this coordinate system, the wave equation (Equation 1.6) reads:

$$\frac{\partial^2 p}{\partial r^2} + \frac{2}{r}\frac{\partial p}{\partial r} = \frac{1}{c^2}\frac{\partial^2 p}{\partial t^2} \tag{1.24}$$

A simple solution of this equation is

$$p(r,t) = \frac{\rho_0}{4\pi r}\dot{Q}\left(t - \frac{r}{c}\right) \tag{1.25}$$

It represents a spherical wave produced by a point source at $r = 0$ with the 'volume velocity' $Q$, which is the rate (in m$^3$/s) at which fluid is expelled by the source. The overdot means partial differentiation with respect to time. Again, the argument $t - r/c$ indicates that any disturbance created by the sound source is propagated outward with velocity $c$, its strength decreasing as $1/r$. Reversing the sign in the argument of $\dot{Q}$ would result in the unrealistic case of an in-going wave.

Now, the only non-vanishing component of the particle velocity is the radial one, which is calculated by applying Equation 1.2 to Equation 1.25:

$$v_r = \frac{1}{4\pi r^2}\left[Q\left(t - \frac{r}{c}\right) + \frac{r}{c}\dot{Q}\left(t - \frac{r}{c}\right)\right] \tag{1.26}$$

If the volume velocity of the source varies according to $Q(t) = \hat{Q}\exp(i\omega t)$, we obtain from Equation 1.25

$$p(r,t) = \frac{i\omega\rho_0}{4\pi r}\hat{Q}\cdot\exp\left[i(\omega t - kr)\right] \tag{1.27}$$

This expression represents — with $k = \omega/c$ — the sound pressure in a harmonic spherical wave. The particle velocity after Equation 1.26 is

$$v_r = \frac{p}{\rho_0 c}\left(1 + \frac{1}{ikr}\right) \tag{1.28}$$

This formula indicates that the ratio of sound pressure and particle velocity in a spherical sound wave depends on the distance $r$ and the frequency $\omega = kc$. Furthermore, it is complex, that is, between both quantities, there is a phase difference. For $kr >> 1$, that is, for distances which are large compared with the wavelength, the ratio $p/v_r$ tends asymptotically to $\rho_0 c$, the characteristic impedance of the medium.

A plane wave is an idealization which does not exist in the real world, at least not in its pure form. However, a sound wave travelling within a rigid-walled tube can come very close to a plane wave if the lateral dimensions of the tube are significantly smaller than the acoustic wavelength. Furthermore, a limited region of a spherical wave may also be considered as a good approximation to a plane wave, provided the distance $r$ from the centre is large compared with all wavelengths involved, that is, $kr >> 1$; see Equation 1.28.

An exactly spherical wave can be generated — at least in principle — by a pulsating sphere, that is, by a sound source with spherical symmetry, the diameter of which varies with time in a regular or irregular manner. Another way to produce spherical waves is to 'simulate' the function of a point source, namely, just to expel or to suck in small amounts of the fluid. This can be achieved, for instance, by a small pulsating sphere or by a loudspeaker mounted into one side of a small airtight box. Most sound sources, however, do not behave as point sources. In these cases, the sound pressure depends not only on the distance $r$ but also on the direction, which can be characterized by a polar angle $\theta$ and an azimuth angle $\phi$ (see Figure 1.3). For distances exceeding a characteristic range, which depends on the sort of sound source and the frequency, the sound pressure is given by

$$p(r,\theta,\phi,t) = \frac{A}{r}\,\Gamma(\theta,\phi)\cdot\exp\left[\mathrm{i}(\omega t - kr)\right] \tag{1.29}$$

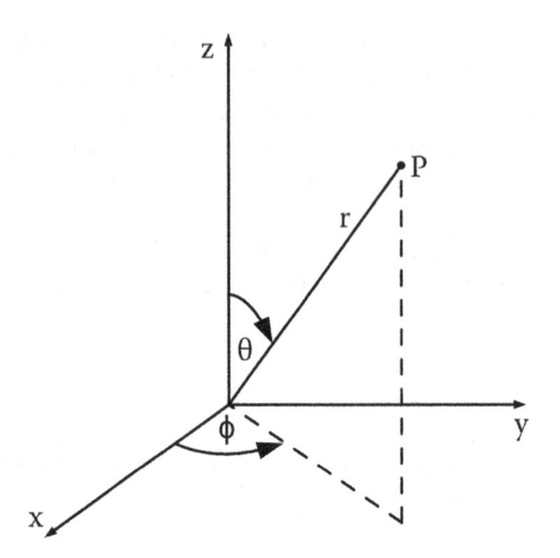


*Figure 1.3* Spherical coordinate system.

where the 'directional factor' $\Gamma(\theta, \phi)$ is normalized so as to make $\Gamma = 1$ for its absolute maximum. $A$ is a constant.

A more general approach to directional sound radiation is to solve the wave equation in spherical coordinates with angular dependence. In contrast to the undirected case (Equation 1.6), the wave equation contains the polar and azimuth angles, $\theta$ and $\phi$, respectively:

$$\frac{\partial^2 p}{\partial r^2}+\frac{2}{r}\frac{\partial p}{\partial r}+\frac{1}{r^2}\left(\frac{1}{\sin\theta}\frac{\partial}{\partial\theta}\left(\sin\theta\frac{\partial}{\partial\theta}\right)+\frac{1}{\sin^2\theta}\frac{\partial^2}{\partial\phi^2}\right)=\frac{1}{c^2}\frac{\partial^2 p}{\partial t^2} \tag{1.30}$$

In this formulation, the solution for the sound pressure is expanded as a Fourier series of radial and angular factors terms. The series expansion of Equation 1.29, for example, is given as

$$p(r,\theta,\phi)=A\sum_{n=0}^{\infty}\sum_{m=-n}^{n}\Gamma_{nm}h_n(kr)Y_n^m(\theta,\phi) \tag{1.31}$$

where $\Gamma_{nm}$ are the equivalent Fourier series coefficients (see Section 1.4 for comparison) of order $n$ and degree $m$. Note that the directivity functions explicit dependency on $(\theta, \phi)$ is now encoded in the order and degree. The function $h_n$ is the Hankel function of the first kind and of the order $n$. $Y$ is a set of orthonormal basis functions, the so-called 'spherical harmonics' (SH); $P$ are the Legendre polynomials functions (Williams 1999). These are defined as:

$$Y_n^m(\theta,\phi)=\sqrt{\frac{(2n+1)}{4\pi}\frac{(n-|m|)!}{(n+|m|)!}}P_n^{|m|}(\cos\theta)\cdot\begin{cases}\cos|m|\phi, & m\geq 0\\ \sin|m|\phi, & m<0\end{cases} \tag{1.32}$$

For the example at hand, the coefficients $\Gamma_{nm}$ can be identified by solving an inverse two-dimensional Fourier transform — also known as spherical harmonic transform — from experimental or numerical data sampled on a spherical grid surrounding the source, that is, using a surrounding spherical microphone array (Williams 1999). This theory of spherical harmonics can be applied to decompose any directional sound radiation pattern or sound reception patterns into the components of these basis functions, yielding a compact representation, which may be used for analysis or further processing, such as interpolation, and even extrapolation, which is quite similar to a frequency spectrum analysis of sound signals (Section 1.4).

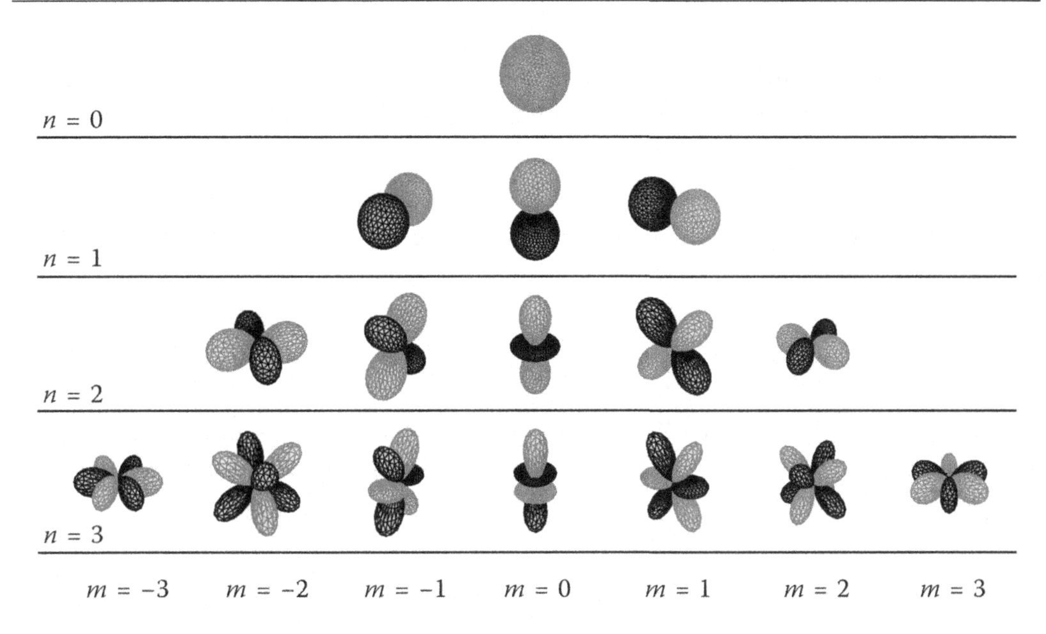


*Figure 1.4* Spherical harmonics.

Apart from the aforementioned example of capturing outgoing sound fields while retaining angular information, the theory can be used to measure incoming sound fields with a compact spherical microphone array or to synthesize sound fields inside a loudspeaker array by a sum of spherical harmonic functions, where only the right mixture, that is, the set of weights of the functions (SH coefficients), has to be determined. For the analysis and synthesis, there are calculation rules expressed in the SH transformation. Examples of application of SH processing are explained in Chapter 8 (on measurement techniques) and Chapter 10 (on prediction methods).

## 1.3 ENERGY DENSITY AND INTENSITY, RADIATION

As shown before, a sound wave is associated with variations in the pressure within a medium and with vibrations of the particles of which it consists. Both imply an increase in the mechanical energy, which the sound wave carries away from the source. The strength of this energy flow is characterized by the 'intensity', sometimes also called the 'energy flux density'. Since the energy transport has a certain direction, we represent the intensity by a vector symbol, $\mathbf{I}$. On the contrary, the energy content of the sound wave is characterized by the energy per unit volume, called the 'energy density', which will be denoted by $w$ in the following.

The sound intensity is defined as the sound energy passing per second through an imagined window of unit area which is perpendicular to the incident sound wave. This picture leads us immediately to a first relation between the intensity and the energy density. In time $dt$, the volume $cdt$ is travelling through the window ($c$ = sound velocity), carrying the energy $wcdt = Idt$, or

$$I = cw \tag{1.33}$$

where $I$ denotes the magnitude of the intensity vector $\mathbf{I}$.

Suppose a force $\mathbf{F}$ is acting on the matter contained in a volume element, enforcing a shift $\mathbf{ds}$ of its boundary. Then the work done by the force is given by the scalar product of both vectors, $\boldsymbol{F}d\boldsymbol{s}$; the rate of work is $\boldsymbol{F}\boldsymbol{v}$. By replacing the force with the force per unit area, that is, with the pressure $p$ (which, by the way, is omnidirectional), we arrive at the energy supplied to the fluid per unit time and area, that is, at the instantaneous intensity:

$$\boldsymbol{I} = p\boldsymbol{v} \tag{1.34}$$

A second relation between the energy density and the intensity is supplied by the principle of energy conservation, which in the present case states that

$$\text{div}\,\boldsymbol{I} = -\frac{\partial w}{\partial t} \tag{1.35}$$

This equation tells us that the outflow of energy from a volume element must be compensated for by a decrease in its energy content. According to the rules of vector analysis, we can write

$$\text{div}\,\boldsymbol{I} = \text{div}(p\boldsymbol{v}) = p\,\text{div}\,\boldsymbol{v} + \boldsymbol{v}\,\text{grad}\,p \tag{1.36}$$

The gradient and the divergence on the right can be expressed by the time derivatives of the particle velocity and of the pressure (see Section 1.1) with the result:

$$\text{div}\,\boldsymbol{I} = -\frac{1}{2\rho_0 c^2}\frac{\partial p^2}{\partial t} - \frac{\rho_0}{2}\frac{\partial |\boldsymbol{v}|^2}{\partial t} \tag{1.37}$$

Or by using Equation 1.35, apart from an additive constant:

$$w = \frac{1}{2\rho_0 c^2}p^2 + \frac{\rho_0}{2}v^2 \quad \text{with } v^2 = |\boldsymbol{v}|^2 \tag{1.38}$$

The first term on the right side represents the potential energy density, whereas the second one is the kinetic energy density.

For plane waves, we can safely replace the intensity vector $\boldsymbol{I}$ with its magnitude $I$. In this case, the sound pressure and the longitudinal component of the particle velocity are related by $p = \rho_0 c v$, and the same holds for a spherical wave at a large distance from the centre ($kr \gg 1$; see Equation 1.28). Hence, we can express the particle velocity in terms of the sound pressure. Then, the energy density and the intensity (according to Equation 1.33) are

$$w = \frac{p^2}{\rho_0 c^2} \quad \text{and} \quad I = \frac{p^2}{\rho_0 c} \tag{1.39}$$

Stationary signals which are not limited in time may be characterized by time averages over a sufficiently long time $t_a$. We introduce the root mean square of the sound pressure by

$$p_{\text{rms}} = \left(\frac{1}{t_a}\int_0^{t_a} p^2 dt\right)^{1/2} = \left(\overline{p^2}\right)^{1/2} \tag{1.40}$$

where the overbar is a shorthand notation indicating time averaging. Then, Equation 1.39 yields

$$\overline{w} = \frac{p_{\text{rms}}^2}{\rho_0 c^2} \text{ and } \overline{I} = \frac{p_{\text{rms}}^2}{\rho_0 c} \tag{1.41}$$

Finally, for a harmonic sound wave with the sound pressure amplitude $\hat{p}$, $p_{rms}$ equals $\hat{p}/\sqrt{2}$, which leads to

$$\overline{w} = \frac{\hat{p}^2}{2\rho_0 c^2} \text{ and } \overline{I} = \frac{\hat{p}^2}{2\rho_0 c} \tag{1.42}$$

The intensity in a spherical wave can also be expressed by using the sound power, $P$:

$$\overline{I} = \frac{P}{4\pi r^2} \tag{1.43}$$

Again, $r$ is the distance from the sound source. By inserting $\hat{p} = \rho_0 \omega \hat{Q}/4\pi r$ from Equation 1.27 into Equation 1.41, one obtains

$$\overline{I} = \frac{\rho_0}{32\pi^2 r^2 c}\hat{Q}^2\omega^2 \text{ and } P = \frac{\rho_0 \hat{Q}^2 \omega^2}{8\pi c} \tag{1.44}$$

If, on the other hand, the power output $P$ of a point source is given, the root mean square of the sound pressure at distance $r$ from the source is

$$p_{rms} = \frac{1}{r}\sqrt{\frac{\rho_0 c P}{4\pi}} \tag{1.45}$$

## 1.4 SIGNALS AND SYSTEMS

Any acoustical signal can be unambiguously described by its time function $s(t)$, where $s$ denotes the sound pressure, the density variations in a sound wave, or a component of particle velocity, for instance. If this function is a sine or cosine function — or an exponential function with imaginary argument — we speak of a harmonic signal, which is closely related to the harmonic waves, as introduced in Section 1.2. Harmonic signals and waves play a key role in acoustics, although real sound signals are almost never harmonic but show a much more complicated time dependence. The reason for this apparent contradiction is the fact that virtually all signals can be considered as superposition of harmonic signals. This is the fundamental statement of the famous Fourier theorem.

The Fourier theorem can be formulated as follows: Let $s(t)$ be a real, non-periodic time function describing, for example, the time variations of the sound pressure or of the volume velocity. We suppose that this function is sufficiently steady (a requirement which is fulfilled in all practical cases), and that the integral $\int_{-\infty}^{\infty}\left[s(t)\right]^2 dt$ has a finite value. Then, the time function can be represented as a superposition of harmonic oscillations with continuously varying frequencies $f$, each of them being represented by an exponential with imaginary argument:

$$s(t) = \int_{-\infty}^{\infty} S(f)\exp(2\pi i f t)\,df \tag{1.46}$$

The amplitudes of these oscillations are given by a function $S(f)$, called the 'spectral function' or the 'amplitude spectrum', or simply the 'spectrum' of the signal $s(t)$. It is related to the function $s(t)$ by

$$S(f) = \int_{-\infty}^{\infty} s(t)\exp(-2\pi i f t)\,dt \tag{1.47}$$

In general, the spectral function is complex:

$$S(f) = |S(f)| \exp[i\psi(f)] \tag{1.48}$$

The absolute value $|S(f)|$ and the function $\psi(f)$ are called the amplitude spectrum and the phase spectrum of the signal $s(t)$. A few examples of time functions and their amplitude spectra are shown in Figure 1.7.

Equation 1.47 is called the Fourier transform of a given function, while Equation 1.46 represents the inverse Fourier transform. It can easily be shown that $S(-f) = S^*(f)$, where the asterisk denotes the transition to the complex conjugate function. $S(f)$ and $s(t)$ are different but equivalent representations of a signal.

The Fourier theorem assumes a somewhat different form if $s(t)$ is a periodic function with the period $T$, that is, if $s(t) = s(t + T)$, then the integral in Equation 1.46 has to be replaced by the 'Fourier series':

$$s(t) = \sum_{n=-\infty}^{\infty} S_n \exp\left(\frac{2\pi int}{T}\right) \tag{1.49}$$

with the coefficients

$$S_n = \frac{1}{T}\int_0^T s(t) \exp\left(\frac{-2\pi int}{T}\right) dt \tag{1.50}$$

where $n$ is an integer. The continuous spectral function $S(f)$ has changed now into a set of discrete 'Fourier coefficients', for which $S_{-n} = S_n^*$. Hence, a periodic signal consists of discrete harmonic vibrations, the frequencies of which are multiples of a fundamental frequency $1/T$. These components are called 'partial oscillations' or 'harmonics', the first harmonic being identical with the fundamental oscillation ($n = 1$).

A very common task is to determine the spectrum of a signal $s(t)$, that is, to perform a spectral analysis. If $s(t)$ is given by a mathematical expression, one will try to compute the integrals in Equation 1.47 or Equation 1.50. If the signal $s(t)$ is obtained by measurement, for instance, as the output voltage of a microphone, a coarse form of spectral analysis can be carried out by passing the signal through a set of band-pass filters with pass bands covering the whole frequency range of interest. Usually, such filters have ratios of the upper and lower cut-off frequencies of 2:1 (octave filters) or 5:4 (third-octave filters).

For a more precise spectral analysis, the continuous function $s(t)$ must be 'sampled', that is, it is replaced with a sequence of samples taken from $s(t)$ at equidistant times $t_n = t_0 + n\Delta t$ with $n = 0, 1, \ldots, N - 1$:

$$s_0, s_1, s_2, \ldots, s_{N-2}, s_{N-1}, s_0, s_1, \ldots$$

The sequence is periodic, that is, after $N$ elements, it is repeated. Then, the Fourier coefficients are given by

$$S_m = \sum_{n=0}^{N-1} s_n \exp\left(-\frac{2\pi inm}{N}\right) \quad (m = 0,1,2\ldots N-1) \tag{1.51}$$

The sequence of these coefficients, which is also periodic with the period $N$, is called the discrete Fourier transform (DFT) of $s_n$. The inverse transformation reads

$$s_n = \frac{1}{N}\sum_{m=0}^{N-1} S_m \exp\left(\frac{2\pi inm}{N}\right) \quad (n = 0,1,2\ldots N-1) \tag{1.52}$$

Nowadays, all these operations are most conveniently performed by means of a digital computer. A particularly efficient procedure of computing spectral functions is the 'fast Fourier transform' (FFT) algorithm (see, for example, work by Bracewell [1986]).

Of course, all the preceding formulae can be written with the angular frequency $\omega = 2\pi f$ instead of the frequency $f$. A real notation of the formulae, using cosine and sine functions instead of exponentials, is also possible. For this purpose, one just has to separate the real parts from the imaginary parts in the preceding equations.

Stationary and aperiodic time functions cannot be analyzed by Equation 1.42 because they are not limited in time, and therefore, the integral would not converge. Often, such processes have a more or less random character. What we can do is to consider a section of the signal with the duration $T_0$. For this, the spectral function $S_{T_0}(f)$ is well defined and can be evaluated in the described way. The 'power spectrum' of the whole signal is then given by

$$W(f) = \lim_{T_0 \to \infty} \frac{1}{T_0} S_{T_0}(f) S_{T_0}^*(f) = \lim \frac{\left|S_{T_0}\right|^2}{T_0} \tag{1.53}$$

(The asterisk denotes the conjugate complex quantity.) The power spectrum, which is an even function of the frequency, does not contain all the information of the original time function, because it is based on the absolute value of the spectral function only, whereas all phase information has been eliminated. Inserted into Equation 1.46, it does not restore the time function but instead yields another important time function, called the 'autocorrelation function' of it:

$$\phi_{ss}(\tau) = \int_{-\infty}^{\infty} W(f) \cdot \exp(2\pi i f \tau) \mathrm{d}f = 2\int_0^{\infty} W(f) \cdot \cos(2\pi f \tau) \mathrm{d}f \tag{1.54}$$

The time variable has been denoted by $\tau$ in order to indicate that it is not identical with the real time. In the usual definition of the autocorrelation function, it appears as a time shift:

$$\phi_{ss}(\tau) = \lim_{T_0 \to \infty} \frac{1}{T_0} \int_{-T_0/2}^{T_0/2} s(t) \cdot s(t+\tau) \mathrm{d}t = \overline{s(t) \cdot s(t+\tau)} \tag{1.55}$$

(As before, the overbar means time averaging.) The autocorrelation function characterises the statistical similarity of a signal at time $t$ and the same signal at a different time $t + \tau$. For signals with finite energy content, time averaging would be meaningless; hence, a modified expression of the autocorrelation function is used:

$$\phi_{ss}(\tau) = \int_0^{\infty} s(t) \cdot s(t+\tau) \mathrm{d}t \tag{1.56}$$

Since $\phi_{ss}$ is the Fourier transform of the power spectrum, the latter is also obtained by inverse Fourier transformation of the autocorrelation function:

$$W(f) = \int_{-\infty}^{\infty} \phi_{ss}(\tau) \cdot \exp(-2\pi i f \tau) \mathrm{d}\tau = 2\int_0^{\infty} \phi_{ss}(\tau) \cdot \cos(2\pi f \tau) \mathrm{d}\tau \tag{1.57}$$

Equations 1.54 and 1.57 are the mathematical expressions of the theorem of Wiener and Khintchine: power spectrum and autocorrelation function are Fourier transforms of each other.

If $s(t + \tau)$ in Equation 1.55 is replaced with $s'(t + \tau)$, where $s'$ denotes a time function different from $s$, one obtains the 'cross-correlation function' of the two signals $s(t)$ and $s'(t)$:

$$\phi_{ss'}(\tau) = \lim_{T_0 \to \infty} \frac{1}{T_0} \int_{-T_0/2}^{T_0/2} s(t) s'(t+\tau) \mathrm{d}t = \overline{s(t) s'(t+\tau)} \tag{1.58}$$

The cross-correlation function provides a measure of the statistical similarity of the two functions $s$ and $s'$. It is closely related to the correlation coefficient

$$\Psi_{ss'} = \frac{\overline{s(t) \cdot s'(t)}}{\sqrt{\overline{s(t)^2} \cdot \overline{s'(t)^2}}} \tag{1.59}$$

which may vary between +1 and −1. If $\Psi = 0$, the functions $s(t)$ and $s'(t)$ are said to be uncorrelated, and the signals they represent are incoherent. It should be noted, however, that a vanishing correlation coefficient is only a necessary condition for incoherence but not a sufficient one.

In a certain sense, a sine or cosine signal can be considered as an elementary signal; it is unlimited in time and steady in all its derivatives, and its spectrum consists of a single line. The counterpart of it is Dirac's delta function $\delta(t)$: it has one single line in the time domain, so to speak, whereas its amplitude spectrum is constant for all frequencies, that is, $S(f) = 1$ for the delta function. This leads to the following representation:

$$\delta(t) = \lim_{f_0 \to \infty} \frac{1}{f_0} \int_{-f_0/2}^{f_0/2} \exp(2\pi i f t) \mathrm{d}f \tag{1.60}$$

The delta function has the following fundamental property:

$$s(t) = \int_{\infty}^{\infty} s(\tau)\delta(t-\tau)\mathrm{d}\tau \tag{1.61}$$

where $s(t)$ is any function of time. Accordingly, any signal can be considered as a close succession of very short pulses, as indicated in Figure 1.5. Especially, for $s(t) = 1$, we obtain

$$\int_{-\infty}^{\infty} \delta(\tau)\mathrm{d}\tau = 1 \tag{1.62}$$

Since the delta function $\delta(t)$ is zero for all $t \neq 0$, it follows from Equation 1.62 that its value at $t = 0$ must be infinite.

Now, consider a linear and time-independent but otherwise unspecified transmission system. Examples of acoustical transmission systems are all kinds of ducts (air ducts, mufflers, wind instruments, etc.) and resonators. Likewise, any two points in an enclosure may be considered as the input and output terminal of an acoustic transmission system. Linearity means that multiplying the input signal with a constant factor results in an output signal which is augmented by the same factor. The properties of such a system are completely characterized by the 'impulse response' $g(t)$, that is, the output signal which is the response to an impulsive input signal represented by the Dirac function $\delta(t)$ (see Figure 1.5). Since the response cannot precede the excitation, the impulse response of any causal system must vanish for $t < 0$. If $g(t)$ is known, the output signal $s'(t)$, with respect to any input signal $s(t)$, can be obtained by replacing the Dirac function in Equation 1.56 with its response, that is, with $g(t)$:

$$s'(t) = \int_{-\infty}^{\infty} s(\tau) g(t-\tau)\mathrm{d}\tau = \int_{-\infty}^{\infty} g(\tau) s(t-\tau)\mathrm{d}\tau \tag{1.63}$$

This operation is known as the convolution of the functions $s$ and $g$. A common shorthand notation of it is

$$s'(t) = s(t) * g(t) = g(t) * s(t) \tag{1.64}$$

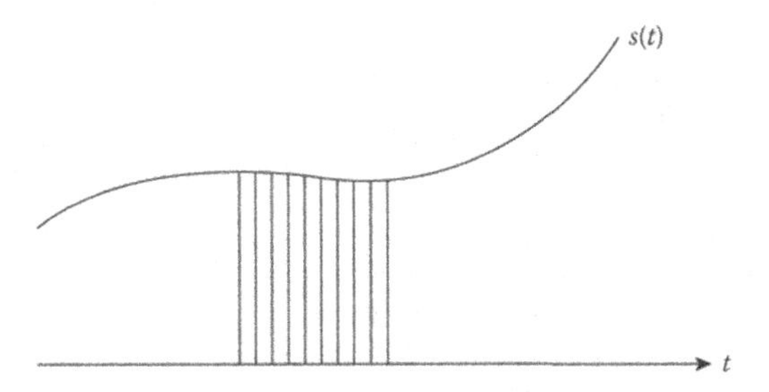


*Figure 1.5* Continuous function as the limiting case of a close succession of short impulses.

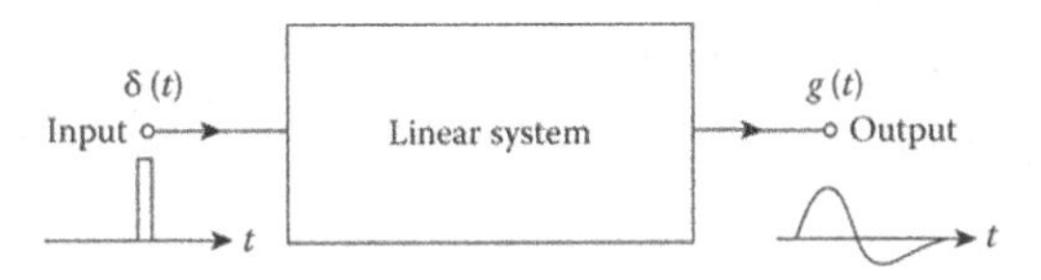


*Figure 1.6* Impulse response of a linear system.

Equation 1.63 has its analogue in the frequency domain, which looks even simpler. Let $S(f)$ be the complex spectrum of the input signal $s(t)$ of our linear system; then, the spectrum of the resulting output signal will be

$$s'(f) = G(f) \cdot S(f) \tag{1.65}$$

The complex function $G(f)$ is the 'transmission function' or 'transfer function' of the system; it is related to the impulse response $g(t)$ by the Fourier transformation:

$$G(f) = \int_{-\infty}^{\infty} g(t) \exp(-2\pi i f t) \mathrm{d}t \tag{1.66}$$

$$g(t) = \int_{-\infty}^{\infty} G(f) \exp(2\pi i f t) \mathrm{d}f \tag{1.67}$$

with $G(-f) = G^*(f)$, since $g(t)$ is a real function. The transfer function $G(f)$ has also a direct meaning: if a harmonic signal with frequency $f$ is applied to a transmission system, its amplitude will be changed by the factor $|G(f)|$, and its phase will be shifted by the phase angle of $G(f)$.

## 1.5 SOUND PRESSURE LEVEL AND SOUND POWER LEVEL

In the frequency range in which our hearing is most sensitive (500–5,000 Hz), the intensity of the threshold of sensation and that of the threshold of pain in hearing differ by about 13 orders of magnitude. For this reason, it would be impractical to characterize the strength of a sound signal by its sound pressure or its intensity. Instead, a logarithmic quantity — the 'sound pressure level' is generally used for this purpose — is defined by

$$L = 20\log_{10}\left(\frac{p_{rms}}{p_0}\right) \text{dB} \tag{1.68}$$

In this definition, $p_{rms}$ denotes the 'root-mean-square' pressure, as introduced in Section 1.3. The quantity $p_0$ is an internationally standardized reference pressure; its value is $2 \times 10^{-5}$ Pa, which corresponds roughly to the normal hearing threshold at 1,000 Hz. The 'decibel' (abbreviated as dB) is not a unit in physical sense but is used rather to recall the preceding level definition. Strictly speaking, $p_{rms}$ as well as the $L$ (sound pressure level) are defined only for stationary sound signals since they both imply an averaging process.

Since there is a simple relation between the sound pressure and the intensity, $L$ can also be expressed by the intensity using Equation 1.41 (second part):

$$L = 10\log_{10}\left(\frac{p_{rms}^2}{p_0^2}\right) \approx 10\log_{10}\left(\frac{I}{I_0}\right)\text{dB} \tag{1.69}$$

with $I_0 = 10^{-12}$ watts/m$^2$. According to Equation 1.68, the strength of two different sound fields or signals can be compared by their level difference:

$$\Delta L = 20\log_{10}\left(\frac{p_{rms_1}}{p_{rms_2}}\right)\text{dB} \tag{1.70}$$

It is often convenient to express the sound power delivered by a sound source in terms of the 'sound power level', defined by

$$L_w = 10\log_{10}\left(\frac{P}{P_0}\right)\text{dB} \tag{1.71}$$

where the reference power is $P_0 = 10^{-12}$ W. Using this quantity, the sound pressure level produced by a point source with power $P$ in the free field can be expressed as follows (see Equation 1.45):

$$L = L_w - 20\log_{10}\left(\frac{r}{r_0}\right) - 11\text{ dB with } r_0 = 1\text{ m} \tag{1.72}$$

## 1.6 SOME PROPERTIES OF HUMAN HEARING

Since the ultimate consumer of all room acoustics is the listener, it is important to consider at least a few facts relating to aural perception. More information may be found in the work by Zwicker and Fastl (1990) or by Blauert (1997), for instance.

One of the most obvious facts of human hearing is that the ear is not equally sensitive to sounds of different frequencies. Generally, the loudness at which a sound is perceived depends, of course, on its objective strength, that is, on the SPL. Furthermore, it depends in a complicated manner on the spectral composition of the sound signal, on its duration, and on several other factors. A widely used measure for the subjective impression of loudness is the 'loudness level', which is SPL of a 1,000 Hz tone, which is heard equally loud as the sound to be characterised. The unit of the loudness level is the 'phon'.

Figure 1.7 presents the contours of equal loudness level for sinusoidal sound signals which are presented to a listener in the form of frontally incident plane waves. The numbers next to the curves indicate the loudness level. The lowest, dashed curve, which corresponds to a loudness level of 3 phons, marks the threshold of hearing. According to this diagram, a pure tone with an SPL of 40 dB and a frequency of 1,000 Hz has, by definition, a loudness level of 40 phons. However, at 100 Hz, its loudness level would be only 24 phons, whereas at 50 Hz it would be almost inaudible.

Using these curves, the loudness level of any pure tone can be determined from its frequency and its SPL. In order to simplify this somewhat-tedious procedure, electrical instruments have been constructed which measure the sound pressure level. Basically, they consist of a calibrated microphone which converts the sound into an electrical signal, amplifiers, and most important, a weighting network, the frequency-dependent attenuation of which approximates the contours of equal loudness. Several weighting functions are in use and have been internationally standardised; the most common of them is the A-weighting curve. Consequently, the result of such a measurement is not the loudness level in phon but the 'A-weighted sound pressure level' in dB(A).

When such an instrument is applied to a sound signal with more complex spectral structure, the result may deviate considerably from the true loudness level. The reason for such errors is the fact that, in our hearing, weak spectral components are partially or completely masked by stronger ones and that this effect is not modelled in the aforementioned sound level metres. Apart from masking in the frequency domain, temporal masking may occur in non-stationary signals. In particular, a strong time-variable signal may mask a subsequent weaker signal. This property is very important in listening in closed spaces, as will be described in more detail in Chapter 7.

One misleading property of the loudness level is its unsatisfactory relation to our subjective perception. In fact, doubling the subjective sensation of loudness does not correspond to twice the loudness level in phon, as should be expected. Instead, it corresponds only to an increase of about 10 phons. This shortcoming is avoided by another loudness scale with 'sone' as a unit. The sone scale is defined in such a way that 40 phons correspond to 1 sone,

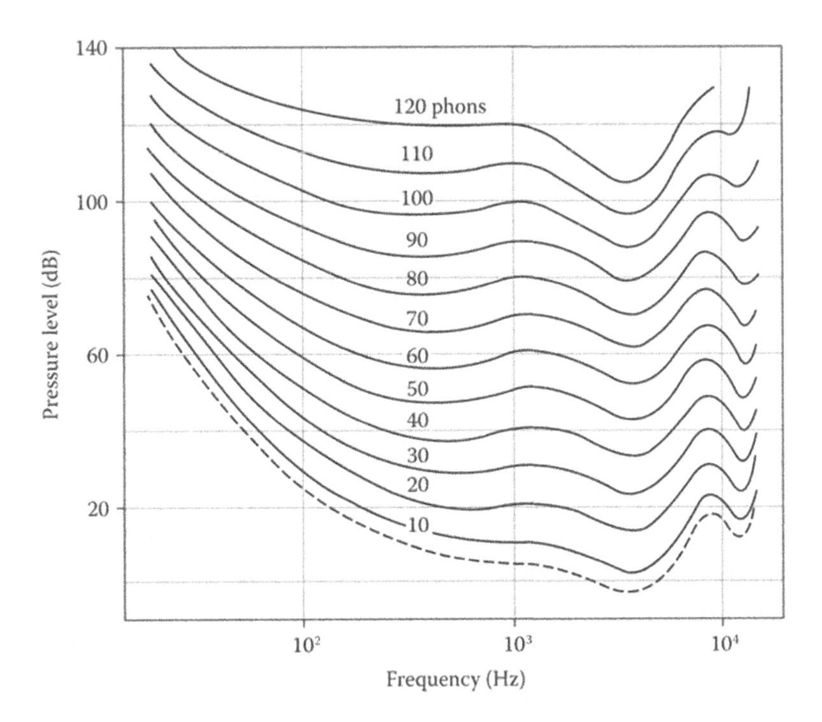


*Figure 1.7* Contours of equal loudness level for frontal sound incidence. The dashed curve corresponds to the average hearing threshold.

and every increase in the loudness level by 10 phons corresponds to doubling the number of sones. Nowadays, instruments as well as computer programs are available by which loudness of almost any type of sound signal can by determined, taking into account the aforementioned masking effects.

Another important property of our hearing is its ability to detect the direction from which a sound wave is arriving, and thus to localise the direction of sound sources. For sound incidence from a lateral direction, it is easy to understand how this effect is brought about: an originally plane or spherical wave is distorted by the human head, by the pinnae and — to a minor extent — by the shoulders and the trunk. This distortion depends on sound frequency and the direction of incidence. As a consequence, the sound signals at both ears show characteristic differences in their amplitude and phase spectrum, or to put it more simply, at lateral sound incidence, one ear is within the shadow of the head but the other is not. The interaural amplitude and phase differences caused by these effects enable our hearing to reconstruct the direction of sound incidence.

Quantitatively, the changes a sound signal undergoes on its way to the entrance of the ear canal can be described by the so-called head-related transfer functions (HRTFs), which characterize the transmission from a very remote point source to the ear canal, for instance, its entrance. Such transfer functions have been measured by many researchers (Blauert 1997). As an example, Figure 1.8 shows HRTFs for seven lateral angles of incidence $\phi$ in the horizontal plane. Frontal sound incidence is defined at $\phi = 0°$; Figure 1.8 (left) presents the magnitude expressed in dB, while Figure 1.8 (right) plots the binaural impulse responses, that is, the inverse Fourier transform of the complex-valued HRTF. By comparing the curves of $\phi = 0°$ and 180°, for instance, the shadowing effect of the head becomes obvious. It should be noted that the HRTF may show individual variations due to differences in the sizes and shapes of human heads, pinnae, and so on. Beyond these differences, they show common and characteristic features.

If the sound source is situated within the vertical symmetry plane, this explanation fails, since then the source produces equal sound signals at both ear canals. But even then, the ear transfer functions show characteristic differences for various elevation angles of the source, and it is commonly believed that the way in which they modify a sound signal enables us to distinguish whether a sound source is behind, above, or in front of our head.

These considerations are valid only for the localization of sound sources in a free sound field. In a closed room, however, the sound field is made up of many sound waves propagating in different directions, and accordingly, matters are more complicated. We shall discuss the subjective effects of more complex sound fields as they are encountered in room acoustics in Chapter 7.

## 1.7 SOUND SOURCES

In room acoustics, we are mainly concerned with three types of sound sources: the human voice, musical instruments, and technical noise sources. (We do not consider loudspeakers here, because they reproduce sound signals originating from other sources. More will be said about loudspeakers in Chapter 11.)

It is a common feature of all these sources that the sounds they produce have a more or less complicated spectral structure — apart from some rare exceptions. In fact, it is the spectral content of speech signals (phonemes) which gives them their characteristics. Similarly, the timbre of musical sounds is determined by their spectra.

The signals emitted by most musical instruments, in particular, by string and wind instruments, including the organ, are nearly periodic. Therefore, their spectra consist of many

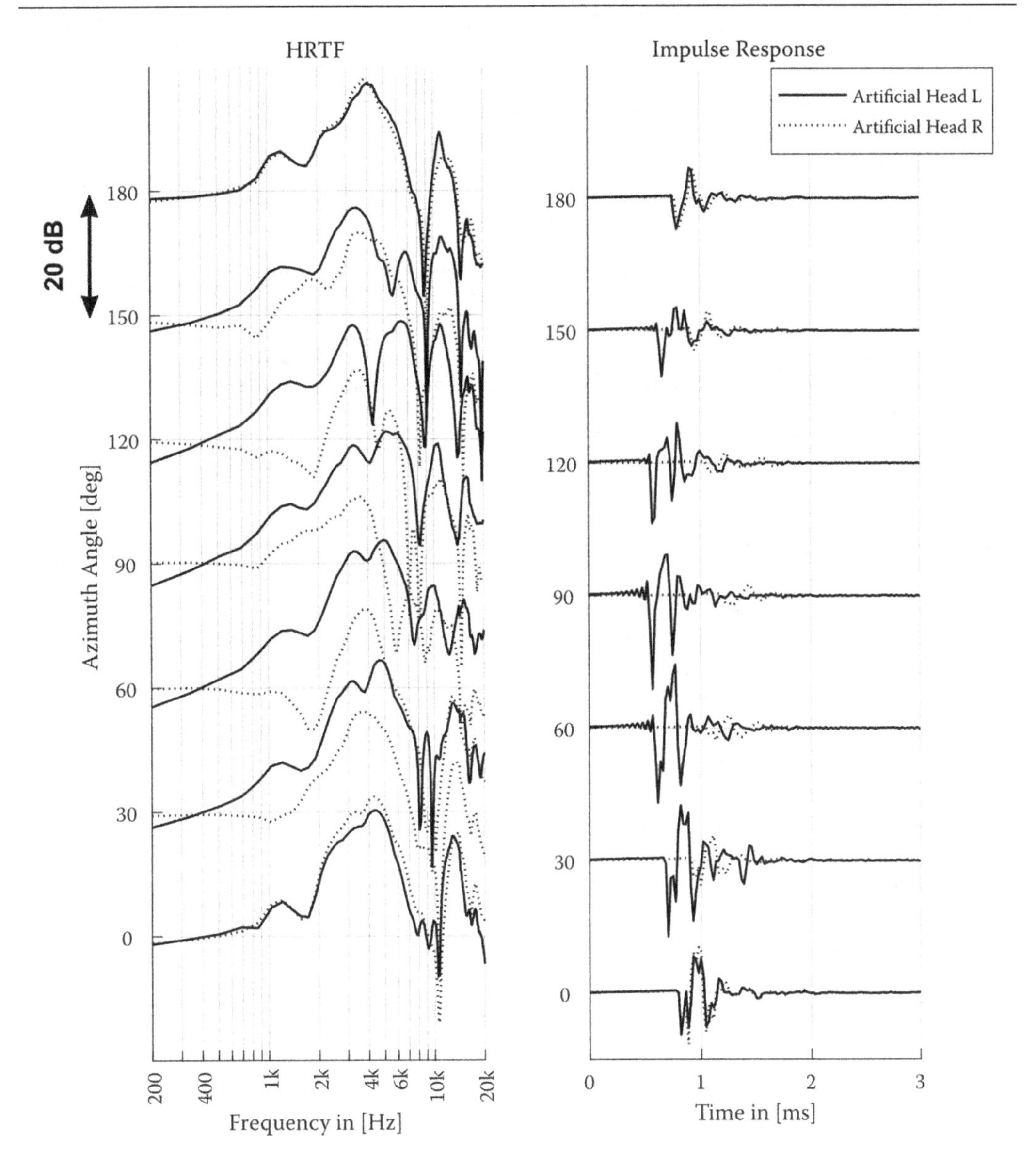


*Figure 1.8* Head-related transfer functions for the left and the right ear for several directions of sound incidence in the horizontal plane for an artificial head: (left) amplitude (logarithmic scale, in dB) and (right) head-related impulse responses.

discrete Fourier coefficients (see Section 1.4), which may be represented by equally spaced vertical lines. The time signals themselves are mixtures of harmonic vibrations with frequencies which are integral multiples of the fundamental frequency $1/T$ ($T$ = period of the signal). The component with the lowest frequency $1/T$ is the fundamental tone, and the higher-order components are called overtones. It is the fundamental which determines what we perceive as the pitch of a tone. This means that our ear receives many harmonic components of quite different frequencies even if we listen to a single musical tone. Likewise, the spectra of many speech sounds, in particular, vowels and voiced consonants, have a line structure. As an

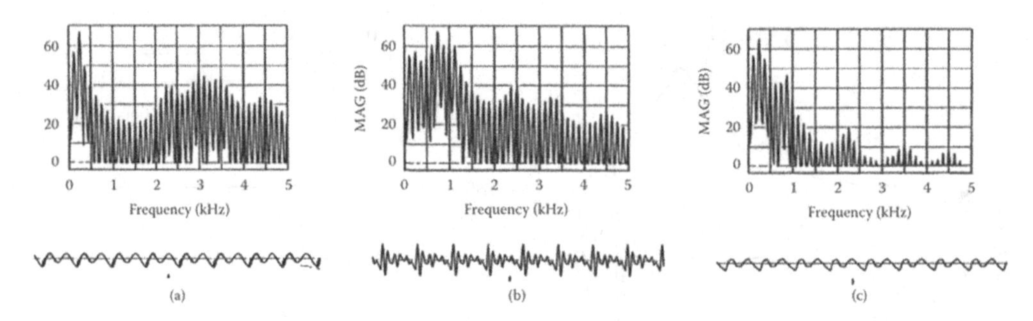


*Figure 1.9* Amplitude spectrum and time function (sound pressure) of vowels (a) /i/, (b) /a/, and (c) /u/. (Since these data have been obtained from experiments, the spectral 'lines' show up as sharp peaks.)

Source: Flanagan (1965).

example, Figure 1.9 presents the time function and the amplitude spectrum of three vowels. There are some characteristic frequency ranges in which the overtones are especially strong. These are called the 'formants' of the vowel.

For normal speech, the fundamental frequency lies between 50 and 350 Hz and is identical to the frequency at which the vocal chords vibrate. The total frequency range of conversational speech may be seen from Figure 1.10, which plots the long-time power spectrum of continuous speech, both for male and female speakers. The high-frequency energy is mainly due to the consonants, for instance, to fricatives such as /s/ or /f/, or to plosives, such as /p/ or /t/. Since consonants are of particular importance for the intelligibility of speech, a room or hall intended for speech, as well as a public address system, should transmit the high frequencies with great fidelity. The transmission of the fundamental vibration, on the other hand, is less important since our hearing is able to reconstruct it if the sound signal is rich in higher harmonics (virtual pitch).

Among musical instruments, large pipe organs have the widest frequency range, with the fundamental tone reaching from 16.5 Hz to more than 8 kHz. The piano follows with a tonal range corresponding to about 7¼ octaves. (One octave corresponds to a frequency ratio of 2:1.) The frequencies of most of the remaining instruments lie somewhere within this range. However, some instruments, especially percussion instruments, produce sound components with even higher frequencies. It should be noted that the given numbers refer to the fundamentals only. Since the spectra of almost all instruments contain higher harmonics which are responsible for the timbre of the sounds they produce, the bandwidth of any acoustical transmission system, whether it is a room or an electroacoustic system, should be able to transmit frequency components with up to 15 kHz at least. Fortunately, it is not the entire frequency range which is the responsibility of the acoustical engineer. At 10 kHz and above, the sound attenuation in air is so dominant that the influence of the boundaries on the propagation of high-frequency sound components can safely be neglected. At frequencies lower than 50 Hz, geometrical considerations are almost useless because of the large wavelengths of the sounds; furthermore, at these frequencies, it is almost impossible to assess correctly the sound absorption by vibrating panels or walls, and hence to control the reverberation. This means that, in this frequency range too, room acoustical design possibilities are very limited. On the whole, it can be stated that the frequency range relevant to room acoustics reaches from 50 to 10,000 Hz, the most important part being between 100 and 5,000 Hz.

The acoustical power output of the sound sources as considered here is relatively low by everyday standards. Table 1.1 lists a few typical data. The human voice generates a sound

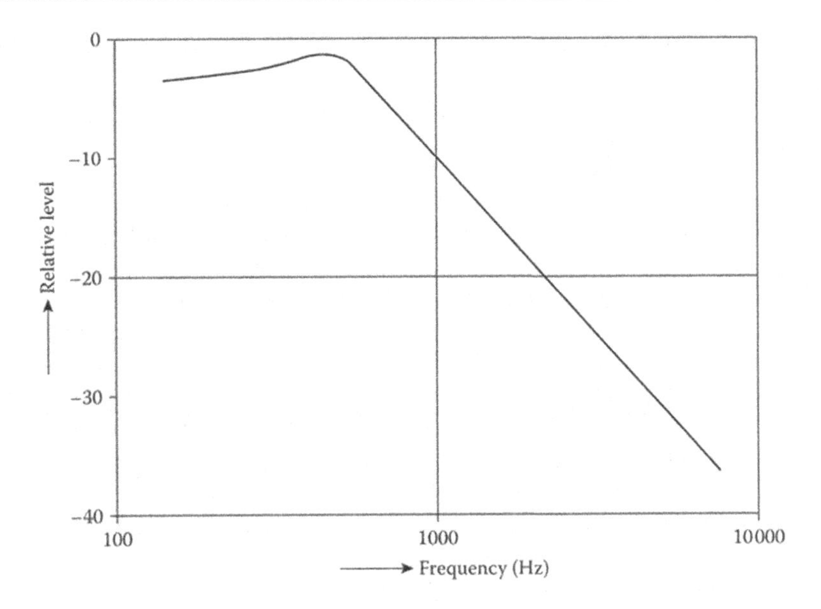


*Figure 1.10* Long-time power spectrum of continuous speech 30 cm from the mouth.

*Table 1.1* Sound power and power level of some sound sources (the data of musical instruments are for playing fortissimo)

| *Source or signal* | *Sound power (mW)* | *Sound power level (dB)* |
|---|---|---|
| Whisper | $10^{-6}$ | 30 |
| Conversational speech | 0.01 | 70 |
| Human voice, maximum | 1 | 90 |
| Violin | 1 | 90 |
| Clarinet, French horn | 50 | 107 |
| Trumpet | 100 | 110 |
| Organ, large orchestra | $10^{4}$ | 130 |

power ranging from 0.001 μW (whispering) to 1,000 μW (shouting); the power produced in conversational speech is of the order of 10 μW, corresponding to a sound power level of 70 dB. The power of a single musical instrument may lie in the range from 10 μW to 100 mW. A full symphony orchestra can easily generate a sound power of 10 W in fortissimo passages. It may be added that the dynamic range of most musical instruments is about 30 dB (woodwinds) to 50 dB (string instruments). A large orchestra can cover a dynamic range of 100 dB.

An important property of the human voice and musical instruments is their directionality, that is, the fact that they do not emit sound with equal intensity in all directions. In speech, this is because of the 'sound shadow' cast by the head. The lower the sound frequency, the less pronounced is the reduction of sound intensity by the head, because with decreasing frequencies, the sound waves are increasingly diffracted around the head. In Figures 1.11a and b, the distribution of the relative SPL for different frequency bands is plotted on a horizontal plane and a vertical plane, respectively. These curves are obtained by filtering out the respective frequency bands from natural speech; the direction denoted by 0° is the frontal direction.

Musical instruments usually exhibit a pronounced directionality because of the dimensions of their sound-radiating surfaces or openings, which, in the interest of high efficiency, are often comparable to or even larger than the wavelengths of the sounds they are to generate. Unfortunately, general statements are almost impossible, since the directional distribution of the radiated sound changes very rapidly, not just from one frequency to the other; it can be quite different for instruments of the same sort but from different manufactures. This is true especially for string instruments, the bodies of which exhibit complicated vibration patterns, particularly at higher frequencies. The radiation from a violin takes place in a fairly uniform way only at frequencies lower than about 450 Hz; at higher frequencies, however, matters become quite involved. For wind instruments, the directional distributions exhibit more common features, since here the sound is not radiated from a curved anisotropic plate with complicated vibration patterns but from a fixed opening, which is very often the end of a horn. The 'directional characteristics of an orchestra' are highly involved, but space is too limited here to discuss this in any detail. For the room acoustician, however, it is important to know that strong components, particularly from the strings but likewise from the piano, the woodwinds, and of course, the tuba, are radiated upwards. For further details, we refer to the exhaustive account of Meyer (2008).

In a certain sense, the sounds from natural sources can be considered as statistical or stochastic signals, and in this context, their autocorrelation function is of interest as it gives some measure of a signal's 'tendency of conservation'. Autocorrelation measurements on speech and music have been performed by several authors (Furdujev 1965). Here, we are reporting results obtained by Ando, who passed various signals through an A-weighting filter (see Section 1.6) and formed their autocorrelation function according to Equation 1.55 with a finite integration time of $T_0$ = 35 s. Two of his results are depicted in Figure 1.12. A characteristic measure of the effective duration of the autocorrelation function is the time delay $\tau_e$, at which its envelope has fallen to one-tenth of its maximum. These values are indicated in Table 1.2 for a few signals. They range from about 10 ms to more than 100 ms.

The variety of possible noise sources is too large to discuss in detail. A common kind of noise in a room is sound intruding from adjacent rooms or from outside through walls, doors, and windows, due to insufficient sound insulation. A typical noise source in halls is the air-conditioning system; some of the noise produced by the machinery propagates in the air ducts and is radiated into the hall through the air outlets.

*Table 1.2* Effective duration of autocorrelation functions of various sound signals

| *Motif* | *Name of piece* | *Composer* | *Eff. duration $\tau_e$, (ms)* |
|---|---|---|---|
| A | Royal Pavane | Gibbons | 127 |
| B | Sinfonietta opus 48, 4th movement (Allegro con brio) | Arnold | 43 |
| C | Symphony No. 102 in B-flat major, 2nd movement (Adagio) | Haydn | 65 |
| D | Siegfried Idyll; bar 322 | Wagner | 40 |
| E | Symphony KV 551 in C major (Jupiter), 4th movement (Molto allegro) | Mozart | 38 |
| F | Poem read by a female | Kunikita | 10 |

Source: Ando (1977).

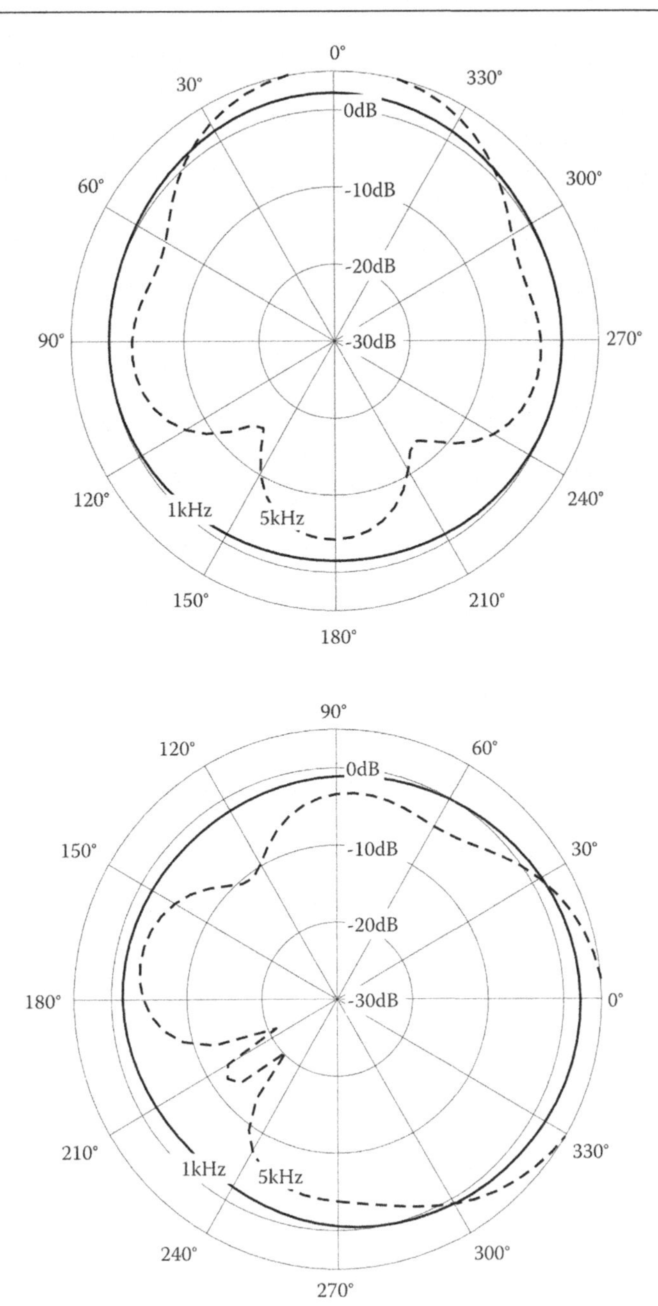


*Figure 1.11* Directional characteristics of a trumpet for two different frequency bands: (top) in the horizontal plane and (bottom) in the vertical plane. The reference musician's view direction is 0,0.

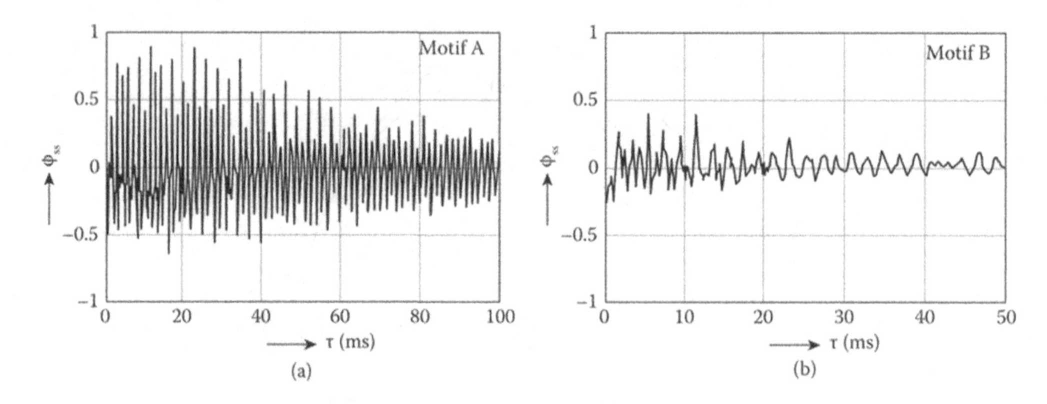


*Figure 1.12* Examples of measured autocorrelation functions: (a) music motif A and (b) music motif B (both from Table 1.2).

Source: Courtesy of Y.J. Ando.

## REFERENCES

Ando Y.J. Subjective preference in relation to objective parameters of music sound fields with a single echo. *J Acoust Soc Am* 1977; 62: 1436.

Blauert J. *Spatial Hearing*. Cambridge, MA: MIT Press, 1997.

Bracewell R.N. *The Fourier Transform and Its Applications*. Singapore: McGraw-Hill, 1986.

Flanagan J.J. *Speech Analysis Synthesis and Perception*. Berlin: Springer-Verlag, 1965.

Furdujev V. Objective evaluation of the acoustics of rooms (in French). Proceedings of the Fifth International Congresses on Acoustics, Liege, 1965, p. 41.

Meyer J. *Acoustics and the Performance of Music*. Berlin: Springer-Verlag, 2008.

Williams E. *Fourier Acoustics: Sound Radiation and Near-Field Acoustical Holography*. London: Academic Press, 1999.

Zwicker E., Fastl H. *Psychoacoustics — Facts and Models*. Berlin: Springer-Verlag, 1990.

Chapter 2

# Reflection and scattering

Up to now, we have dealt with sound propagation in a medium which was unbounded in every direction. In contrast to this simple situation, room acoustics is concerned with sound propagating in enclosures where the sound-conducting medium is bounded on all sides by walls, ceiling, and floor. These boundaries usually reflect a certain fraction of the sound energy impinging on them. Another fraction of the energy is 'absorbed', that is, it is extracted from the sound field inside the room, either by conversion into heat or by being transmitted to the outside by the walls. It is just this combination of the numerous reflected components which is responsible for what is known as 'the acoustics of a room' and also for the complexity of the sound field in a room.

Before we discuss the properties of such involved sound fields, we shall consider in this chapter the process which is fundamental for their occurrence: the reflection of a plane sound wave by a single wall or surface. In this context, we shall encounter the concepts of wall impedance and absorption coefficient, which are of special importance in room acoustics. The sound absorption by a wall will be dealt with mainly from a formal point of view, whereas the discussion of the physical causes of sound absorption and of the functional principles of various absorbent arrangements will be discussed in Chapter 6.

Strictly speaking, the simple laws of sound reflection to be explained in this chapter hold only for unbounded walls. Any free edge of a reflecting wall or panel scatters some sound energy in all directions. The same happens when a sound wave hits any other obstacle of limited extent, such as a pillar or a listener's head, or when it arrives at a basically plane wall which has an irregular surface. Since scattering is a common phenomenon in room acoustics, we shall briefly deal with it in this chapter.

Throughout this chapter, we shall assume that the incident, undisturbed wave is a plane wave. In reality, however, all waves originate from a sound source and are therefore spherical waves or superpositions of spherical waves. The reflection of a spherical wave from a plane wall is highly complicated, unless we assume that the wall is rigid. Section 2.4 discusses more on this. A comprehensive discussion of the exact theory and its various approximations may be found in the work by Sommerfeld. For our present discussion, it may be sufficient to assume that the sound source is not too close to the reflecting wall or to the scattering obstacle so that the curvature of the wave fronts can be neglected without too much error.

## 2.1 REFLECTION FACTOR, ABSORPTION COEFFICIENT, AND WALL IMPEDANCE

If a plane wave strikes a plane and uniform wall of infinite extent, in general, a part of the sound energy will be reflected back in the form of a secondary wave, with an amplitude and

DOI: 10.1201/9781003389873-2

phase which differ from those of the incident wave. Both waves interfere with each other and form a 'standing wave', at least partially.

The changes in amplitude and phase which take place during the reflection of a wave are expressed by the complex reflection factor

$$R = |R| \exp(i\chi)$$

of the wall. Its absolute value $|R|$, also called the 'modulus' or the 'magnitude' of the reflection factor, and its phase angle $\chi$ depend on the frequency and the direction of the incident wave.

According to Equation 1.34, the intensity of a plane wave is proportional to the square of the pressure amplitude. Therefore, the intensity of the reflected wave is smaller by a factor $|R|^2$ than that of the incident wave; the fraction $1-|R|^2$ of the incident energy is lost during reflection. This quantity is called the 'absorption coefficient' of the wall:

$$\alpha = 1 - |R|^2 \tag{2.1}$$

For a wall with zero reflectivity ($R = 0$), the absorption coefficient has its maximum value 1. Then, the wall is said to be totally absorbent or 'matched to the sound field'. If $R = 1$ (in-phase reflection, $\chi = 0$), the wall is 'rigid' or 'hard'; in the case of $R = -1$ (phase reversal, $\chi = \pi$), the wall is 'soft'. In both cases, there is no sound absorption ($\alpha = 0$). Soft walls, however, are very rarely encountered in room acoustics and only in limited frequency ranges.

The acoustical properties of a wall surface — as far as they are of interest in room acoustics — are completely described by the reflection factor for all angles of incidence and for all frequencies. Another quantity which is even more closely related to the physical behaviour of the wall and to its construction is based on the motion of the medium particles next to the wall. It is called the wall impedance and is defined by

$$Z = \left( \frac{p}{v_n} \right)_{\text{surface}} \tag{2.2}$$

where $v_n$ denotes the velocity component normal to the wall. For non-porous walls that are excited into vibration by the sound field, the normal component of the particle velocity is identical to the velocity of the wall vibration. Like the reflection factor, the wall impedance is generally complex and a function of the angle of sound incidence. Frequently, the 'specific acoustic impedance' is used, which is the wall impedance divided by the characteristic impedance of the air:

$$\zeta = \frac{Z}{\rho_0 c} \tag{2.3}$$

The reciprocal of the wall impedance is the 'wall admittance', and the reciprocal of $\zeta$ is called the 'specific acoustic admittance' of the wall.

As explained in Section 1.2, any complex quantity can be represented in a rectangular coordinate system (see Figure 1.2). This holds also for the wall impedance. In this case, the length of the arrow corresponds to the magnitude of $Z$, while its inclination angle is the phase angle of the wall impedance:

$$\mu = \arg(Z) = \arctan\left( \frac{\operatorname{Im} Z}{\operatorname{Re} Z} \right) = \arctan\left( \frac{\operatorname{Im} \zeta}{\operatorname{Re} \zeta} \right) \tag{2.4}$$

If the frequency changes, the impedance will usually change as well, and also the length and inclination of the arrow representing it. The curve connecting the tips of all arrows is called the 'locus of the impedance in the complex plane'.

## 2.2 REFLECTION OF PLANE WAVES

In this section, we consider a plane and smooth wall and a plane sound wave impinging on it at an angle θ, which may have any value between 0° and 90°. Without loss of generality, we can assume that the wall normal and the wave normal of the incident wave lie in the x–y plane of a rectangular coordinate system. The situation is depicted in Figure 2.1. The incident wave and the wave reflected from the plane surface are indicated by bold arrows. The dashed path will be explained in the next section.

The sound pressure in the incident wave can be described by Equation 1.23 setting $\alpha = \theta$, $\beta = \pi/2 - \theta$, $\gamma = \pi/2$:

$$p_i(x,y) = \hat{p}_0 \exp\left[-ik(x\cos\theta + y\sin\theta)\right] \tag{2.5}$$

(In this and the following expressions, we omit, for the sake of simplicity, the factor $\exp(i\omega t)$, which is understood to be common to all sound pressures and particle velocities.)

In the incident wave, the medium particles oscillate in the direction of sound propagation, and their velocity is obtained by dividing the pressure by the characteristic impedance $\rho_0 c$. Hence, the component of their velocity normal to the wall is

$$(v_i)_x = \frac{\hat{p}_0}{\rho_0 c}\cos\theta \cdot \exp\left[-ik(x\cos\theta + y\sin\theta)\right] \tag{2.6}$$

To arrive at a corresponding representation of the reflected wave, one has to note that the latter travels in the opposite direction with respect to the $x$-axis (not to the $y$-axis); therefore, the sign of $x$ in Equations 2.5 and 2.6 must be reversed (see Section 1.2). Furthermore,

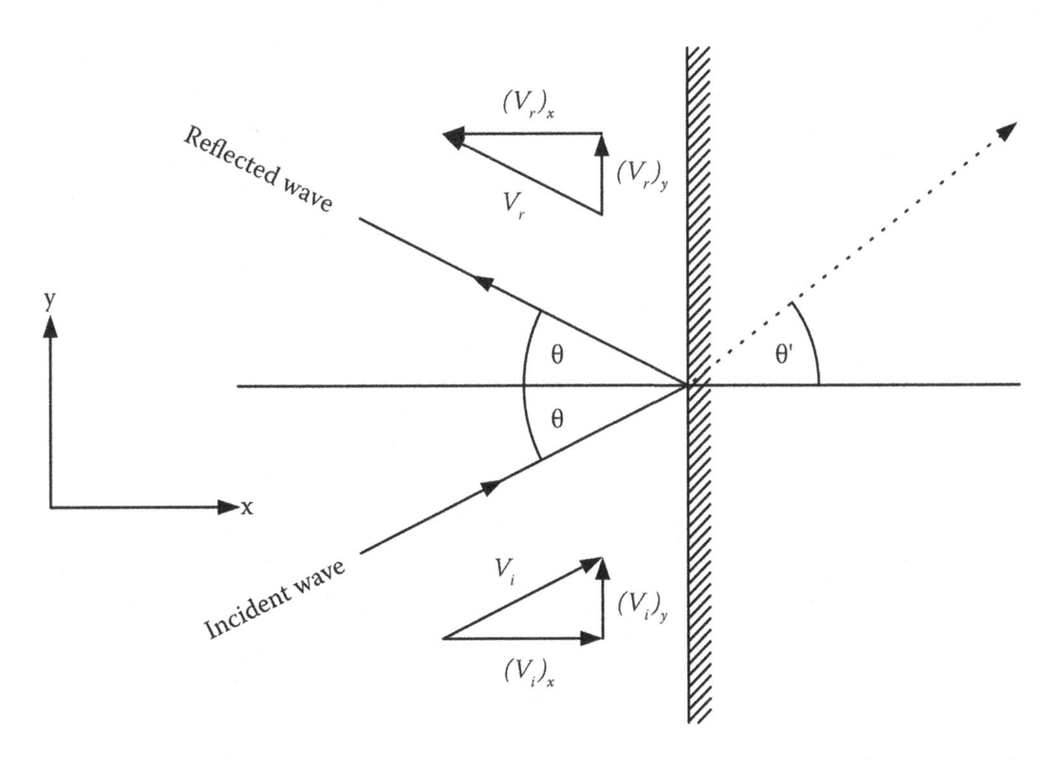


*Figure 2.1* Sound reflection from a plane wall. The dashed line marks the path of a wave intruding into the medium on the right of the boundary.

the reflection reduces the amplitude and changes the phase of the wave; both changes are described by the complex reflection factor $R$. The sign of the particle velocity is also reversed since $p/v$ has opposite signs for positive- and negative-going waves. So we obtain for the reflected wave:

$$p_r(x,y) = R\hat{p}_0 \exp\left[-ik(-x\cos\theta + y\sin\theta)\right] \tag{2.7}$$

$$(v_r)_x = \frac{-R\hat{p}_0}{\rho_0 c}\cos\theta \cdot \exp\left[-ik(-x\cos\theta + y\sin\theta)\right] \tag{2.8}$$

Again, the wave normal of the reflected wave includes the angle $\theta$ with the wall normal. This is the law of 'specular' reflection well known from optical mirrors.

By setting $x = 0$ in Equations 2.6 through 2.8, and by dividing $(p_i + p_r)$ by $((v_i)_x + (v_r)_x)$, we obtain

$$Z = \frac{\rho_0 c}{\cos\theta} \cdot \frac{1+R}{1-R} \tag{2.9}$$

or, after solving for $R$:

$$R = \frac{Z\cos\theta - \rho_0 c}{Z\cos\theta + \rho_0 c} = \frac{\zeta\cos\theta - 1}{\zeta\cos\theta + 1} \tag{2.10}$$

According to Equation 2.1, the absorption coefficient of the reflecting surface is

$$\alpha = \frac{4\mathrm{Re}(\zeta)\cos\theta}{|\zeta|^2\cos^2\theta + 2\mathrm{Re}(\zeta)\cos\theta + 1} \tag{2.11}$$

In Figure 2.2, this relation is represented graphically for normal sound incidence ($\theta = 0$). The diagram presents the circles of constant absorption coefficient in the complex $\zeta$ plane, that is, abscissa and ordinate in Figure 2.2 are the real and imaginary parts of the specific wall impedance, respectively. As $\alpha$ increases, the circles contract towards the point $\zeta = 1$, which corresponds to a perfect match of the wall to the medium. For grazing incidence, that is, for $\theta \to 90°$, the absorption of the wall vanishes as cos 90° is zero.

A special case is a surface the impedance of which is independent of the direction of sound incidence. This applies if the normal component of the particle velocity at any wall element depends only on the sound pressure at that element and not on the pressure at neighbouring elements. Walls or surfaces with this property are referred to as 'locally reacting'. For such boundaries, the only angle dependence of the absorption coefficient is that of the cosine function in Equation 2.11. Figure 2.3 plots the absorption coefficient of locally reacting surfaces with different characteristic impedances.

In practice, surfaces with local reaction are rather the exception than the rule. They are encountered whenever the material behind the insonified surface is unable to propagate waves or vibrations in a direction parallel to the surface. The reason for this behaviour may be some anisotropy, as, for instance, that of the Rayleigh model, which will be described in Section 6.5. Examples of surfaces with extended reaction are panels that can vibrate under the influence of an impinging sound wave; the vibrations of neighbouring elements are coupled to each other by the bending stiffness.

Another example of a boundary with extended reaction is the surface of a homogeneous and isotropic medium, which completely fills the right half-space in Figure 2.1 ($x > 0$). In this case, there will not be just one secondary wave but two of them: one of them will intrude into the medium at the right; its direction differs from that of the incident wave. In Figure 2.1, it is

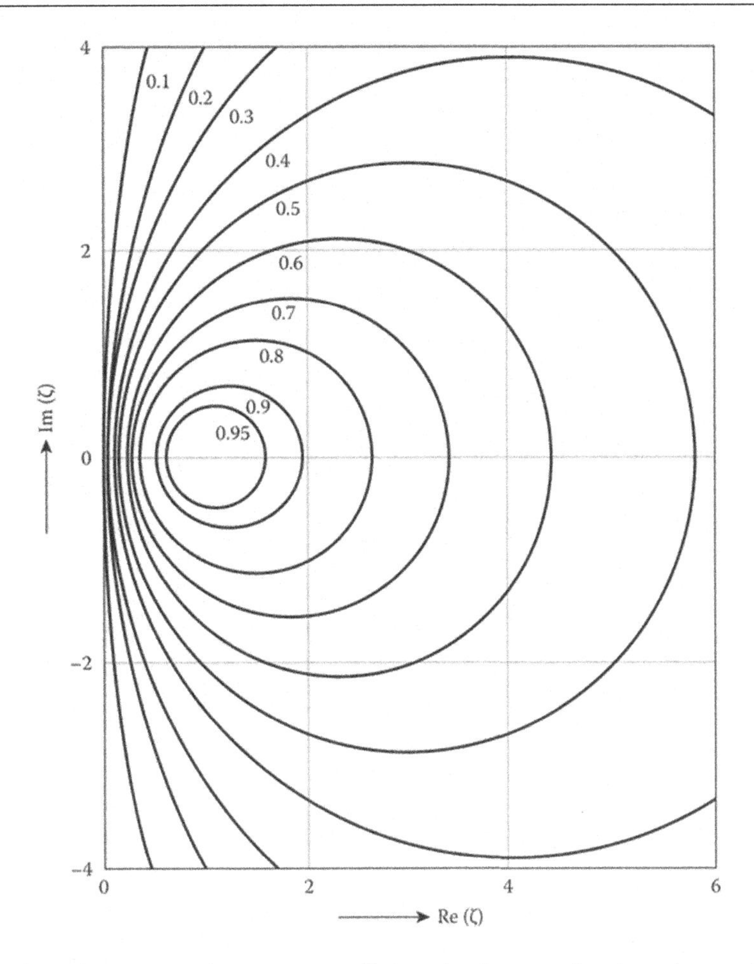


*Figure 2.2* Circles of constant absorption coefficient in the complex impedance plane, for normal sound incidence (θ = 0). The numbers next to the circles denote the absorption coefficient.

indicated by the dashed line. The acoustic properties of the medium at the right are characterized by its characteristic impedance $Z$ and by the complex wave number $k'$. We suppose that the material is of the porous type, such as glass wool or Rockwool. Under certain conditions (see Section 6.5), both quantities are related by

$$\frac{Z'}{\rho_0 c} = \frac{k'}{k} \quad \text{with } k = \frac{\omega}{c} \tag{2.12}$$

For a quantitative treatment, we complete Equations 2.5 through 2.8 by two expressions representing the sound pressure and the particle velocity (x-component) of the wave transmitted into the right half-space:

$$p_t(x,y) = T\hat{p}_0 \exp\left[-ik'(x\cos\theta' + y\sin\theta')\right] \tag{2.13}$$

$$(v_t)_x = \frac{T\hat{p}_0}{Z'}\cos\theta' \cdot \exp\left[-ik'(-x\cos\theta' + y\sin\theta')\right] \tag{2.14}$$

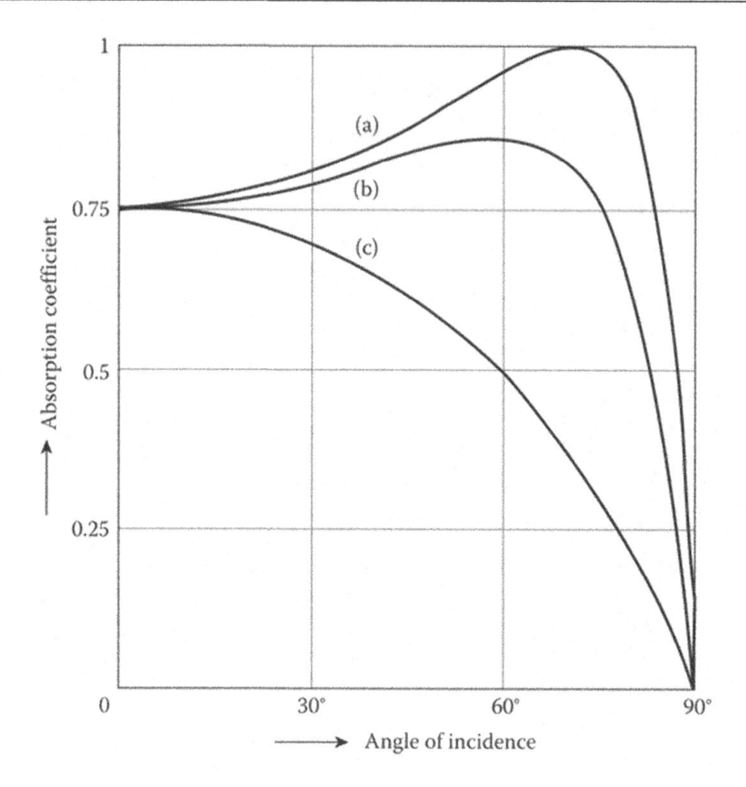


*Figure 2.3* Absorption coefficient of a locally reacting surface, as a function of the angle of incidence. The specific impedance is (a) $\zeta$ = 3, (b) $\zeta$ = 1.5 + 1.323i, and (c) $\zeta$ = 1/3.

In these equations, $T$ denotes the complex transmission factor, which is to express the changes the amplitude and the phase undergo when the wave enters the half-space $x > 0$. Its definition is analogous to that of the reflection factor, namely, as the complex factor by which the amplitude and the phase of the wave are changed when it enters the new medium.

The reflection factor $R$ and the transmission factor $T$ can be evaluated from Equations 2.5 through 2.8 and Equations 2.13 and 2.14. At first, we require that the expressions for the incident, the reflected and the transmitted wave, must show the same periodicity with respect to the $y$-direction. This leads immediately to

$$k\sin\theta = k'\sin\theta'$$

or

$$\frac{\sin\theta'}{\sin\theta} = \frac{k}{k'} \tag{2.15}$$

For lossless media, that is, if $k'$ is real, this is the Snell's law of refraction known from optics. In the present case, however, it should be noted that both $k'$ and $\theta'$ are complex quantities.

Next, we observe that the mentioned wave expressions must fulfil two boundary conditions, namely, that the total sound pressure and the total normal component of the particle velocity have the same value at both sides of the boundary $x = 0$, or

$$p_i + p_r = p_t \text{ and } (v_i)_x = (v_t)_x$$

This yields the relations

$$1+R=T \text{ and } \frac{\cos\theta}{\rho_0 c}(1-R)=\frac{\cos\theta'}{Z'}T$$

from which we obtain

$$R=\frac{Z'\cos\theta-\rho_0 c\cos\theta'}{Z'\cos\theta+\rho_0 c\cos\theta'} \tag{2.16}$$

This result agrees with Equation 2.10, apart from the factors cos $\theta'$ in the nominator and the denominator. Thus, we expect that the absorption coefficient calculated with Equation 2.16 does not dramatically differ from that calculated under the assumption of local reaction. Therefore, in most cases, it is sufficient to use the simpler Equation 2.11 with constant (specific) wall impedance $\zeta$. This has the additional advantage that it allows a simple formulation of the boundary conditions at the walls of a room (see Section 3.1).

We conclude this section with a short description of the sound field in front of a reflecting wall. The superposition of the incident and the reflected wave results in what is called a standing wave. The total pressure amplitude is obtained by adding Equations 2.5 and 2.7 and forming the absolute value of the sum:

$$|p(x)|=\hat{p}_0\sqrt{1+|R|^2+2|R|\cos(2kx\cos\theta+\chi)} \tag{2.17}$$

Similarly, the amplitude of the particle velocity is found from Equations 2.6 and 2.8:

$$|v(x)|=\frac{\hat{p}_0}{\rho_0 c}\sqrt{1+|R|^2-2|R|\cos(2kx\cos\theta+\chi)} \tag{2.18}$$

The time dependence of the pressure and the velocity is taken into account simply by a factor $\exp(i\omega t)$, which was omitted in the earlier expressions.

According to these equations, the pressure amplitude and the velocity amplitude in the standing wave vary periodically between the maximum values

$$p_{\max}=\hat{p}_0(1+|R|) \text{ and } v_{\max}=\frac{\hat{p}_0}{\rho_0 c}(1+|R|) \tag{2.19}$$

and the minimum values

$$p_{\min}=\hat{p}_0(1-|R|) \text{ and } v_{\min}=\frac{\hat{p}_0}{\rho_0 c}(1-|R|) \tag{2.20}$$

But in such a way that each maximum of the pressure amplitude coincides with a minimum of the velocity amplitude, and vice versa. The distance between two maxima is $(\lambda/2)/\cos\theta$, in particular, at normal incidence, half the acoustic wavelength. Figure 2.4 shows the situation for normal sound incidence ($\theta = 0$). So by measuring the pressure amplitude as a function of $x$, we can evaluate the wavelength. More importantly, the magnitude and the phase angle of the reflection factor can be evaluated. This is the basis of a standard method for measuring the impedance and the absorption coefficient of wall materials (see Section 8.6).

The common factor $\exp(-iky\cdot\sin\theta)$ in Equations 2.5 through 2.8 tells us that the 'standing' wave field is traveling parallel to the reflecting surface with a velocity

$$c_y=\frac{\omega}{k_y}=\frac{\omega}{k\sin\theta}=\frac{c}{\sin\theta} \tag{2.21}$$

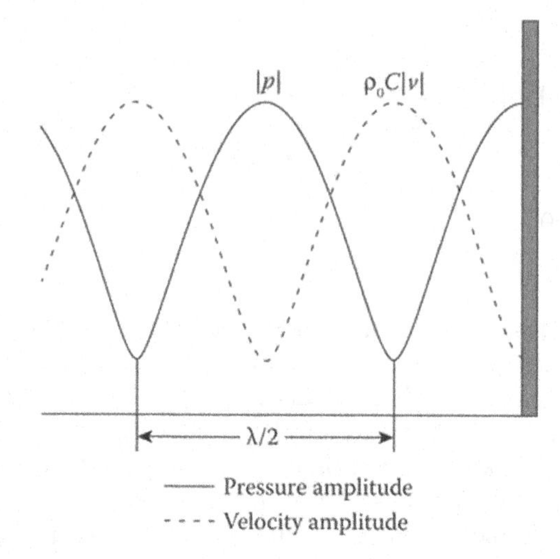


*Figure 2.4* Standing wave in front of a plane surface with the real reflection factor $R = 0.7$, normal sound incidence.

## 2.3 A FEW EXAMPLES

The purpose of this section is to make the reader familiar with the concepts developed in the preceding section and — at the same time — to describe some arrangements which are the basis of frequently used sound absorbers. If not stated otherwise, normal incidence of the primary plane wave is assumed in this section.

First, we consider a layer of some homogeneous material of thickness $d'$ (see Figure 2.5). We characterize it acoustically by its characteristic impedance $Z_0'$ and the angular wave number $k'$; both quantities may be complex. At its right boundary, the layer is loaded with the impedance $Z_r$. A plane sound wave arriving from the left will excite two plane waves within the layer travelling in opposite directions. Hence, the sound pressure and the particle velocity within the layer are given by

$$p(x) = A\exp(-ik'x) + B\exp(ik'x)$$

and

$$Z'v(x) = A\exp(-ik'x) + B\exp(ik'x)$$

with the two unknown constants $A$ and $B$. Dividing both expressions by each other leads to

$$Z_l = \frac{A+B}{A-B} \cdot Z' \text{ at } x = 0 \tag{2.22}$$

and

$$Z_r = \frac{A\exp(-ik'd') + B\exp(ik'd')}{A\exp(-ik'd') - B\exp(ik'd')} \cdot Z' \text{ at } x = d' \tag{2.23}$$

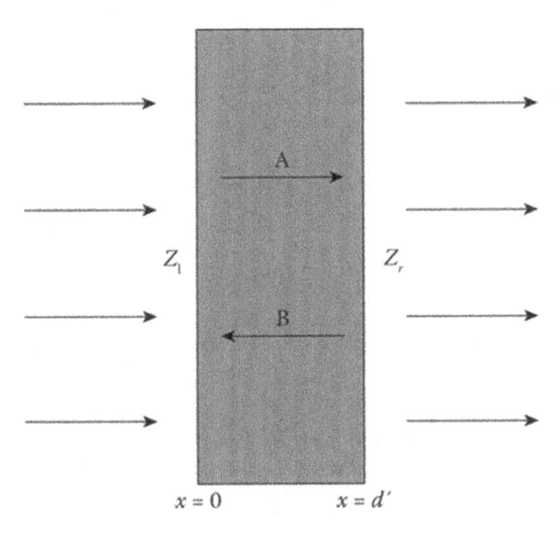


*Figure 2.5* Sound transmission through a layer (thickness $d'$) of a homogeneous material specified by its characteristic impedance $Z'$ and its propagation constant $k'$.

The fraction $B/A$ can be eliminated from both expressions. Finally, we express the exponentials by trigonometric functions using Equation 1.19 and arrive after some simple operations at the wall impedance, which acts on the incident wave at the left boundary:

$$Z_l = \frac{Z_r + iZ'\tan(k'd')}{Z' + iZ_r\tan(k'd')} \cdot Z' \tag{2.24}$$

This relation tells us that the impedance $Z_r$ with which the layer is loaded at its rear side is transformed into another impedance $Z_l$ appearing at its front side. It is very useful since it can be modified and combined in many ways. If, for instance, the layer is backed by a rigid plane ($Z_r \to \infty$), its wall impedance at the front side ($x = 0$) is

$$Z_l = \frac{Z'}{i\tan(k'd')} = -iZ' \cdot \cot(k'd') \tag{2.25}$$

This formula describes the wall impedance of the simplest type of a sound absorber: a porous layer fixed immediately in front of a hard wall. We shall discuss it at greater length in Section 6.6. If the values for air are inserted into this expression ($Z' = \rho_0 c$ and $k' = k$), the result is the 'wall impedance' of an air layer of thickness $d'$:

$$Z_a = -i\rho_0 c \cdot \cot(kd') \tag{2.26}$$

Of course, the 'wall' is only fictive in this case; from Equation 2.11 (with $\theta = 0$), it follows immediately that its absorption coefficient is zero. Nevertheless, this formula is quite useful. For $kd' < \pi/2$ (i.e. for $d' < \lambda/4$), the impedance $Z_a$ of the air cushion is that of an elastic spring.

As a second example, we consider an example of a thin porous layer — some curtain or blanket — which is stretched or hung in front of a rigid wall at distance $d$. The $x$-axis is normal to the layer and the wall; accordingly, the former has the coordinate $x = 0$, while the solid wall is located at $x = d$ (see Figure 2.6).

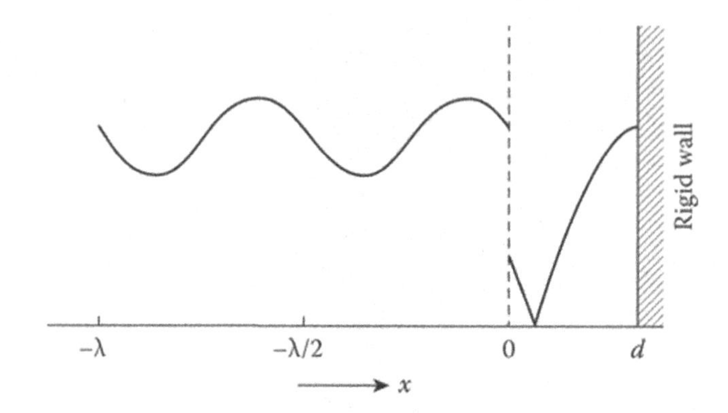


*Figure 2.6* Standing wave in front of a thin porous layer (dashed vertical line), situated at a distance $d$ from a rigid wall. The plotted curve is the pressure amplitude for $r_s = \rho_0 c$ and $d/\lambda = 5/16$.

First, we assume that the porous layer is fixed or so heavy that it will not vibrate under the influence of a sound wave incident from the left. Any pressure difference between the two sides of the layer forces an airstream through the pores with the flow velocity $v_s$. The latter is related to the pressures $p$ and $p'$ in front and behind the layer by

$$r_s = \frac{p - p'}{v_s} \tag{2.27}$$

where $r_s$ is the flow resistance of the porous layer. We assume that this relation is valid for a steady flow of air as well as for an alternating flow. The unit of flow resistance is 1 Pa · s · m$^{-1}$ = 1 kg · m$^{-2}$ · s$^{-1}$. Another commonly used unit is the Rayl (1 Rayl = 10 kg m$^{-2}$ s$^{-1}$).

Because of the conservation of matter, the particle velocities at both sides of the layer must be equal to each other and equal to the mean flow velocity through the layer:

$$v(0) = v'(0) = v_s \tag{2.28}$$

Then, it follows from Equation 2.27 that the wall impedance of the arrangement in Figure 2.6 is

$$Z = \frac{p}{v_s} = r_s + \frac{p'}{v_s} \tag{2.29}$$

The last term on the right-hand side of this equation represents the impedance $Z_a$ of the air cushion between the fabric and the rigid wall (see Equation 2.26, with $d' = d$). Thus, Equation 2.29 reads

$$Z = r_S - i\rho_0 c \cdot \cot(kd) \tag{2.30}$$

Hence, the wall impedance of the complete arrangement is just the flow resistance of the porous layer plus the impedance of the air space behind it.

In the complex plane of Figure 2.2, this wall impedance can be represented as a vertical line at a distance $r_s/\rho_0 c$ from the imaginary axis. Increasing the wave number or the frequency is equivalent to going repeatedly from $-i\infty$ to $+i\infty$ on that line. As can be seen from the circles of constant absorption coefficient, the latter has a maximum whenever $Z$ is real, that is, whenever the depth $d$ of the air space is an odd multiple of $\lambda/4$. Introducing $\zeta = Z/\rho_0 c$ from

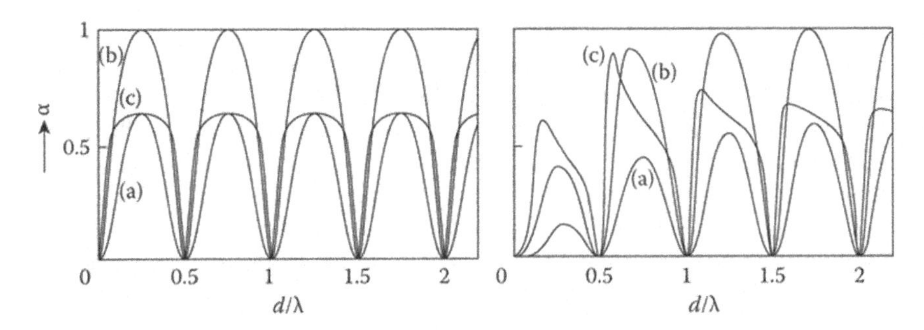


*Figure 2.7* Absorption coefficient of a porous layer backed with an air cushion of thickness $d$ (see Figure 2.6): (a) $r_s = 0.25\rho_0 c$, (b) $r_s = \rho_0 c$, and (c) $r_s = 4\rho_0 c$. A: $r_s d/M_p' c = 0$ (i.e. the layer is kept at rest); B: $r_s d/M_p' c = 4$.

Equation 2.30 into Equation 2.11 (with $\theta = 0$) yields the absorption coefficient of a porous layer in front of a rigid wall:

$$\alpha(f) = \frac{4r_s'}{(r_s'+1)^2 + \cot^2(2\pi fd/c)} \tag{2.31}$$

with $r_s' = r_s/\rho_0 c$. In the left part of Figure 2.7, the content of this formula is plotted as a function of the frequency parameter $d/\lambda = fd/c$ for three different values of the flow resistance. The curves show the expected periodicity.

In reality, a porous membrane will not remain at rest in a sound field, as was assumed before, but will vibrate as a whole since its mass is finite. Then, the total particle velocity in the plane $x = 0$ of Figure 2.6 consists of two components. One of them, $v_m$, is the velocity of the layer which is set into motion by the pressure difference between its faces:

$$p - p' = M_p' \frac{\partial v_m}{\partial t} = i\omega M_p' \cdot v_m \tag{2.32}$$

where $M_p'$ is the specific mass of the membrane, that is, its mass per unit area. The other contribution to the total velocity, named $v_s$, is due to the airflow forced through its pores according to Equation 2.27. Hence, the quantity $t_s$ in Equation 2.27 must be replaced by $Z_r$, which is defined as the ratio of the pressure difference between both sides of the layer, $p - p'$, and the total velocity $v_m + v_s$, which is

$$Z_r = \left(\frac{1}{r_s} + \frac{1}{i\omega M_p'}\right)^{-1} = \frac{r_s}{1 - i\omega_s/\omega} \tag{2.33}$$

with the characteristic frequency $\omega_s = r_s/M_p'$. Accordingly, $r_s$ in Equation 2.30 — but not in Equation 2.31 — must be replaced by $Z_r$. It is left for the readers to work out a modified formula corresponding to Equation 2.31. The right diagram in Figure 2.7 demonstrates the influence of the finite mass on the absorption coefficient of the arrangement. It is negligible for $\omega >> \omega_s$.

In practical applications, it may be advisable to provide for a varying distance between the porous fabric and the rigid wall in order to smooth out the irregularities of the absorption coefficient. This can be achieved by hanging or stretching the fabric in pleats, as is usually done with draperies.

In the next example, we consider an arrangement similar to that shown in Figure 2.6, however, with an additional thin layer of some impervious material. This layer has the specific mass $M'$ and is placed immediately at the left side of the porous one without touching it. For the sake of simplicity, we assume that the porous layer cannot vibrate as a whole, that is, $\omega M'_p >> r_s$ in Equation 2.33; thus, we have $Z_r \approx r_s$. Clearly, the velocity $v_m$ of the impervious membrane must be equal to the velocity $v_s$ of the air flowing through the porous layer, $v_m = v_s$. However, the pressure differences $\delta p$ generating these motions are different. For the porous layer, we have $(\delta p)_s = r_s v_s$ after Equation 2.27. The motion of the impervious layer is controlled by its mass according to Equation 2.32; hence, the pressure difference between its faces is $(\delta p)_m = i\omega M' v_s$. The total pressure difference, divided by the velocity $v_s$, is

$$\frac{(\delta p)_s + (\delta p)_m}{v_s} = r_s + i\omega M'$$

Finally, the impedance $Z_a = -i\rho_0 c \cot(kd)$ of the air cushion must be added to this expression, as in Equation 2.30. In many practical applications, the product $kd$ will be much smaller than unity; hence, we can use the approximation $\cot(kd) \approx 1/kd = c/\omega d$. Thus, the wall impedance of the whole arrangement is

$$Z = r_s + i\left(\omega M' - \frac{\rho_0 c^2}{\omega d}\right) \tag{2.34}$$

In the complex plane, this impedance is represented as a vertical line with the distance $r_s$ from the imaginary axis (see Figure 2.8a). But in contrast to the preceding example, the locus moves only once from $-i\infty$ to $+i\infty$ if the frequency is varied from zero to infinity. When it crosses the real axis, the absolute value of the wall impedance reaches its minimum. Since $Z = p/v_s$, a given sound pressure will then cause a particularly high velocity of the impervious sheet. This effect is known as resonance. According to Equation 2.34, this will occur at the angular frequency:

$$\omega_0 = \left(\frac{\rho_0 c^2}{M'd}\right)^{1/2} \tag{2.35}$$

And $f_0 = \omega_0/2\pi$ is the 'resonance frequency' of the system.

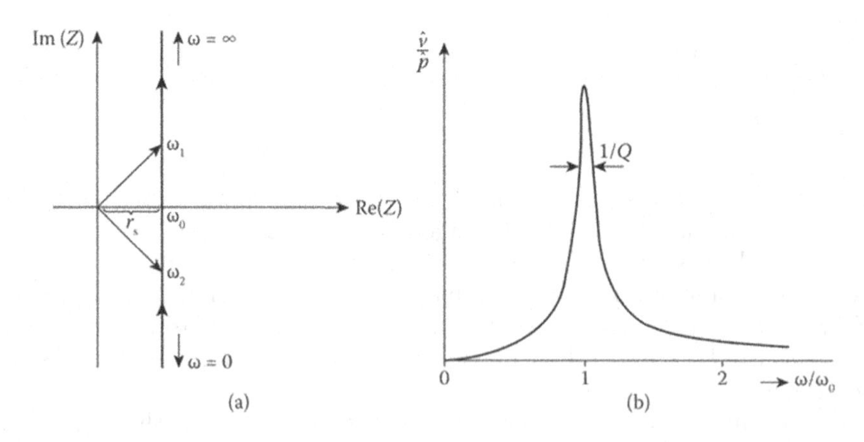


*Figure 2.8* Resonance system: (a) locus of the wall impedance in the complex plane and (b) magnitude of the ratio of velocity/pressure as a function of the driving frequency.

In Figure 2.8b, a typical resonance curve is depicted, that is, the ratio of the velocity amplitude and the pressure amplitude as a function of the sound frequency. This ratio is equal to the magnitude of the wall admittance, $1/|Z|$:

$$\left|\frac{v}{p}\right| = \frac{\omega}{M'\left[\left(\omega^2 - \omega_0^2\right)^2 + 4\delta^2\omega^2\right]^{1/2}} \tag{2.36}$$

where the 'damping constant'

$$\delta = \frac{r_s}{2M'} \tag{2.37}$$

has been introduced. The resonance maximum is $|p|/2M'\delta$.

Here we assume that the damping constant $\delta$ is small compared with $\omega_0$. Then,

$$\omega_1 = \omega_0 + \delta \text{ and } \omega_2 = \omega_0 - \delta$$

are the angular frequencies for which the phase angle of the wall impedance becomes ±45°. Furthermore, at these frequencies, the velocity amplitude has dropped by a factor $\sqrt{2}$ below its maximum. The difference $\Delta\omega = \omega_1 - \omega_2$ is the 'half power bandwidth' or, divided by the resonance angular frequency $\omega_0$, the 'relative half-width' of the resonance. The 'quality factor', or '$Q$'-factor, of the system is defined as the reciprocal of the latter:

$$\frac{\Delta\omega}{\omega_0} = \frac{1}{Q} = \frac{2\delta}{\omega_0} = \frac{r_s}{\omega_0 M'} \tag{2.38}$$

Each of these quantities may be used as a measure for the width of a resonance curve. Practical resonance absorbers as used in rooms, for instance, for the control of reverberation time, will be described in Section 6.3.

In our last example, we return to the absorptivity of a cloth or fabric with a thin layer of porous material. However, in contrast to the arrangement considered earlier, we now imagine that the material is not placed in front of a rigid wall but is hanging freely in a sound field. As before, we characterize the acoustical behaviour of the fabric by its complex flow resistance $Z_r$ according to Equation 2.33, which comprises the friction of air moving within the pores as well as the mass inertia of the material. The sound pressure and the normal component of the particle velocity of an incident plane wave are given by Equations 2.5 and 2.6, with $\theta$ denoting the angle of sound incidence (Figure 2.1). Similarly, the wave reflected from the blanket is represented by Equations 2.7 and 2.8. Additionally, a third plane wave will be transmitted into the space behind the fabric. Since the medium at both sides is the same, the transmitted wave will continue its travel in the same direction as the incident wave without being refracted, and also, the characteristic impedance remains unaltered. Hence, Equations 2.13 and 2.14 describing the transmitted wave must be modified:

$$p_t(x,y) = T\hat{p}_0 \exp\left[-ik(x\cos\theta + x\sin\theta)\right] \tag{2.39}$$

$$(v_t)_x = \frac{T\hat{p}_0}{\rho_0 c} \cdot \cos\theta \cdot \exp\left[-ik(x\cos\theta + x\sin\theta)\right] \tag{2.40}$$

At the plane $x = 0$, the sound pressures and particle velocities have to fulfil the following boundary condition:

$$(v_i)_x + (v_r)_x = (v_t)_x$$

That is, the normal component of the particle velocity is the same on both sides. And according to the definition of $Z_r$ as the pressure difference between both sides of the curtain divided by the normal component of the flow velocity, we have

$$\frac{p_i + p_r - p_t}{(v_t)_x} = Z_r$$

From both conditions, the unknown quantities $R$ and $T$ are obtained by inserting Equations 2.5 to 2.8 and Equations 2.39 and 2.40 (with $x = y = 0$) into these conditions, with the result:

$$R(\theta) = \frac{\zeta_r \cos\theta}{2 + \zeta_r \cos\theta} \tag{2.41}$$

$$T(\theta) = \frac{2}{2 + \zeta_r \cos\theta} \tag{2.42}$$

With $\zeta_r = Z_r/\rho_0 c$.

In contrast to the situation considered in Section 6.3, the curtain is not a boundary of the room. Therefore, the sound energy transmitted through it is not converted into heat but remains still in the sound field. Hence, the energy dissipated within the porous material is obtained by subtracting both the reflected and the transmitted sound energy from the incident one. Accordingly, we replace Equation 2.1 by a slightly different definition of the absorption coefficient:

$$\alpha = 1 - |R|^2 - |T|^2 \tag{2.43}$$

Inserting Equations 2.41 and 2.42 yields

$$\alpha(\theta) = \frac{4\mathrm{Re}(\zeta_r)\cos\theta}{|\zeta_r \cos\theta + 2|^2} = \frac{\mathrm{Re}(\zeta_r)\cos\theta}{(|\zeta_r|/2)^2 \cos^2\theta + \mathrm{Re}(\zeta_r)\cos\theta + 1} \tag{2.44}$$

Of particular interest is the absorption of a curtain hanging freely in a closed room with a diffuse sound field so that it is exposed to sound waves arriving from all directions. This question is discussed in Section 2.5.

## 2.4 REFLECTION OF SPHERICAL WAVES

Up to now we dealt exclusively with the reflection of plane waves, although this wave type is highly idealized, as has been discussed in Section 1.2. A much more realistic wave type is the spherical wave as usually produced by a point source. At large distances from the source, the curvature of its wave fronts will become so small that it can be treated as a plane wave without too much error. According to Equation 1.24, this will be the case when $kr >> 1$ ($r$ = distance, $k = \omega/c$). In general, the reflection of a spherical wave from a plane wave of arbitrary impedance is much more complicated than that of a plane wave. The only exception is reflection from a rigid plane; in this case, the reflected wave is also spherical and seems to originate from the mirror image of the original source with respect to the plane.

To illustrate the situation, we consider in Figure 2.9 a point source $S$ which emits a harmonic spherical wave. We are interested in the amplitude and the phase of the sound signal received in a point $R$, which is located at some distance from it. Both points are situated over

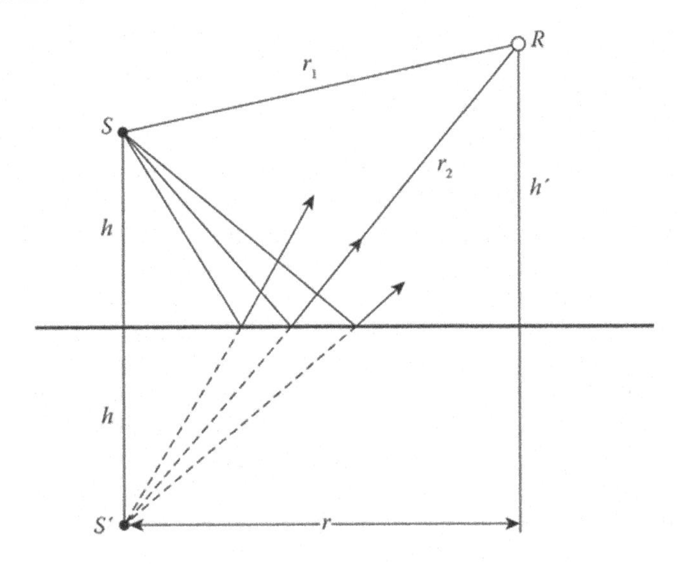


*Figure 2.9* Reflection of a spherical wave from a plane surface. $S'$ is the mirror image of S with respect to the plane, and $r$ is the horizontal distance of S and $S'$.

a horizontal plane boundary which is assumed to be of the locally reacting type, that is, its impedance does not depend on the direction of the sound incidence. Their heights above that boundary are $h$ and $h'$, respectively. Moreover, Figure 2.9 shows the reflection of a few sound rays. (The 'sound ray' is a useful notion employed in geometric acoustics which will be dealt with in greater detail in Chapter 4.) Exactly one of these rays will reach the receiver $R$. It is evident that all rays emerging from $S$ hit the reflecting surface under different angles in contrast to those of a plane wave which would be parallel, making everywhere the same angle with the boundary plane.

According to Figure 2.9, all reflected rays appear to originate from one point $S'$, which is symmetrical to the real sound source $S$ with respect to the boundary. $S$' is usually referred to as the image source or virtual source of $S$ with respect to the reflecting plane. Sources of this kind play a major role in room acoustics (see Sections 4.1 and 9.8). In fact, they facilitate the construction of the ray path connecting a real sound source with a given receiving point via one reflection. Figure 2.9 shows such a path. Its length is denoted by $r_2$, while $r_1$ denotes the length of the direct, that is, the unreflected component. The total sound pressure in $R$ is composed of the contributions $p_0$ and $p_r$ of the incident and the reflected ray, respectively:

$$p = p_0 + p_r$$

According to Equation 1.27, the direct component is given by

$$p_0 = \frac{i\omega\rho_0}{4\pi}\hat{Q} \cdot \frac{\exp(-ikr_1)}{r_1}$$

(As usual, we have omitted the time factor $\exp(i\omega t)$.) The most obvious way to calculate the sound pressure of the reflected component by the 'geometric approximation' is

$$p_r \approx (p_r)_{geo} = \frac{i\omega\rho_0}{4\pi}\hat{Q} \cdot \frac{\exp(-ikr_2)}{r_2} \cdot R(\theta_0) \tag{2.45}$$

where $R(\theta_0)$ is the reflection factor of the boundary for the incidence angle $\theta_0$. Equation 2.45 is not the exact solution of the problem, because the reflection factor $R(\theta)$ is defined for plane waves only but not for spherical waves. Nevertheless, it is a useful and simple approximation.

Without performing any calculations, we can draw a few useful conclusions from Equation 2.10, that is, the relationship between the impedance of the reflecting surface and its reflection factor. If the wall impedance of the reflecting surface is relatively high, the angle dependence of its reflection factor is not very pronounced; in the limit of $|Z| \to \infty$, it can be neglected. Furthermore, Equation 2.10 tells us that the reflection factor $R$ does not show much variation in the vicinity of $\theta_0 = 0$ (see also Figure 2.3); hence, it will be nearly constant if both points $S$ and $R$ are situated one above the other. The situation is different if both points are close to the reflecting surface. Accordingly, we expect that at grazing incidence, the geometric approximation fails because of the strong angle dependence of the reflection factor.

An exact expression of the sound pressure $p_r$ of the reflected wave can be obtained by employing the 'Sommerfeld identity' (Sommerfeld 1964), which represents a spherical wave $\exp(-ikr)/r$ as the superposition of infinitely many cylindrical waves with different wavelengths, each of them being described by a zero-order Bessel function $J_0$. On the basis of this identity, the following expression can be derived:

$$p_r = -\frac{\omega^2 \rho_0}{4\pi c}\hat{Q} \cdot \int_C R(\theta) J_0(kr \cdot \sin\theta) \cdot \exp\left[ik(h+h')\cos\theta\right]\sin\theta d\theta \tag{2.46}$$

It should be noted that the angle $\theta$ in this expression may be complex. This is accounted for by choosing an integration path $C$ running first from 0 to $\pi/2$ along the real $\theta$-axis, and then from this point parallel to the imaginary axis to $\pi/2 + i\infty$.

A comprehensive discussion of this integral and its numerical solution has been presented by F. P. Mechel in his book on sound absorbers (Mechel 1989). At present, we shall not dive deeper into this complicated matter. Instead, we report on a more recent publication by Suh and Nelson (1999), who calculated the correct pressure $p_r$ of the reflected wave by numerical integration of Equation 2.46 and compared this with the pressure $(p_r)_{geo}$ according to Equation 2.45, again assuming local reaction of the reflecting boundary. For this purpose, these authors calculated the relative difference between both the results:

$$\Delta = \frac{\left|p_r - (p_r)_{geo}\right|}{|p_r|} \cdot 100\% \tag{2.47}$$

Generally, this quantity shows a monotonic increase with the angle of incidence $\theta_0$. As a 'figure of merit', we can consider the angle $\theta_{max}$, which is the upper limit of the range, in which the relative difference according to Equation 2.47 remains below 1%. This limit is shown in the last column of Table 2.1 for three different surfaces, the characteristic impedances of which are listed in the second column. The third column contains the distance $r$ measured in wavelengths. Generally, the accuracy of the approximation (2.45) seems to be sufficient for practical purposes as long as the angle of incidence is below 45°. However, its use may become problematic for larger incidence angles, in particular, for grazing sound incidence.

## 2.5 RANDOM SOUND INCIDENCE

In a closed room, the typical sound field does not consist of a single plane wave but is composed of many such waves, each with its own particular amplitude, phase, and direction. To find the effect of a wall on such a complicated sound field, we ought, of course, to consider

*Table 2.1* Maximum angles of incidence $\theta_{max}$ for $\Delta$ < 1%

| *Material* | *Specific impedance,* $\zeta$ | $r(\lambda)$ | $\theta_{max}$ |
|---|---|---|---|
| 1 | 5.00 + 11.00i | 2 | 30° |
| | | 5 | 45° |
| | | 10 | 53° |
| 2 | 1.00 − 2.83i | 2 | 11° |
| | | 5 | 31° |
| | | 10 | 37° |
| 3 | 0.59 + 0.57i | 2 | 5° |
| | | 5 | 12° |
| | | 10 | 26° |

Source: Adapted from Suh and Nelson (1999).

the reflection of each wave separately and then to add all sound pressures, taking regard of their phases.

With certain assumptions, we can resort to some simplifications that allow general statements on the effect of a reflecting wall. If there are numerous waves incident on the wall, the phases of which are randomly distributed, one can neglect all phase relations and the interference effects caused by them. Then, the components are called incoherent. In this case, the total energy at some point can be calculated just by adding the energies of the components, which are proportional to the squares of the sound pressures:

$$p_{\text{rms}}^2 = \sum_n (p_{\text{rms}})^2 \quad \text{or} \quad I = \sum_n I_n \tag{2.48}$$

Which means, the total intensity is the sum of all component intensities.

Here we make the additional assumption that the intensities of the incident sound waves are uniformly distributed over all possible directions; hence, each solid angle element carries the same energy per second. In this case, we speak of 'random sound incidence', and the sound field associated with it is said to be isotropic or 'diffuse'. In room acoustics, the diffuse sound field plays the role of a standard field, and the reader will frequently encounter it in this book.

For a quantitative treatment, it is convenient to use a spherical polar coordinate system as depicted in Figure 2.10. Its origin is the centre of a wall element $dS$; the wall normal is its polar axis. We consider an element of solid angle $d\Omega$ around a direction which is determined by the polar angle $\theta$ and the azimuth angle $\phi$. Expressed in these angular coordinates, the solid angle element is $d\Omega = \sin\theta \, d\theta \, d\phi$.

First, we calculate how the square of the sound pressure amplitude depends on the distance from the wall, which, for the moment, is assumed to be perfectly rigid ($R = 1$). A wave hitting this wall under an angle $\theta$ gives rise to a standing wave with the squared pressure amplitude, according to Equation 2.17:

$$|p|^2 = 2\hat{p}_0^2\left[1 + \cos(2kx\cos\theta)\right] \tag{2.49}$$

By averaging this expression over all directions of incidence, that is, over the solid angle $2\pi$, we obtain

$$|p|_{av}^2 = \frac{1}{2\pi}\iint_{2\pi} |p|^2 d\Omega = 2\hat{p}_0^2 \frac{1}{2\pi}\int_0^{2\pi} d\phi \int_0^{\pi/2} \left[1 + \cos(2kx\cos\theta)\right]\sin\theta \, d\theta$$

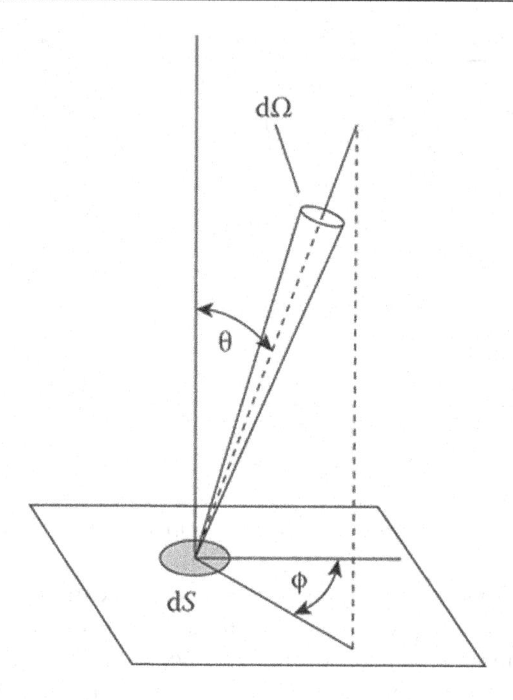


*Figure 2.10* Spherical polar coordinates.

or

$$|p|_{av}^2 = 2\hat{p}_0^2\left[1 + \frac{\sin 2kx}{2kx}\right] \tag{2.50}$$

This quantity, divided by $|p_\infty|^2 = 2\hat{p}_0^2$, is plotted in Figure 2.11 (solid curve) as a function of the distance $x$ from the wall. Next to the wall, the square pressure fluctuates, as it does in every standing wave. With increasing distance, however, these fluctuations fade out and the square pressure approaches a constant limiting value which is half of that immediately on the wall. Accordingly, the sound pressure level close to the wall would surpass that measured far away by 3 dB. For the same reason, the sound absorption of an absorbent surface adjacent and perpendicular to a rigid wall is higher near the edge than at a distance of several wavelengths from the wall.

When the sinusoidal excitation signal is replaced with random noise of limited bandwidth, the pressure distribution is obtained by applying a second averaging process to Equation 2.50, namely, over the frequency band. As an example, the dashed curve of Figure 2.11 plots the result of averaging $|p|_{av}^2$ over an octave band, that is, a frequency band with $f_2 = 2f_1$, where $f_1$ and $f_2$ denote the lower and the upper limiting frequencies of the band. Here, the typical wavelength $\lambda$ corresponds to the frequency $\sqrt{f_1 f_2} = f_1\sqrt{2}$. Now, the standing wave has levelled out for all distances exceeding $x > 0.5\lambda$, but there is still a pronounced increase in sound pressure if the wall is approached. This increase of $|p|_{av}^2$ to twice its far distance value is obviously caused by the fact that the reflecting surface enforces certain phase relations between all impinging and reflected waves. In any case, we can conclude that in a diffuse sound field, phase effects are limited to a relatively small range next to the walls, which is of the order of half a wavelength.

Next, we derive an important rule according to which absorption coefficients are averaged over all directions, assuming random sound incidence as before. If we consider a wall element

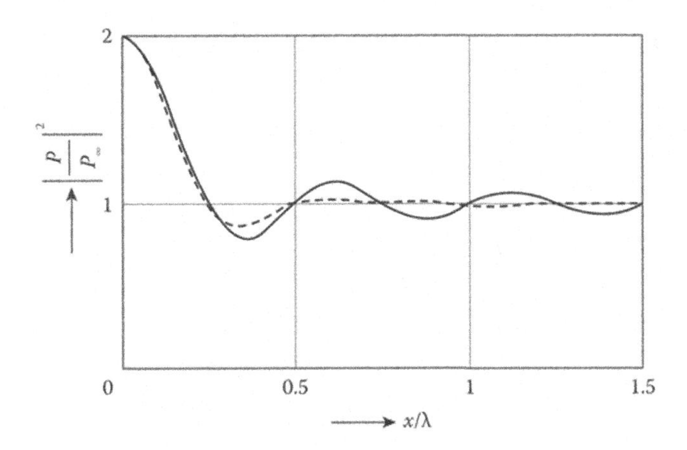


*Figure 2.11* Squared and normalized sound pressure amplitude in front of a rigid wall at random sound incidence: solid line represents sine tone, and dashed line represents random noise of octave bandwidth.

with area d$S$. Its projection in the direction $\phi$, $\theta$ is d$S$ cos $\theta$ (see Figure 2.10). If $I$ denotes the intensity of the sound arriving from that direction, $I \cdot \cos\theta \cdot \mathrm{d}S\,\mathrm{d}\Omega$ is the sound energy falling per second on d$S$ from the solid angle d$\Omega$. By integrating this over all solid angle elements, assuming $I$ independent of $\phi$ and $\theta$ (diffuse sound field), we obtain the total energy influx per second at d$S$:

$$E_\mathrm{i} = I\mathrm{d}S\int_0^{2\pi}\mathrm{d}\phi\int_0^{\pi/2}\cos\theta\sin\theta\mathrm{d}\theta = \pi I\mathrm{d}S \tag{2.51}$$

From the energy $I\cos\theta\,\mathrm{d}S\,\mathrm{d}\Omega$, the fraction $\alpha(\theta)$ is absorbed; thus, the totally absorbed energy per second is

$$E_\mathrm{a} = I\mathrm{d}S\int_0^{2\pi}\mathrm{d}\phi\int_0^{\pi/2}\alpha(\theta)\cos\theta\sin\theta\mathrm{d}\theta = 2\pi I\mathrm{d}S\int_0^{\pi/2}\alpha(\theta)\cos\theta\sin\theta\mathrm{d}\theta \tag{2.52}$$

By dividing both expressions, we get the absorption coefficient for random or uniformly distributed incidence:

$$\alpha_\mathrm{uni} = \frac{E_\mathrm{a}}{E_\mathrm{i}} = 2\int_0^{\pi/2}\alpha(\theta)\cos\theta\sin\theta\mathrm{d}\theta = \int_0^{\pi/2}\alpha(\theta)\sin(2\theta)\mathrm{d}\theta \tag{2.53}$$

This expression is often referred to as the 'Paris formula' in the literature.

We apply this integral to Equation 2.11, which expresses the angle dependence of a locally reacting surface with the specific wall impedance $\zeta = \xi + i\eta$. The integration can readily be carried out with the result:

$$\alpha_\mathrm{uni} = \frac{8\xi}{|\zeta|^2}\cdot\left[1+\frac{\xi^2-\eta^2}{\eta|\zeta|^2}\cdot\arctan\left(\frac{\eta}{1+\xi}\right)-\frac{\xi}{|\zeta|^2}\cdot\ln\left(1+2\xi+|\zeta|^2\right)\right] \tag{2.54}$$

The content of this expression is depicted in Figure 2.12 in the form of curves of constant absorption coefficient $\alpha_\mathrm{uni}$ in a coordinate system, the abscissa and the ordinate of which are the phase angle $\mu = \arctan(\eta/\xi)$ and the absolute value $|\zeta|$ of the specific impedance, respectively. The absorption coefficient has its absolute maximum 0.951 for the real impedance $\zeta = 1.567$. Thus, in a diffuse sound field, a locally reacting boundary can never be totally absorbent.

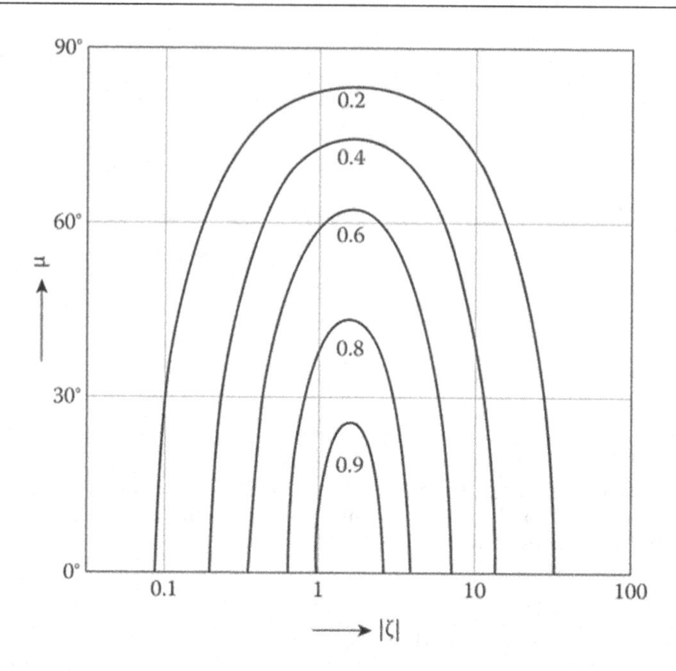


*Figure 2.12* Contours of constant absorption coefficient $\alpha_{uni}$ of locally reacting surfaces at random sound incidence. The abscissa is the magnitude, and the ordinate is the phase angle μ of the specific impedance ζ.

In the last example of Section 2.3, we discussed the absorption of a freely hanging curtain and derived the following expression for its absorption coefficient:

$$\alpha = \frac{\xi_r \cos\theta}{\left(|\zeta_r|/2\right)^2 \cos^2\theta + \xi_r \cos\theta + 1} \tag{2.55}$$

Here, θ is the angle under which an incident plane wave arrives at the porous sheet, the acoustic properties of which are given by its complex flow resistance after Equation 2.33:

$$Z_r = \frac{r_s}{1 - i\omega_s/\omega} = \rho_0 c \zeta_r$$

$\xi_r$ and $\eta_r$ are the real and the imaginary parts of the specific impedance $\zeta_r$. Obviously, Equation 2.55 is very similar to Equation 2.11, which describes the angle dependence of a locally reacting surface. Hence, we can calculate the Paris average of the absorption coefficient (2.55) by using Equation 2.54 after a few replacements, $\xi$ with $\xi_r/2$, $\zeta$ with $\zeta_r/2$, and $\eta$ with $\eta/2$.

In Figure 2.13a, $\alpha_{uni}$ of the curtain is plotted as a function of the frequency ratio $f/f_s = \omega/\omega_s$ for various values of the flow resistance $r_s$. Far below the characteristic frequency $f_s$, the absorption coefficient is very small, as at these frequencies the fabric nearly completely follows the vibrations imposed by the sound field. With increasing frequency, the inertia of the curtain becomes more and more relevant, leading to an increasing motion of the air inside the pores of the fabric. For high frequencies, the porous layer stays practically at rest, and the absorption coefficient becomes frequency-independent. This limiting value is plotted in Figure 2.13b as a function of the flow resistance $r_s$. It has a maximum $\alpha_{uni} = 0.951$ at $r_s = 3.135\rho_0 c$. This discussion shows that freely hanging curtains, large flags, and so on may considerably add to the absorption in a room and may well be used to control its reverberation.

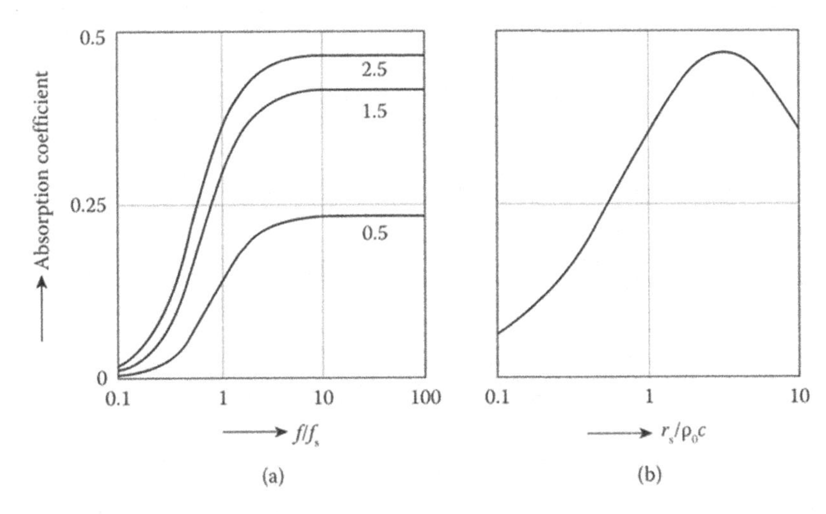


*Figure 2.13* Absorption coefficient of a freely hanging curtain, random sound incidence: (a) as a function of frequency, parameter is $r_s/\rho_0 c$, and (b) high-frequency limit of the absorption coefficient as a function of $r_s/\rho_0 c$.

## 2.6 REFLECTION FROM FINITE-SIZED PLANE SURFACES

So far, we have considered sound reflection from plane walls of infinite extension. If a reflecting wall has finite dimensions with a free boundary, the latter will become the origin of an additional sound wave when it is irradiated with sound. This additional wave is brought about by diffraction and, hence, may be referred to as a 'diffraction wave'. It spreads more or less in all directions.

The simplest example is diffraction by a semi-infinite wall, that is, a rigid plane with one straight edge, as depicted in Figure 2.14. If this wall is exposed to a plane sound wave at normal incidence, one might expect that it reflects some sound into a region A, while another region B, the 'shadow zone', remains completely free of sound. This would, indeed, be true if the acoustical wavelength were vanishingly small. In reality, however, a diffraction wave originating from the edge of the wall modifies this picture. Behind the wall, that is, in region B, there is still some sound intruding into the shadow zone. And in region C, the plane wave is disturbed by interferences with the diffraction wave. On the whole, there is a steady transition from the undisturbed, that is, the primary sound wave, to the field in the shadow zone. This is shown in Figure 2.14, where the squared sound pressure in a plane parallel to the diffracting half-plane is depicted. Of course, the extension of this transition depends on the angular wave number $k$ and the distance $d$. A similar effect occurs at the upper boundary of region A with the reflected wave.

If the 'wall' is a reflector of limited extension, for example, a freely suspended panel, the line source from which the diffraction wave originates is wound around the edge of the reflector, so to speak. As an example, Figure 2.15a shows a rigid circular disc with radius $a$, irradiated from a point source S. We consider the sound pressure at point $P$. Both $P$ and $S$ are situated on the middle axis of the disc at distances $R_1$ and $R_2$ from its centre, respectively. In Figure 2.15b, the squared sound pressure of the reflected wave in $P$ is plotted as a function of the disc radius as calculated from the approximation

$$|p|^2 = p_{\max}^2 \sin^2\left(\frac{ka^2}{2\breve{R}}\right) \text{with} \frac{1}{\breve{R}} = \frac{1}{2}\left(\frac{1}{R_1} + \frac{1}{R_2}\right) \tag{2.56}$$

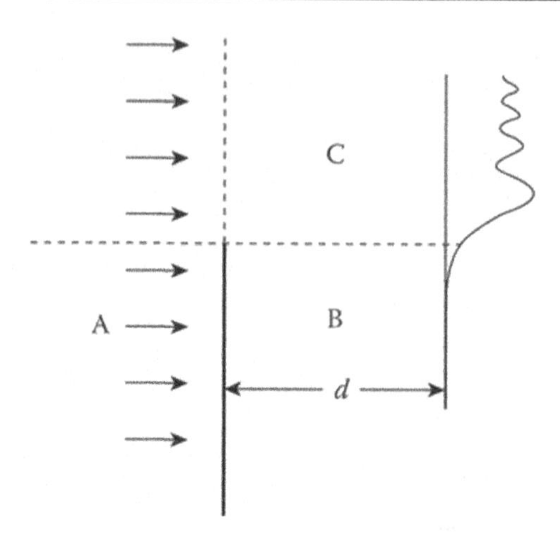


*Figure 2.14* Diffraction of a plane wave from a rigid half-plane. The diagram shows the squared sound pressure amplitude across the boundary B–C *(kd* = 100).

Source: Morse and Ingard (1968).

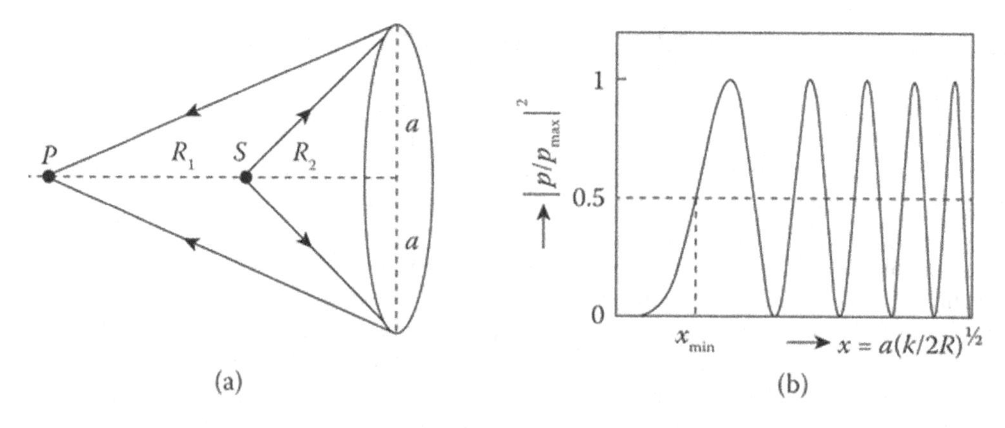


*Figure 2.15* Sound reflection from a circular disc: (a) arrangement ($S$ = point source, $P$ = observation point) and (b) squared sound pressure amplitude of reflected wave.

which is valid for distances that are large compared to the disc radius $a$. For very small discs — or for very low frequencies — the reflected sound is negligibly weak, since the primary sound wave is nearly completely diffracted around the disc, and the obstacle is virtually not present. With increasing disc radius or frequency, the pressure in $P$ grows rapidly; for higher values of the dimensionless frequency parameter $x = a\left(k/2\breve{R}\right)^{1/2}$, it shows strong fluctuations. The latter are caused by interferences between the sound reflected specularly from the disc and the diffraction wave originating from its rim.

A simple way to explain these fluctuations is by drawing a set of concentric circles on the disc which have the same centre as the disc itself. The radii $\rho_n$ of these circles are chosen in such a way that the length of the path connecting $S$ with $P$ (see Figure 2.15a) via a point on the circle with radius $\rho_n$ point exceeds $R_1 + R_2$ by an integral multiple of half the wavelength. Expressed in formulae

$$\sqrt{R_1^2 + \rho_n^2} + \sqrt{R_2^2 + \rho_n^2} - R_1 - R_2 = n\frac{\lambda}{2}$$

or, since for $\rho_n << R_{1,2}$,

$$\sqrt{R_{1,2}^2 + \rho_n^2} \approx R_{1,2} + \frac{\rho_n^2}{2R_{1,2}}, \text{ and so on.}$$

and so on.

$$\rho_n^2 \approx n\lambda\left(\frac{1}{R_1} + \frac{1}{R_2}\right)^{-1} = \frac{n\lambda\breve{R}}{2} \tag{2.57}$$

Each of these circles separates two zones on the plane called 'Fresnel zones' (Figure 2.16), which contribute to opposite signs to the sound pressure in $P$. As long as the disc radius $a$ is smaller than the radius of the first zone, that is, $a \leq \sqrt{\lambda\breve{R}/2}$, the contributions of all disc points have the same sign. With increasing disc radius (or decreasing wavelength), additional Fresnel zones will enter the disc from its rim, each of them lessening the effect of the preceding one.

We consider the reflection from the disc as significant if $|p|^2$ for the first time equals its average value, which is half its maximum value. This is the case if the argument of the sine in Equation 2.56 is $\pi/4$. This condition defines a minimum frequency $f_{min}$, above which the disc can be considered as an efficient reflector:

$$f_{min} = \frac{c\breve{R}}{4a^2} \approx 85\frac{\breve{R}}{a^2}\text{Hz} \tag{2.58}$$

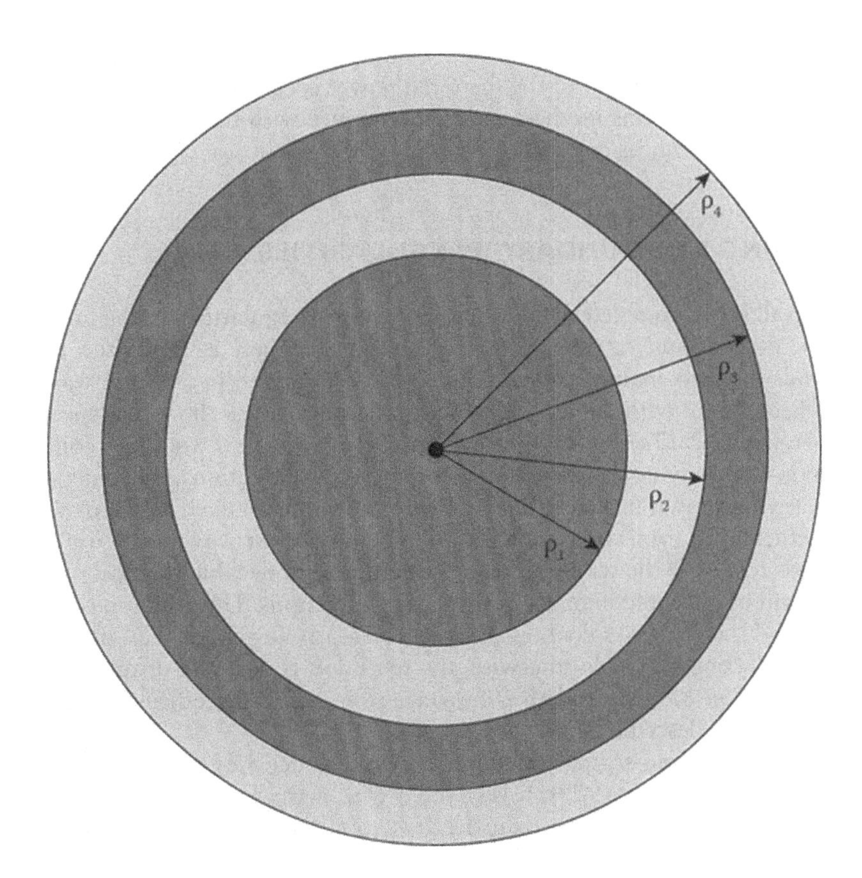


*Figure 2.16* The first four Fresnel zones on a disc with the radius $a = \rho_4$ (after Equation 2.57).

(In the second version, $\breve{R}$ and $a$ are expressed in metres.) A circular panel with a diameter of 1 m, for instance, viewed from a distance of 5 m ($R_1 = R_2 = 5$ m), reflects an incident sound wave at frequencies above 1,700 Hz. For lower frequencies, its effect is much smaller.

Similar considerations applied to a rigid strip with the width $h$ yield for the minimum frequency of geometrical reflections

$$f_{\min} = 185\frac{\breve{R}}{h^2}\,\text{Hz} \tag{2.59}$$

($\breve{R}$ and $h$ in metres) with the same meaning of $\breve{R}$ as in Equation 2.56. If the reflector is tilted by an angle $\theta$ against the primary sound wave, $a$ in Equation 2.56 and $h$ in Equation 2.59 must be multiplied with a factor $\cos\theta$.

Generally, any obstacle or surface of limited extension distorts a primary sound wave by diffraction unless its dimensions are very small compared to the wavelength. One part of the diffracted sound is scattered more or less in all directions. For this reason, this process is also referred to as 'sound scattering'. (The role of sound scattering by the human head in hearing has already been mentioned in Section 1.6.)

The scattering efficiency of a body is often characterized by its 'scattering cross section', defined as the ratio of the total power scattered $P_s$ and the intensity $I_0$ of the incident wave:

$$Q_s = \frac{P_s}{I_0} \tag{2.60}$$

If the dimensions of the scattering body are small compared to the wavelength, $P_s$, and hence $Q_s$, is very small. In the opposite case of short wavelengths, the scattering cross section of the obstacle approaches twice its visual cross section, that is, $2\pi a^2$ for a sphere or a circular disc with radius $a$. Then, one-half of the scattered power is concentrated into a narrow beam behind the obstacle and forms its shadow by interference with the primary wave, while the other half is deflected from its original direction.

## 2.7 SCATTERING BY BOUNDARY IRREGULARITIES

Very often, a wall is not completely smooth but contains irregularities in the form of coffers, bumps, plastic decorations, or other projections. The way these irregularities influence the reflected sounds depends mainly on their dimensions, measured in acoustic wavelengths. If they are small compared with the wavelength, they do not disturb the wall's 'specular' reflection at all (see Figure 2.17a). In the opposite case, that is, if they are large compared with the wavelength, they may be treated as plane or curved wall sections, reflecting the incident sound specularly as shown in Figure 2.17c. There is an intermediate range of wavelengths, however, in which each projection adds a scattered component wave to the reflected sound field (see Figure 2.17b). If the wall structure is irregular, a noticeable fraction of the incident sound energy will be scattered in many non-specular directions. This fraction is characterized by the product $s(1 - \alpha)$, where $s$ denotes the so-called scattering coefficient. If $s = 1$, we speak of a 'diffusely reflecting surface'; otherwise, the reflection is partially diffuse — or specular (for $s = 0$). In Section 8.8, methods for measuring the scattering efficiency of acoustically rough surfaces will be described.

As an example of a sound-scattering boundary, we consider the ceiling of a particular concert hall (Meyer and Kuttruff 1959). It is covered with many bodies made of gypsum, such as pyramids and spherical segments; their depth is about 30 cm on the average. Figure 2.18 shows the directional distribution of the sound reflected from that ceiling, measured at a frequency of 1,000 Hz, with normally incident sound waves; the plotted quantity is the sound

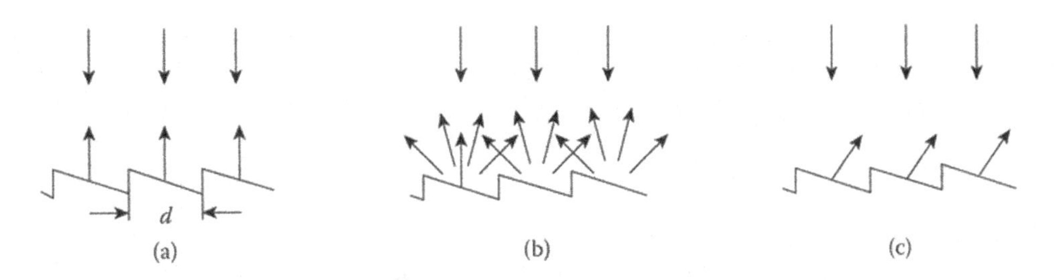


*Figure 2.17* Scattering by wall irregularities: (a) $d << \lambda$, (b) $d \approx \lambda$, and (c) $d >> \lambda$ ($d$ = typical dimension of irregularities).

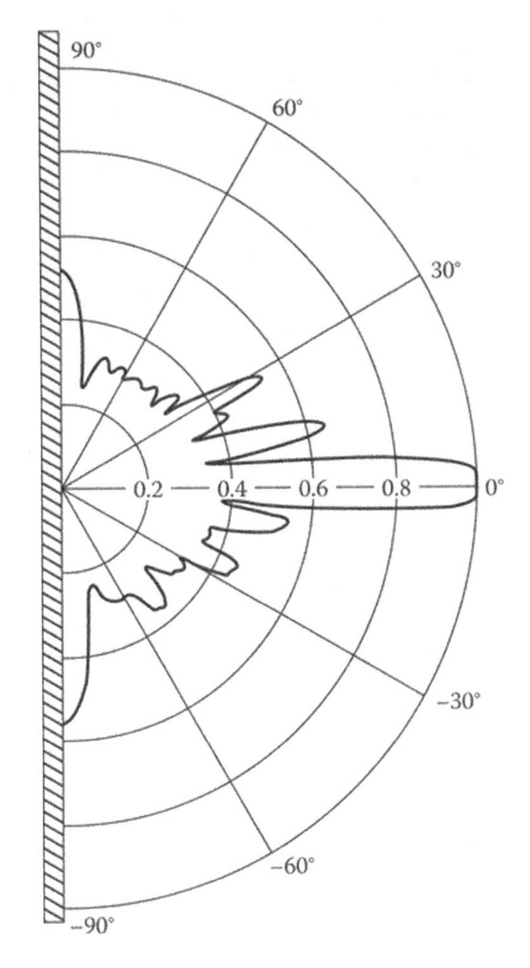


*Figure 2.18* Polar diagram of a highly irregular ceiling. The plotted quantity is the sound pressure amplitude.

Source: Meyer and Kuttruff (1959).

pressure amplitude. (This measurement has been carried out on a scale model of the ceiling.) The pronounced maximum at 0° corresponds to the specular component, which is still of considerable strength.

The occurrence of diffuse or partially diffuse reflections is not restricted to walls with a geometrically structured surface; they may also be produced by walls with non-uniform impedance. To understand this, we return to Figure 2.14 and imagine that the dotted vertical line marks a totally absorbing wall. This would not change the structure of the sound field at the left side of the wall, that is, the disturbance caused by the diffraction wave in region A. Therefore, we can conclude that any change in wall impedance creates a diffraction wave.

A practical example of this kind are walls lined with relatively thin panels which are mounted on a rigid framework. At the points where the lining is fixed, the panel is very stiff and cannot react to the incident sound field. Between these points, however, the lining is more compliant because it can perform bending vibrations, particularly if the frequency of the exciting sound is close to the resonance frequency of the lining (see Equation 2.35). Scattering will be even stronger if adjacent partitions are tuned to different resonance frequencies by variations in the panel masses or the depths of the air space behind them.

Now, we consider a plane wall subdivided into parallel strips with equal width $d$ and with different reflection factors $R_n = |R_n| \exp(i\chi_n)$, as shown in Figure 2.19. We assume that $d$ is noticeably smaller than the wavelength. A plane wave hitting the wall at normal incidence will excite all strips with about equal amplitude and phase, and each of them will react to it by emitting a secondary wave or wavelet. The sound pressure far from the wall is obtained by summation of all these contributions:

$$p(\vartheta) \propto \sum_n |R_n| \exp\left[i(\chi_n - nkd\sin\vartheta)\right] \tag{2.61}$$

By varying $|R_n|$ and $\chi_n$ in a suitable way, the specular reflection can more or less be destroyed, and its energy will be scattered into non-specular directions instead.

Next, our goal is to optimize the diffusion of the reflected sound with simple means. For this purpose, it is assumed that the magnitude of all reflection factors in Equation 2.61 is 1. In principle, randomness of the angular distribution of the reflected sound could be affected by arranging for phase angles $\chi_n$ distributed randomly within the interval from 0 to $2\pi$. To achieve complete randomness of the scattered sound in this way, however, would require a very large number of elements.

A similar effect can be reached with 'pseudorandom sequences' of phase angles. If these sequences are periodic, arrangements of this kind act as phase gratings, with the grating constant $Nd$ if $N$ denotes the number of elements within one period. As with optical gratings, constructive interference of the wavelets reflected from corresponding elements will occur if the condition

$$\sin\vartheta_m = m\frac{2\pi}{Nkd} = m\frac{\lambda}{Nd}\left(m < \frac{Nd}{\lambda}\right) \tag{2.62}$$

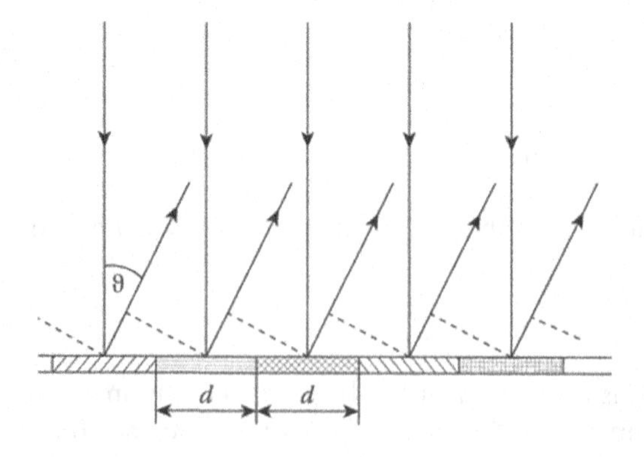


*Figure 2.19* Sound reflection from strips with equal widths but with different reflection factors, arranged in a plane.

is fulfilled. (The limitation of $m$ accounts for the fact that the value of the sine function cannot exceed unity.) Inserting $\sin\vartheta_m$ from Equation 2.62 into Equation 2.61 yields the far-field sound pressure in the $m$th diffraction order:

$$p_m = \sum_{n=0}^{N-1} \exp(i\chi_n)\exp\left(-\frac{2\pi imn}{N}\right) \tag{2.63}$$

$p_m$ is the discrete Fourier transform of the sequence $\exp(i\chi_n)$, as may be seen by comparing this expression with Equation 1.46. Hence, a uniform distribution of the reflected energy over all diffraction orders can be achieved by finding phase shifts $\chi_n$, for which the power spectrum of $\exp(i\chi_n)$ is flat.

It was M. R. Schroeder's idea to select the phase angles according to certain periodic number sequences which are known to have the required spectral properties, and to realize them in the form of properly corrugated surfaces (Schroeder 1979). They are believed to improve the acoustics of concert halls and recording studios by creating 'lateral sound waves', which are known to be relevant for good music acoustics (see Section 9.3).

Probably, the best-known kind of 'pseudostochastic diffusers', as they are also called, is based on a number theoretical scheme named 'quadratic residues'. Suppose that the period $N$ is a prime number. Then, one sequence with the desired properties is the so-called Gauss sequence $\exp(-i2\pi n^2/N)$. Since phases are insensitive for added multiples of $2\pi$, $n^2$ can be reduced to $n^2 \bmod N$:

$$\chi_n = \frac{2\pi}{N}\left(n^2 \mathrm{mod} N\right)(n = 0,1,\ldots,N-1) \tag{2.64}$$

(The modulo operation $A \bmod N$ yields the remainder after division of a number $A$ by an integer $N$.) Other useful sequences exploit the properties of primitive roots of prime numbers or the index function. Diffusers of the latter kind suppress the specular reflection completely. (Schroeder 1985)

The phase shifts $\chi_n$ according to Equation 2.64 are generated by a periodic series of equally spaced wells which have different depths $h_n$ and are separated by thin and rigid partitions. A sound wave hitting such an arrangement will excite secondary waves in the wells. Each of these waves travels towards the rigid bottom of the well. When the reflected wave reappears at the opening, it will have gained a phase shift $\chi_n = 2kh_n = 4\pi(h_n/\lambda)$; hence, the required depths are

$$h_n = \frac{\chi_n}{2k} = \frac{\lambda_d}{4\pi}\chi_n \tag{2.65}$$

$\lambda_d$ is a free design parameter of the diffuser, the 'design wavelength'. The diffuser works optimally for the 'design frequency' $f_d = c/\lambda_d$ and integral multiples of it. On the other hand, there are critical frequencies at which no scattering takes place at all. This occurs when all depths are integral multiples of the acoustical wavelength.

A second design parameter is the width $d$ of the wells. If it is too small, the number of allowed diffraction orders after Equation 2.62 may become very small or — at low frequencies — even zero, that is, there will be no or no effective scattering of the reflected sound. Furthermore, narrow wells show increased losses due to the viscous and thermal boundary layer on their walls. If, on the other hand, the wells are too wide, not only the fundamental wave mode but also higher-order wave modes may be excited inside the wells, resulting in a more involved sound field. Practical diffusers have well widths of a few centimetres. Figure 2.20 shows a section of a quadratic residue diffuser (QRD) with $N = 7$. It goes without saying that such a diffuser works not only for normal incident sound but also for waves arriving from oblique directions.

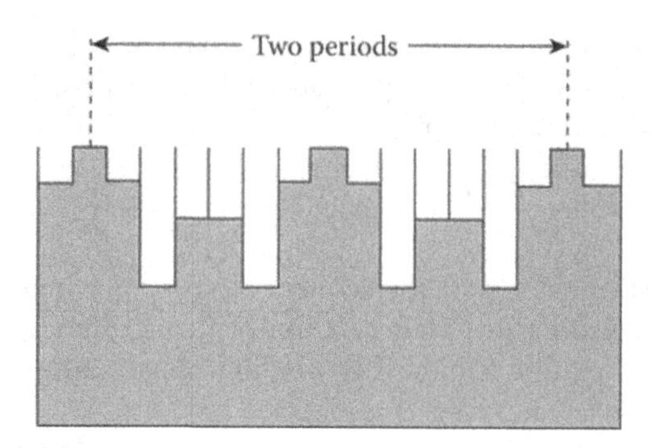


*Figure 2.20* Quadratic residue diffuser with $N = 7$.

Real diffusers have a finite extension, of course. As a consequence, the scattered sound energy is not concentrated in discrete directions but in lobes of finite widths around the angles $\vartheta_m$ of Equation 2.62. Figure 2.21 presents a polar diagram showing the angular distribution of the scattered sound pressure amplitude for a QRD consisting of two periods with $N = 7$; its design frequency is 285 Hz.

The concept of Schroeder diffusers is easily applied to two-dimensional structures which scatter the incident sound into all diffraction orders within the half-space. They consist of periodic arrays of parallel and rigid-walled channels. For instance, a two-dimensional QRD can be constructed by choosing the phase angles of the reflection factors according to

$$\chi_{nm} = \frac{2\pi}{N} \cdot \left(n^2 + m^2\right) \mathrm{mod} N \left(n, m = 0, 1, \ldots, N-1\right) \tag{2.66}$$

The array is periodic in both directions with the prime number $N$.

The explanation of pseudorandom diffusers as presented here is only qualitative since it neglects all losses and assumes all wells behave independently. In reality, the orifices of the wells or channels are coupled to each other by local airflows that tend to equalize local pressure differences. A more rigorous treatment of pseudorandom diffusers starts from the spatial Fourier expansion of the scattered sound field. For a more detailed description of this method, the reader is referred to the works by Mechel (1989) and Cox and D'Antonio (2004).

Fujiwara and Miyajima (1992) observed that pseudostochastic diffusers show unexpectedly high absorptivity. This effect is probably caused by the aforementioned equalizing airflows. Depending on the sound frequency, these flows may assume relatively high velocities which are not associated with radiation into the far field but are confined to the surface of the diffuser and to the interior of the wells. However, they lead to high viscous and thermal losses inside the channels (Kuttruff 1994). As pointed out by Mechel (1985), additional losses are probably caused by the fact that the local flows are forced to go around the sharp edges of adjacent troughs. In Figure 2.22, the absorption coefficient of a QRD with a design frequency of 285 Hz is plotted as a function of the frequency. It is negligible at low frequencies but shows a marked rise well below the design frequency and remains at a relatively high level, showing several distinct maxima and minima.

As was shown by Fujiwara and Miyajima (1992) and by Mechel (1985), relatively high absorptivities are not a peculiarity of pseudostochastic structures, as discussed earlier, but occur for any collection of wells or tubes with different lengths, the openings of which are close to each other. An example is a pipe organ, which shows remarkable sound absorption,

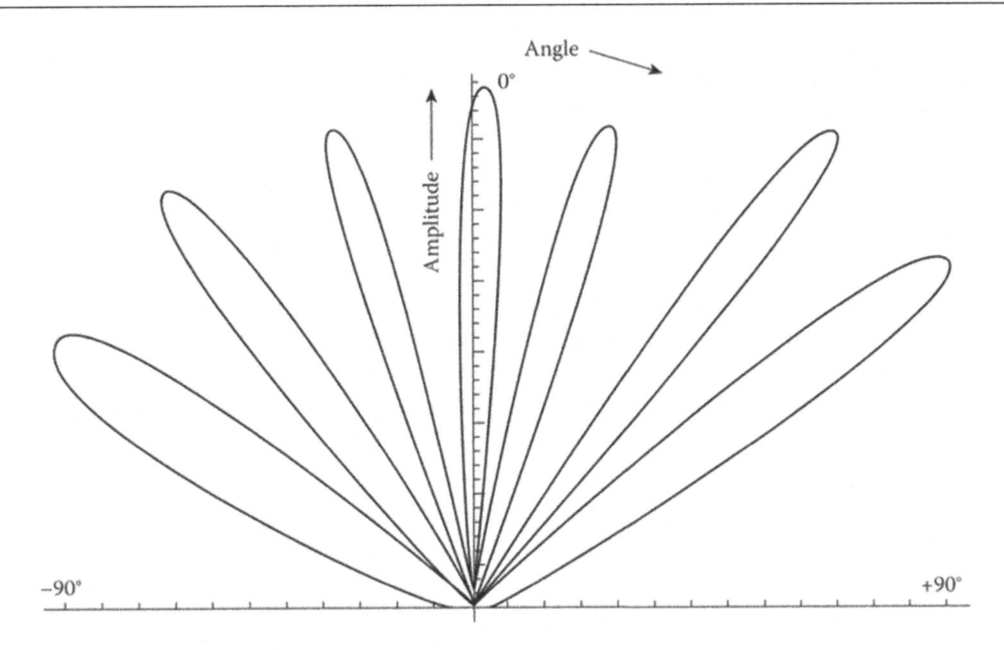


*Figure 2.21* Scattering diagram of a quadratic residue diffuser according to Figure 2.20: for two periods, the spacing $d$ is $\lambda/2$.

Source: Schroeder (1986).

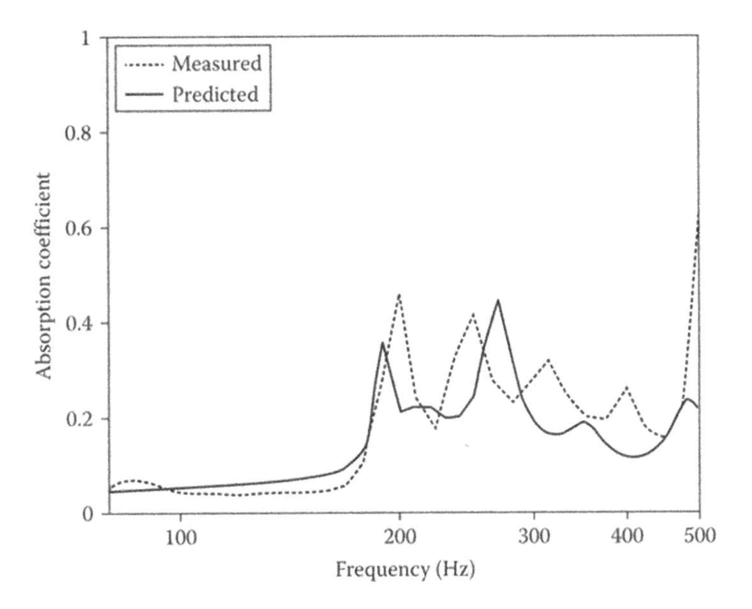


*Figure 2.22* Absorption coefficient of a quadratic residue diffuser with $N$ = 7 (see Figure 2.20), made of aluminium.

Source: Fujiwara and Miyajima (1992).

although no porous materials whatsoever are used in its construction. According to Meyer (2008), the absorption coefficient of an organ, related to the area of its prospect, is as high as about 0.55–0.60 in the frequency range from 125 to 4,000 Hz and has at least some influence on the reverberation time of a concert hall or a church.

The phase grating diffusers, as invented by M. R. Schroeder, are certainly based on an ingenious concept. Nevertheless, they suffer from the fact that the scattered energy is concentrated in a number of grating lobes which are separated by large minima. This is caused by the periodic repetition of a base element. One way to overcome this disadvantage is to use aperiodic number sequences. Another method is to combine two different base schemes — for instance, QR diffusers of different lengths or different 'polarities' — according to an aperiodic binary sequence, for instance, a Barker code. These and other possibilities are discussed in the work by Cox and D'Antonio (2004).

## REFERENCES

Cox T.J., D'Antonio P. *Acoustic Absorbers and Diffusers*. London: Spon Press, 2004.

Fujiwara K., Miyajima M. Absorption characteristics of practically constructed Schroeder diffusers of quadratic-residue type. *Appl Acoust* 1992; *35*: 149.

Kuttruff H. Sound absorption by pseudostochastic diffusers (Schroeder diffusers). *Appl Acoust* 1994; *42*: 215.

Mechel F.P. *Sound Absorbers (in German)*. Stuttgart: S. Hirzel Verlag, 1989.

Mechel F.P. The wide-angle diffuser — a wide-angle absorber? *Acustica* 1985; *81*: 379.

Meyer E., Kuttruff H. On the acoustics of the new Beethovenhalle in Bonn (in German). *Acustica* 1959; *9*: 465.

Meyer J. *Acoustics and the Performance of Music*. Berlin: Springer-Verlag, 2008.

Morse P.M., Ingard K.U. *Theoretical Acoustics*. New York: McGraw-Hill, 1968.

Schroeder M.R. Binaural dissimilarity and optimum ceilings for concert halls: More lateral sound diffusion. *J Acoust Soc Am* 1979; *65*: 958.

Schroeder M.R. Phase gratings with suppressed specular reflection. *Acustica* 1985; *81*: 364.

Schroeder M.R. *Number Theory in Science and Communication*, 2nd edn. Berlin: Springer Verlag, 1986.

Sommerfeld A. *Partial Differential Equations in Physics*. New York: Academic Press, 1964.

Suh J.S., Nelson P.A. Measurement of the transient response of rooms and comparison with geometrical acoustic models. *JASA* 1999; *105*: 230.

Chapter 3

# Sound waves in a room

In Chapter 2, we saw the laws which a plane sound wave obeys upon reflection from a single plane wall and how this reflected wave is superimposed on the incident one. Now, we shall try to obtain some insight into the complicated distribution of sound pressure or sound energy in a room which is enclosed on all sides by walls.

We could try to describe the resulting sound field by means of a detailed calculation of all the reflected sound components and by finally adding them together, that is to say, by a manifold application of the reflection laws which we dealt with in Chapter 2. Since each wave which has been reflected from wall A will be reflected from walls B, C, D, and so on and will arrive eventually once more at wall A, this procedure leads only asymptotically to a final result, not to mention the avalanche-like growth of required calculations. Nevertheless, this method is highly descriptive, and therefore, it is frequently applied in a much simplified form in geometrical room acoustics. We shall return to it in Chapter 5.

In this chapter, we shall choose a different way of tackling our problem which will lead to a solution in closed form — at least a formal one. This advantage is paid for by a higher degree of abstraction, however. Characteristics of this approach are certain boundary conditions which have to be set up along the room boundaries and which describe mathematically the acoustical properties of the walls, the ceiling, and the other surfaces. Then, solutions to the wave equations are sought which satisfy these boundary conditions. This method is the basis of what is frequently called 'the wave theory of room acoustics'.

It will turn out that this method in its exact form, too, can only be applied to highly idealized cases with reasonable effort. The rooms with which we are concerned in our daily life, however, are more or less irregular in shape, partly because of the furniture, which forms part of the room boundary. Rooms such as concert halls, theatres, or churches deviate from their basic shape because there are balconies, galleries, pillars, columns, and many other sorts of wall irregularities, not to mention the persons attending an event in the room. Then, even the formulation of boundary conditions may turn out to be quite involved, and the solution of a given problem requires extensive numerical calculations. Therefore, the immediate application of the wave theory to practical problems in room acoustics is very limited. Nevertheless, the wave theory offers the most reliable and appropriate description from a physical point of view, and therefore, it is essential for a more than superficial understanding of sound propagation in enclosures. For the same reason, we should keep in mind the results of the wave theory when we are applying more simplified methods, in order to keep our ideas in perspective. Finally, the wave theoretical description is the basis of the very powerful 'finite element method' of calculating the sound field in an enclosure.

DOI: 10.1201/9781003389873-3

## 3.1 FORMAL SOLUTION OF THE WAVE EQUATION

The starting point for an exact description of the sound field in a room is again the wave equation (Equation 1.6), which will be used here in a time-independent form. That is to say, we assume, as earlier, a harmonic time law for the pressure, the particle velocity, and so on, with an angular frequency ω. Then the equation, known as the Helmholtz equation, reads

$$\Delta p + k^2 p = 0 \quad \text{with} \quad k = \frac{\omega}{c} \tag{3.1}$$

Furthermore, we assume that the room under consideration has locally reacting walls and ceiling, the acoustical properties of which are completely characterized by a wall impedance $Z$, which depends on the coordinates and the frequency, but not on the angle of sound incidence.

According to Equation 1.2, the velocity component normal to any wall or boundary is

$$\mathrm{v}_n = -\frac{1}{i\omega\rho_0}(\operatorname{grad} p)_n = \frac{i}{\omega\rho_0}\frac{\partial p}{\partial n} \tag{3.2}$$

The symbol $\partial/\partial n$ denotes partial differentiation in the direction of the outward normal to the wall. We replace $v_n$ by $p/Z$ (see Equation 2.2) and obtain

$$Z\frac{\partial p}{\partial n} + i\omega\rho_0 p = 0 \tag{3.3}$$

or, using the specific impedance $\zeta = Z/\rho_0 c$:

$$\zeta\frac{\partial p}{\partial n} + ikp = 0 \tag{3.4}$$

Now, it can be shown that the wave equation has non-zero solutions fulfilling the boundary condition (3.3) or (3.4) only for particular discrete values of $k$, called 'eigenvalues' (Morse and Feshbach 1953; Morse and Ingard 1968). In the following text, we shall frequently distinguish these quantities from each other by a single index number $n$ or $m$, though it would be more adequate, in principle, to use a trio of subscripts because of the three-dimensional nature of the problem. Each eigenvalue $k_n$ is associated with a solution $p_n(\mathrm{r})$, named 'eigenfunction' or 'characteristic function'. Here, $r$ is used as an abbreviation for the three spatial coordinates, for instance, $x$, $y$, $z$. It represents a three-dimensional standing wave, a 'normal mode' of the room. Whenever the boundary or a part of it has non-zero absorption, both the eigenfunctions and the eigenvalues are complex. Sometimes, it may happen that two or more eigenfunctions belong to the same eigenvalue — an example for this is the cubical room. In this case, we speak of degenerate eigenvalues.

At this point, we need to comment on the wave number $k$ in the boundary condition (3.3) or (3.4). Implicitly, it is also contained in ζ, since the specific wall impedance depends in general on the frequency $\omega = kc$ except in the limiting case of a rigid boundary, that is, of $\zeta \to \infty$. Hence, both the eigenfunctions $p_n$ and the eigenvalues $k_n$ are frequency-dependent. The eigenfunctions are mutually orthogonal, which means that

$$\iiint_V p_n(r)p_m(r)\,\mathrm{d}V = \begin{cases} K_n & \text{for } n = m \\ 0 & \text{for } n \neq m \end{cases} \tag{3.5}$$

where the integration has to be extended over the whole volume $V$ enclosed by the walls. Here, $K_n$ is a constant with the dimension $Pa^2\ m^3 = N^2\ m^{-1}$.

If all the eigenvalues and eigenfunctions were known, we could — at least in principle — evaluate any desired acoustical property of the room, for instance, its response to arbitrary sound sources, either constant or time-variable ones, its reverberation, the spatial distribution of the energy density, and so forth. Suppose the sound sources are distributed continuously over the room according to a density function $q(\mathrm{r})$, where $q(\mathrm{r})\mathrm{d}V$ is the volume velocity of a volume element $\mathrm{d}V$ at $r$. Furthermore, we assume that all source elements are operating at the same driving frequency $\omega$. By adding $\rho_0 q(\mathrm{r})$ to the right-hand side of Equation 1.4, it is easily seen that the Helmholtz equation (Equation 3.1) has to be modified into

$$\Delta p + k^2 p = -i\omega\rho_0 q(\mathrm{r}) \tag{3.6}$$

Since the eigenfunctions form a complete and orthogonal set of functions, we can expand the source function in a series of $p_n$:

$$q(r) = \sum_n C_n p_n(r) \quad \text{with} \quad C_n = \frac{1}{K_n}\iiint_V p_n(r) q(r)\mathrm{d}V \tag{3.7}$$

Where the summation is extended over all possible combinations of subscripts. In the same way, the solution $p_\omega(\mathrm{r})$, which we are looking for, can be expanded in eigenfunctions:

$$p(\mathrm{r}) = \sum_n D_n p_n(\mathrm{r}) \tag{3.8}$$

Our problem is solved if the unknown coefficients $D_n$ are expressed by the known coefficients $C_n$. For this purpose, we insert both series into Equation 3.6:

$$\Sigma D_n\left(\Delta p_n + k^2 p_n\right) = -i\omega\rho_0\,\Sigma C_n p_n$$

Now, $\Delta p_n = -k_n^2 p_n$ (see Equation 3.1). Using this relation and equating term by term in the earlier equation, we obtain

$$D_n = i\omega\rho_0 \frac{C_n}{k_n^2 - k^2} \tag{3.9}$$

The final solution assumes a particularly simple form if the sound source is a point source with the volume velocity $Q$, located at the arbitrary point $r_0$. Then, the source function is represented mathematically by a delta function:

$$q(\mathrm{r}) = Q\ \delta(\mathrm{r} - \mathrm{r}_0)$$

Or in Cartesian coordinates:

$$q(x,y,z) = Q\ \delta(x - x_0)\cdot\delta(y - y_0)\cdot\delta(z - z_0)$$

Because of Equation 1.56, the coefficient $C_n$ in Equation 3.7 is then given by

$$C_n = \frac{1}{K_n} Q p_n(r_0)$$

Using this relation and Equations 3.9 and 3.8, we finally find the sound pressure in a room, which is excited by a point source emitting a sine signal with the angular frequency $\omega$:

$$p_\omega(r) = i\omega Q \rho_0 \sum_n \frac{p_n(r)p_n(r_0)}{K_n\left(k_n^2 - k^2\right)} \tag{3.10}$$

This function is also called the 'Green's function' of the room. It is interesting to note that it is symmetric in the coordinates of the sound source and of the point of observation. Hence, if we put the sound source at $\boldsymbol{r}$ instead of $\boldsymbol{r}_0$, we observe the same sound pressure at point $\boldsymbol{r}_0$ as we did before at $\boldsymbol{r}$, when the sound source was at $\boldsymbol{r}_0$. Thus, Equation 3.10 is the mathematical expression of the famous reciprocity theorem, which can be applied sometimes with advantage to measurements in room acoustics.

As mentioned before, the eigenvalues are, in general, complex quantities. Putting

$$k_n = \frac{\omega_n}{c} + i\frac{\delta_n}{c} \tag{3.11}$$

and assuming that $\delta_n << \omega_n$, we obtain from Equation 3.10

$$p_\omega(r) = \rho_0 c^2 \omega Q \sum_n \frac{p_n(r)p_n(r_0)}{\left[2\delta_n\omega_n + i\left(\omega^2 - \omega_n^2\right)\right]K_n} \tag{3.12}$$

This expression is the transfer function of the room between the two points $\boldsymbol{r}$ and $\boldsymbol{r}_0$. Each term of this sum represents a 'resonance' of the room, since the sound pressure amplitude assumes a maximum when the driving frequency $\omega$ comes close to $\omega_n$. Therefore, the corresponding frequencies $f_n = \omega_n/2\pi$ are often called the 'resonance frequencies' of the room. Another commonly used name is 'eigenfrequencies'. The $\delta_n$ will turn out to be 'damping constants' (see Equation 2.37).

If the sound source is not emitting a sinusoidal signal but instead a signal which is composed of several spectral components, then $Q = Q(\omega)$ can be considered as its spectral function, and we can represent the source signal as a Fourier integral (see Section 1.4):

$$s(t) = \frac{1}{2\pi}\int_{-\infty}^{+\infty} Q(\omega)\exp(i\omega t)\,\mathrm{d}\omega$$

Since the response to the spectral component with angular frequency $\omega$ is just given by Equations 3.10 or 3.12, the sound pressure at the point $\boldsymbol{r}$ as a function of time is

$$p(r,t) = \frac{1}{2\pi}\int_{-\infty}^{+\infty} p_\omega(r)\exp(i\omega t)\,\mathrm{d}\omega$$

where the constant volume flow $Q$ in the formula for $p_\omega$ has to be replaced by $Q(\omega)$.

## 3.2 NORMAL MODES IN A RECTANGULAR ROOM WITH RIGID BOUNDARIES

In order to put some life into the abstract formalism outlined in the preceding section, we consider a room with parallel pairs of walls, the pairs being perpendicular to each other. It will be referred to in the following as a 'rectangular room'. In practice, rooms with exactly this shape do not exist. On the other hand, many concert halls or other halls, churches, lecture rooms, and so on are much closer in shape to the rectangular room than to any other of simple geometry, and so the results obtained for strictly rectangular rooms can be at least qualitatively applied to many rooms encountered in practice. Therefore, our example not only is intended for the elucidation of the theory discussed earlier but also has some practical bearing as well.

Our room is assumed to extend from $x = 0$ to $x = L_x$ in the $x$-direction, and similarly from $y = 0$ to $y = L_y$ in the $y$-direction, and from $z = 0$ to $z = L_z$ in the $z$-direction (see Figure 3.1). As far as the properties of the wall are concerned, we start with the simplest case, namely, that of all the walls being rigid. That is to say that, at the surface of the walls, the normal components of the particle velocity must vanish.

In Cartesian coordinates, the Helmholtz equation (Equation 3.1) may be written as

$$\frac{\partial^2 p}{\partial x^2} + \frac{\partial^2 p}{\partial y^2} + \frac{\partial^2 p}{\partial z^2} + k^2 p = 0 \tag{3.13}$$

The variables can be separated, which means that we can compose the solution of three factors:

$$p(x,y,z) = p_x(x) \cdot p_y(y) \cdot p_z(z)$$

each of them depending only on one of the space variables. If this product is inserted into the Helmholtz equation, the latter splits up into three ordinary differential equations. The same is true for the boundary conditions. For instance, $p_x$ must satisfy the equation

$$\frac{\mathrm{d}^2 p_x}{\mathrm{d}x^2} + k_x^2 p_x = 0 \tag{3.14}$$

together with the boundary condition

$$\frac{\mathrm{d}p_x}{\mathrm{d}x} = 0 \quad \text{for } x = 0 \quad \text{and} \quad x = L_x \tag{3.15}$$

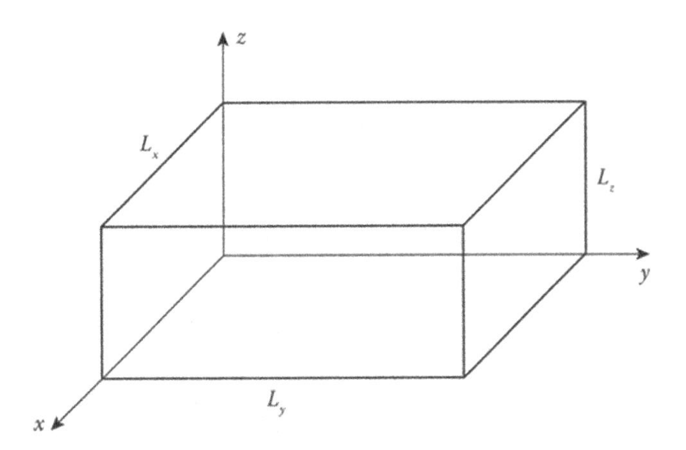


*Figure 3.1* Rectangular room.

Similar equations hold for $p_y(y)$ and $p_z(z)$; the newly introduced constants $k_x$, $k_y$, and $k_z$ are related by

$$k_x^2 + k_y^2 + k_z^2 = k^2 \tag{3.16}$$

The general solution of Equation 3.14 can be written as

$$p_x(x) = A_1 \cos(k_x x) + B_1 \sin(k_x x) \tag{3.17}$$

The constants $A_1$ and $B_1$ are used for adapting this solution to the boundary conditions (3.15). So it is seen immediately that we must put $B_1 = 0$, since only the cosine function possesses the horizontal tangent at $x = 0$ required by Equation 3.15. For the occurrence of a horizontal tangent, too, at $x = L_x$, we must have $\cos(k_x L_x) = \pm 1$; thus, $k_x L_x$ must be an integral multiple of $\pi$. The constant $k_x$ must therefore assume one of the values

$$k_x = \frac{n_x \pi}{L_x} \tag{3.18}$$

$n_x$ being a non-negative integer. Similarly, we obtain for the allowed values of $k_y$ and $k_z$

$$k_y = \frac{n_y \pi}{L_y} \tag{3.19}$$

$$k_z = \frac{n_z \pi}{L_z} \tag{3.20}$$

By inserting these values into Equation 3.16, one arrives at the eigenvalues of the wave equation:

$$k_{n_x n_y n_z} = \pi \left[ \left( \frac{n_x}{L_x} \right)^2 + \left( \frac{n_y}{L_y} \right)^2 + \left( \frac{n_z}{L_z} \right)^2 \right]^{1/2} \tag{3.21}$$

The eigenfunctions associated with these eigenvalues are simply obtained by multiplying the three cosines, each of which describes the dependence of the pressure on one coordinate:

$$p_{n_x n_y n_z}(x, y, z) = C \cdot \cos\left( \frac{n_x \pi x}{L_x} \right) \cdot \cos\left( \frac{n_y \pi y}{L_y} \right) \cdot \cos\left( \frac{n_z \pi z}{L_z} \right) \tag{3.22}$$

where $C$ is an arbitrary constant. This formula represents a 'normal mode' of the room, which can be conceived as a three-dimensional standing wave. The pressure amplitude is zero at all points, at which at least one of the cosines becomes zero. This occurs for all values of $x$ which are odd integers of $L_x/2n_x$, and for the analogous values of $y$ and $z$. So these points of vanishing sound pressure form three sets of equidistant planes, called 'nodal planes', which are perpendicular to one another. The numbers $n_x$, $n_y$, and $n_z$ indicate the numbers of nodal planes perpendicular to the $x$-axis, the $y$-axis, and the $z$-axis, respectively. On both sides of a nodal plane, the instantaneous sound pressures have opposite signs. (For non-rectangular rooms, the surfaces of vanishing sound pressure are curved. They are referred to as 'nodal surfaces'.)

In Figure 3.2, the sound pressure distribution in the plane $z = 0$ is depicted for $n_x = 3$ and $n_y = 2$. The loops are curves of constant pressure amplitude, namely, for $|p/p_{max}| = 0.25$, 0.5

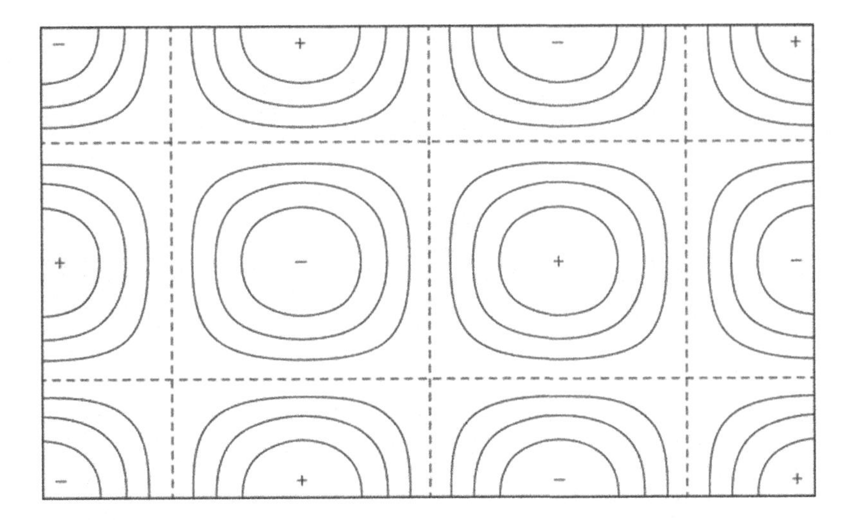


*Figure 3.2* Sound pressure distribution in the plane z = 0 of a rectangular room for $n_x = 3$ and $n_y = 2$.

and 0.75. The intersections of vertical nodal planes with the plane $z = 0$ are indicated by dotted lines.

The eigenfrequencies corresponding to the eigenvalues of Equation 3.21, which are real because of the particular boundary condition (3.15), are given by

$$f_{n_x n_y n_z} = \frac{c}{2\pi} k_{n_x n_y n_z} \tag{3.23}$$

In Table 3.1, the lowest 20 eigenfrequencies (in Hz) of a rectangular room with dimensions $L_x = 4.7$ m, $L_y = 4.1$ m, and $L_z = 3.1$ m are listed for $c = 340$ m/s, together with the corresponding combinations of subscripts, which indicate the structure of the mode.

By employing the relation cos x = $(e^{ix} + e^{-ix})/2$ (see Equation 1.19), Equation 3.22 can be rewritten in the following form:

$$p_{n_x n_y n_z} = \frac{c}{8} \Sigma \exp\left[\pi i \left( \pm \frac{n_x}{L_x} x \pm \frac{n_y}{L_y} y \pm \frac{n_z}{L_z} z \right)\right] \tag{3.24}$$

*Table 3.1* Eigenfrequencies of a rectangular room with dimensions 4.7 × 4.1 × 3.1 m³ (in Hz)

| $f_n$ | $n_x$ | $n_y$ | $n_z$ | $f_n$ | $n_x$ | $n_y$ | $n_z$ |
|---|---|---|---|---|---|---|---|
| 36.17 | 1 | 0 | 0 | 90.47 | 1 | 2 | 0 |
| 41.46 | 0 | 1 | 0 | 90.78 | 2 | 0 | 1 |
| 54.84 | 0 | 0 | 1 | 99.42 | 0 | 2 | 1 |
| 55.02 | 1 | 1 | 0 | 99.80 | 2 | 1 | 1 |
| 65.69 | 1 | 0 | 1 | 105.79 | 1 | 2 | 1 |
| 68.75 | 0 | 1 | 1 | 108.51 | 3 | 0 | 0 |
| 72.34 | 2 | 0 | 0 | 109.68 | 0 | 0 | 2 |
| 77.68 | 1 | 1 | 1 | 110.05 | 2 | 2 | 0 |
| 82.93 | 0 | 2 | 0 | 115.49 | 1 | 0 | 2 |
| 83.38 | 2 | 1 | 0 | 116.16 | 3 | 1 | 0 |

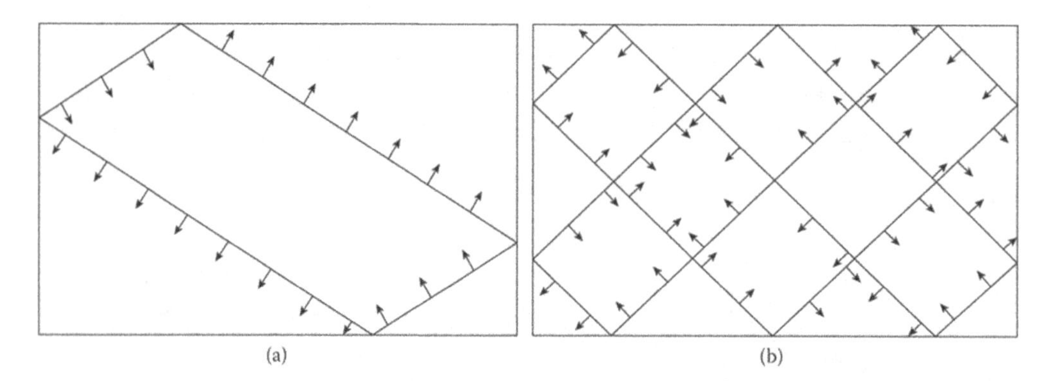


*Figure 3.3* Plane wave fronts creating standing waves in a rectangular room: (a) $n_x$:$n_y$ = 1:1 and (b) $n_x$:$n_y$ = 3:2.

wherein the summation has to be extended over the eight possible combinations of signs in the exponent. Each of these eight terms — multiplied by the usual time factor exp($i\omega t$) — represents a plane travelling wave (see Equation 1.23), whose direction of propagation is defined by the angles $\beta_x$, $\beta_y$, and $\beta_z$, which the wave normal makes with the coordinate axes, where

$$\cos\beta_x : \cos\beta_y : \cos\beta_z = \left(\pm\frac{n_x}{L_x}\right) : \left(\pm\frac{n_y}{L_y}\right) : \left(\pm\frac{n_z}{L_z}\right) \tag{3.25}$$

with

$$\cos^2\beta_x + \cos^2\beta_y + \cos^2\beta_z = 1$$

If one of the three characteristic integers $n_i$, for instance $n_z$, equals zero, then the corresponding angle ($\beta_z$ in this example) is 90°; the propagation takes place perpendicularly to the respective axis, that is, parallel to all planes which are perpendicular to that axis. The corresponding vibration pattern is frequently referred to as a 'tangential mode'. If there is only one non-zero integer $n$, the propagation is parallel to one of the coordinate axes, that is, parallel to one of the room edges. Then, we are speaking of an 'axial mode'. Modes with all integers different from zero are called 'oblique modes'. In Figure 3.3, two combinations of two-dimensional plane waves are shown which correspond to two different eigenfunctions.

## 3.3 NUMBER AND DENSITY OF EIGENFREQUENCIES

We can get an illustrative survey on the number and type of normal modes as well as the directions of the plane waves producing them by the following geometrical representation: We interpret $k_x$, $k_y$, and $k_z$ as Cartesian coordinates in a three-dimensional $k$-space. Each of the allowed values of these coordinates, as given by Equations 3.18 to 3.20, corresponds to a point in this space. These points form a rectangular point lattice in the first octant of our $k$-space (see Figure 3.4). (Negative values obviously do not yield additional eigenvalues, since Equation 3.21 is not sensitive to the signs of the characteristic integers $n_x$.) The lattice points corresponding to tangential and axial modes are situated on the coordinate planes and on the axes, respectively. The vector pointing from the origin of the coordinate system to a certain lattice point has — according to Equation 3.25 — the same direction as one of the plane waves of which the associated mode is made up (see Equation 3.24).

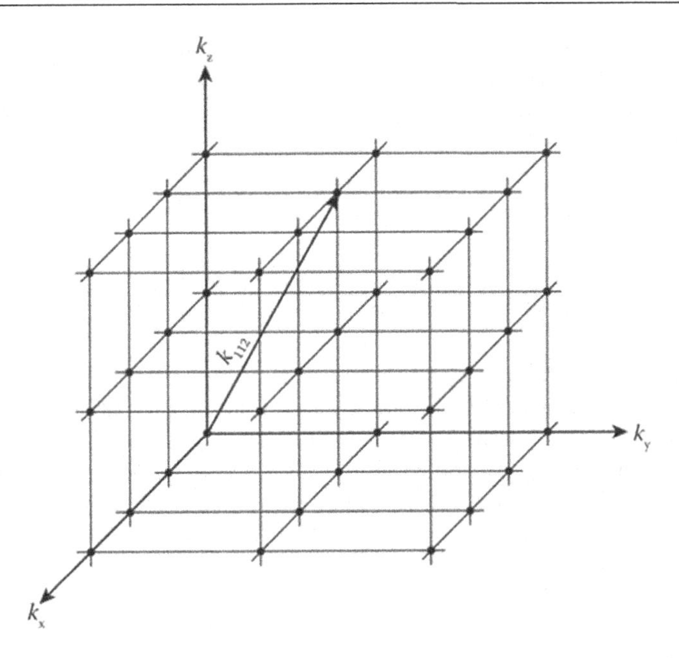


*Figure 3.4* Eigenvalue lattice in the k-space for a rectangular room. The arrow pointing from the origin to an eigenvalue point indicates the direction of one of the eight wave planes which the corresponding mode consists of (see Figure 3.3); its length is proportional to the eigenvalue.

This representation allows a simple estimate of the number of eigenfrequencies to be expected between the frequency 0 and some other given frequency $f$. Regarded geometrically, Equation 3.16 represents a spherical surface in the $k$-space with radius $k$, enclosing a 'volume' $4\pi k^3/3$. Of this volume, however, only the portion situated in the first octant is of interest, that is, the volume $\pi k^3/6$. On the other hand, the distances between one certain lattice point and its nearest neighbours in the three coordinate directions are $\pi/L_x$, $\pi/L_y$, and $\pi/L_z$. The $k$-'volume' per lattice point is therefore $\pi^3/L_xL_yL_z = \pi^3/V$, where $V$ is the geometrical volume of the room under consideration. Now, we are ready to write down the number of lattice points inside the first octant up to radius $k$, which is equivalent to the number of eigenfrequencies from 0 to an upper limit $f = kc/2\pi$:

$$N_f = \frac{\pi k^3/6}{\pi^3/V} = \frac{Vk^3}{6\pi^2} = \frac{4\pi}{3} V \left(\frac{f}{c}\right)^3 \tag{3.26}$$

The average density of eigenfrequencies on the frequency axis, that is, the number of eigenfrequencies per Hz at the frequency $f$, is

$$\frac{dN_f}{df} = 4\pi V \frac{f^2}{c^3} \tag{3.27}$$

Hence, the average spacing of adjacent eigenfrequencies is

$$\langle \delta f_n \rangle = \left(\frac{dN_f}{df}\right)^{-1} = \frac{c^3}{4\pi V f^2} \tag{3.28}$$

A more rigorous derivation of $N_f$ must account for the fact that the lattice points situated on one of the coordinate planes, representing tangential modes, are shared by two adjacent octants. Hence, only half of them have been accounted for so far. Similarly, all lattice points on the coordinate axes related to axial modes are common to four octants; therefore, only one-fourth of them are contained in Equation 3.26, and their contribution to the total number of modes is too small by a factor 4.

The number of all lattice points corresponding to tangential modes can be calculated in about the same way which led us to Equation 3.26. The result is

$$N_{tan} = \frac{1}{4\pi}k^2\left(L_xL_y + L_yL_z + L_zL_x\right) = \frac{Sk^2}{8\pi} = \frac{\pi S}{2}\left(\frac{f}{c}\right)^2$$

Where we introduced the total area of all walls, $S = 2(L_xL_y + L_yL_z + L_zL_x)$. Thus, one correction term to Equation 3.26 reads $(\pi S/4)\,(f/c)^2$. Likewise, the total number of lattice points situated on one of the coordinate axes within the first octant is $kL/4\pi = Lf/2c$, with $L = 4(L_x + L_y + L_z)$ denoting the sum of all edge lengths of the room. However, we should keep in mind that half of these points are already contained in the earlier expression, while another quarter of them have been, as mentioned, counted in Equation 3.26. Hence, the correction term due to the axial modes is $Lf/8c$, and the corrected expression of the number of modes with eigenfrequencies up to a frequency $f$ is

$$N_f = \frac{4\pi}{3}V\cdot\left(\frac{f}{c}\right)^3 + \frac{\pi}{4}S\cdot\left(\frac{f}{c}\right)^2 + \frac{L}{8}\cdot\frac{f}{c} \tag{3.29}$$

It can be shown that in the limiting case $f \to \infty$, Equation 3.26 is valid not only for rectangular rooms but also for rooms of arbitrary shape. This is not too surprising, since any enclosure can be conceived as being composed of many (or even infinitely many) rectangular rooms. For each of them, Equation 3.26 yields the number $N_i$ of eigenfrequencies. Since this equation is linear in $V$, the total number of eigenfrequencies is just the sum of all $N_i$.

We bring this section to a close by applying Equations 3.26 and 3.27 to two simple examples. The rectangular room the eigenfrequencies of which are listed in Table 3.1 has a volume of 59.7 m$^3$. For an upper frequency limit of 116 Hz, Equation 3.26 indicates 10 eigenfrequencies as compared with the 20 listed in Table 3.1. Using the more accurate formula (3.29), we obtain 21 eigenfrequencies. That means we must not neglect the corrections due to tangential and axial modes when dealing with such small rooms at low frequencies.

As a second example, we consider a rectangular room with dimensions 50 m × 24 m × 14 m whose volume is 16,800 m$^3$. (This might be a large concert hall, for instance.) In the frequency range from 0 to 10,000 Hz, there are, according to Equation 3.26, about $1.7 \times 10^9$ eigenfrequencies. At 1,000 Hz, the number of eigenfrequencies per hertz is about 5,200; thus, the average distance of two eigenfrequencies on the frequency axis is less than 0.0002 Hz. These figures underline the practical impossibility of evaluating the sound field even in a moderately sized room by calculating normal modes.

## 3.4 NON-RIGID WALLS

In this section, we are still dealing with rectangular rooms. But now we consider a room the walls of which are not completely rigid. This means that the normal components of particle velocity may have non-vanishing values along the boundary. Accordingly, we have to replace the boundary condition we applied before by the more general condition of Equation 3.3 or 3.4. As in the preceding section, the solution of the wave equation consists of three factors, $p_x$, $p_y$, and $p_z$, each of which depends on one spatial coordinate only. If the specific wall impedance is constant over each wall pair, the boundary condition for $p_x$ reads, under the assumption of locally reacting boundaries:

$$\begin{aligned} \zeta_x \frac{dp_x}{dx} &= ikp_x \quad \text{for} \quad x = 0 \\ \zeta_x \frac{dp_x}{dx} &= -ikp_x \quad \text{for} \quad x = L_x \end{aligned} \tag{3.30}$$

Analogous conditions can be set up for the walls perpendicular to the $y$-axis and the $z$-axis.

Again, the general solution for $p_x$ is given by Equation 3.17. However, for the present purpose, it is more useful to write this solution in its complex version, according to Equations 1.19, with arbitrary constants $C_1$ and $D_1$:

$$p_x(x) = C_1\exp(-ik_x x) + D_1\exp(ik_x x) \tag{3.31}$$

By inserting it into the boundary conditions, we obtain two linear and homogeneous equations for $C_1$ and $D_1$. These equations have a non-vanishing solution only if the determinant of their coefficients is zero. This leads to the following transcendental equation:

$$\exp(ik_x L_x) = \pm\frac{k_x\zeta_x - k}{k_x\zeta_x + k} \tag{3.32}$$

In general, this equation must be solved numerically. Once the allowed values of $k_x$ have been determined, the ratio of the two constants $C_1$ and $D_1$ can be calculated, and we are ready to write down the $x$-component of the solution, apart from a common constant.

At this point, the reader should remember that the exponential function with imaginary argument is a periodic function with the period $2\pi$, that is,

$$\exp(iz) = \exp(iz + i2\pi) = \exp(iz + in_x\pi)$$

where $n_x$ is any even number. Furthermore, the plus or minus sign in Equation 3.32 can be accounted for by admitting not only even numbers $n_x$ in the last exponential but also any integer number, since $\exp(i\pi) = -1$. Thus, to obtain the complete solution of Equation 3.32, a factor $\exp(i\pi n_x)$ must be inserted somewhere:

$$\exp(ik_x L_x) = \frac{k_x\zeta_x - k}{k_x\zeta_x + k}\cdot\exp(i\pi n_x) \tag{3.33}$$

From now on, we restrict the discussion to enclosures with nearly rigid boundaries, that is, to the limiting case $|\zeta| >> 1$. Then, we expect that the eigenvalues and eigenfunctions are not very different from those of the rigid-walled room. By taking the logarithm of Equation 3.33, we obtain:

$$k_x L_x = -i\cdot\ln\left(\frac{k_x\zeta_x - k}{k_x\zeta_x + k}\right) + n_x\pi$$

The logarithm can be approximated by expanding it into a power series for $k/k_x\,\zeta_x$ and truncating this series after its first term:

$$k_x L_x \approx n_x\pi + i\cdot\frac{2k}{k_x\zeta_x}$$

Since the second term in this equation is small compared to the first one, we can replace $k_x$ in the second term with $n_x\,\pi/L_x$:

$$k_x \approx \frac{n_x \pi}{L_x} + i\frac{2k}{\pi n_x \zeta_x} \tag{3.34}$$

Comparing this result with Equation 3.18 confirms our expectation that the allowed values of $k_x$ are not much different from those of the hard-walled room. With increasing 'order' $n_x$ of the mode, the difference becomes even smaller.

Suppose that the wall is reactive, that is, it is free of absorption. Then, its specific impedance is purely imaginary, and the correction term is real. If Im $\zeta$ is positive, which indicates that the motion of the wall is mass-controlled, then the allowed value is higher than in the rigid-walled room. Conversely, a compliant wall, that is, a wall with the impedance of a spring (Im $\zeta < 0$), will lower the allowed value $k_x$.

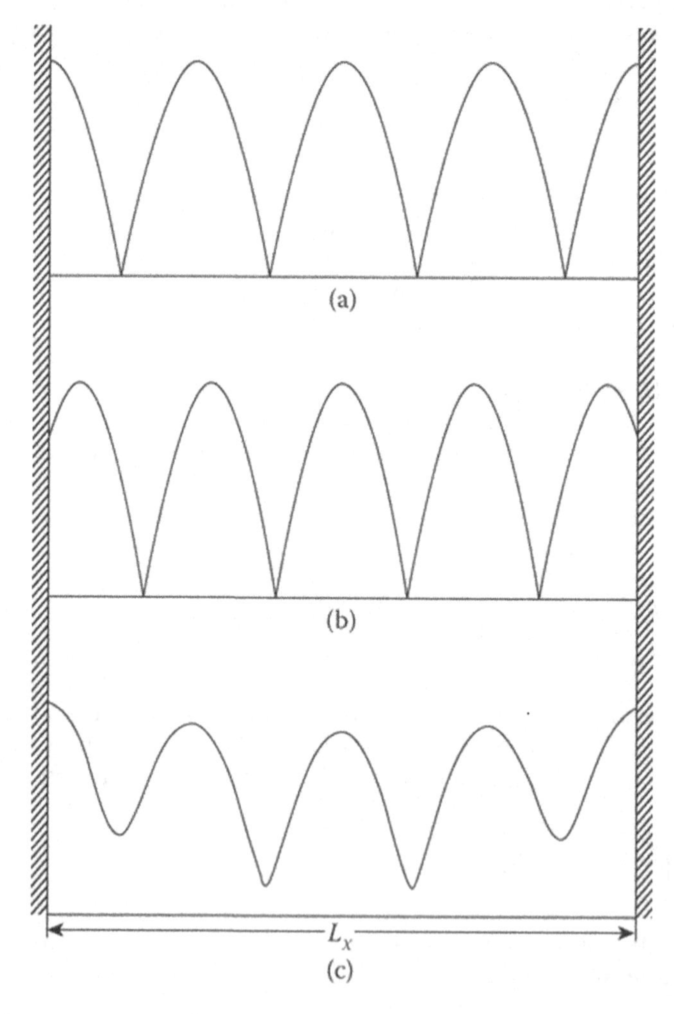


*Figure 3.5* One-dimensional normal mode: pressure distribution for $n_x = 4$: (a) $\zeta = \infty$, (b) $\zeta = i$, and (c) $\zeta = 2$.

Whenever the specific impedance has a non-vanishing real part which indicates wall losses, $k_x$ is complex. If the allowed values of $k_x$ are denoted by $k_{xn_x}$, the eigenvalues of the original differential equation are given as earlier by (see Equation 3.16)

$$k_{n_x n_y n_z} = \left( k_{xn_x}^2 + k_{yn_y}^2 + k_{zn_z}^2 \right)^{1/2} \tag{3.35}$$

In Figure 3.5, the absolute value of the $x$-dependent factor of a certain eigenfunction is represented for three cases: for rigid walls ($\zeta_x = \infty$), for mass-loaded walls with no energy loss ($\zeta_x$, $= i$), and for walls with real impedance. In the second case, the standing wave is simply shifted together, but its shape remains unaltered. On the contrary, in the third case of lossy walls, there are no longer exact nodes, and the pressure amplitude is different from zero at all points. This can easily be understood by keeping in mind that the walls dissipate energy, which must be supplied by waves travelling towards the walls; thus, a pure standing wave is not possible. This situation is comparable to a standing wave in front of a single plane with a reflection factor less than unity, as shown in Figure 2.4.

## 3.5 NUMERICAL SOLUTION OF THE WAVE EQUATION: THE FINITE ELEMENT METHOD

The preceding sections should have shown that it is pointless to search after a comprehensive picture of the sound field in a room — well-suited for practical calculations — by solving the wave equation. One reason for this is the tremendous number of eigenvalues and eigenfrequencies which must be computed for this purpose. Another one is the fact that these quantities and functions are usually complex because of the losses which are present in every real room. And finally, the shapes of most real rooms are so complicated that a simple formulation of boundary conditions is nearly impossible, let alone the great diversity of wall materials. Therefore, when it comes to compute the sound field in practical situations, we have to resort to numerical methods from which we can obtain approximate solutions, the accuracy of which depends on the computational expenditure we are ready to spend.

A particularly efficient and flexible numerical procedure is the finite element method (FEM). It has proved very useful in many disciplines of physics and engineering, for instance, in elasticity, heat conduction, electrical and magnetic fields, and so on. In room acoustics, it can be used to determine the eigenfunctions, or to calculate the steady-state or the transient sound field in a room. The basic idea is to desist from the original goal, namely, to compute the complex sound pressure amplitudes in all room points and instead to search an approximate solution which is exact in $N$ previously assigned points, so-called nodes, while the sound pressure at all other room points is obtained by interpolation. This is achieved by transforming the partial differential equation into a set of linear algebraic equations from which the correct nodal sound pressure amplitudes $p_n$ ($n = 1, 2, \ldots, N$) can be determined.

The first step to arrive at such a set of equations is to subdivide the interior of the room into a large number of small, however, finite-volume elements, for instance, into tetrahedrons or parallelepipeds. This procedure is often called 'creating a mesh'. Each of these elements, which may vary in shape, size, and orientation, is defined by a certain number of 'nodes'. For tetrahedral elements, for instance, the simplest choice of node positions would be their four vertices. In this case, linear interpolation is sufficient to determine the sound pressure at any point of an element.

There are several ways to derive the aforementioned set of linear equations. For the following, the 'method of weighted residuals' has been chosen for calculating the steady-state response of a room. We follow here the very clear description of this method given by Astley (2007). The boundaries of the room are assumed to be of the locally reacting type and, hence, can be characterized by their specific impedance $\zeta$, which may vary from one location to another. Furthermore, we use the simplest possible shape of elements, namely, that of tetrahedrons, which carry one node at each of its vertices. So each element carries four nodes, and each node connects four elements.

We start with the Helmholtz equation (3.6), that is, we assume excitation of the room with a harmonic sound signal the angular frequency of which is $\omega = kc$. Multiplying this equation with some continuous and differentiable function $v(\mathrm{r})$, called 'test function', and integrating the result over the room volume $V$ lead us to:

$$\iiint_V \left[\Delta p + k^2 p + i\omega\rho_0 q(r)\right] v(r)\,\mathrm{d}V = 0 \tag{3.36}$$

If this equation is required to be valid for any test function $v$, we can conclude that the expression within the bracket is zero, which leads us back to Equation 3.6. Any approximate solution $\tilde{p}$, inserted into Equation 3.36, would yield a non-vanishing value of the bracket, the so-called 'residuum'. Thus, the test function acts as a weighting function, which is to distribute this residuum in a suitable way over the whole domain $V$.

Before searching for a suitable approximation $\tilde{p}$, we apply Green's integral theorem,

$$\iiint_V v\Delta p \mathrm{d}V = -\iiint_V \mathrm{grad}\, v \cdot \mathrm{grad}\, p \mathrm{d}V + \iint_S v\frac{\partial p}{\partial n}\,\mathrm{d}S \tag{3.37}$$

to Equation 3.36. $S$ denotes the area of the room boundary, and the point in the second integral indicates scalar multiplication of both gradient functions. At the same time, the derivative in the integral over $S$ is expressed by the specific wall impedance according to the boundary condition (3.4). The result of this transformation is

$$\iiint_V \left(\mathrm{grad}\, v \cdot \mathrm{grad}\, p - vpk^2\right)\mathrm{d}V + ik\iint_S vp\frac{\mathrm{d}S}{\zeta} = i\omega\rho_0 \iiint_V vq\;\,\mathrm{d}V \tag{3.38}$$

This formula contains not only the Helmholtz equation (3.6) but also the boundary condition (3.4).

The next step is the discretization of Equation 3.38. For this purpose, we approximate the complex sound pressure amplitude by the following series:

$$p(r) \approx \tilde{p}(r) = \sum_{n=1}^{N} p_n \Lambda_n(r) \tag{3.39}$$

The functions $\Lambda_n$ (r), called shape functions, are defined by the requirement to be unity at the $n$th node and to vanish at all other nodes:

$$\Lambda_n(\mathbf{r}_m) = \begin{cases} 1 & \text{for} \quad m = n \\ 0 & \text{for} \quad m \neq n \end{cases}$$

where $\mathbf{r}_m$ indicates the position on the $n$th node. For all other points within the $n$th element, the shape function $\Lambda_n$ is obtained by linear interpolation.

To illustrate the shape functions, let us have a look at the analogous one-dimensional problem (Figure 3.6). In this case, the elements are straight-line segments separated by nodes which are aligned along the coordinate axis. The shape functions resemble the Greek character $\Lambda$; they all have the height 1, and each of them connects two elements with each other. The approximation (3.39) is obtained by 'modulating' the peaks of the triangles according to

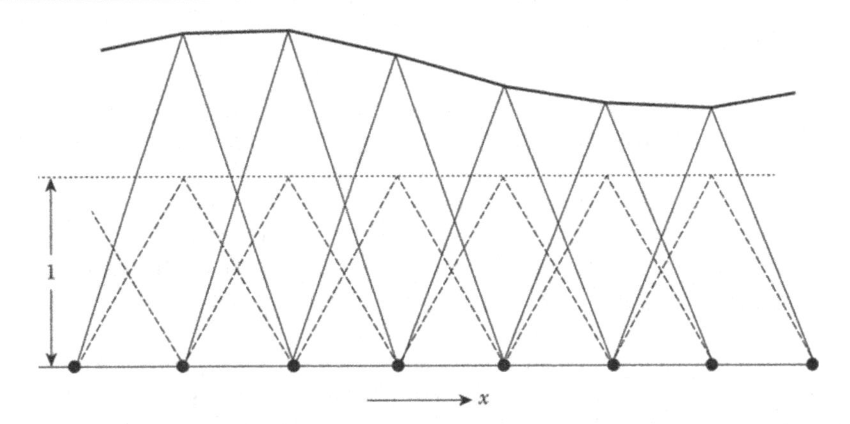


*Figure 3.6* Shape functions (dashed lines) for one-dimensional elements of equal size. The approximate solution is a piecewise linear function shown as the thick line at the top obtained by adding the modified shape functions (solid lines).

the nodal values $p_n$ and to add them. The result is a piecewise linear function (dashed line; see Figure 3.6b). For two-dimensional problems, the process is analogous; the simplest elements are triangles in a plane, each of which is defined by three nodes (the vertices of the triangle). Accordingly, each shape function looks like a tent. The three-dimensional case is a bit more difficult to imagine, but the principle of discretization is the same.

Now we substitute Equation 3.39 into Equation 3.38; at the same time, we identify the test function $v$ successively with the shape functions $\Lambda_1, \Lambda_2, \ldots, \Lambda_N$. The results of both operations are $N$ linear equations,

$$\sum_{n=1}^{N}\left(K_{mn} - k^2 M_{mn} + ikC_{mn}\right)\cdot p_n = ik\rho_0 c Q_m \qquad (m = 1,2,\ldots,N) \tag{3.40}$$

with the abbreviations

$$K_{mn} = \iiint \mathrm{grad}\,\Lambda_m \cdot \mathrm{grad}\Lambda_n \mathrm{d}V, \quad M_{mn} = \iiint \Lambda_m \Lambda_n \mathrm{d}V$$

$$C_{mn} = \iint \Lambda_m \Lambda_n \frac{\mathrm{d}S}{\xi}, \qquad Q_m = \iiint q\Lambda_m \mathrm{d}V$$

which can be solved with standard methods to determine the nodal pressures $p_n$.

The coefficients $K_{mn}$ may be conceived as the components of a quadratic and symmetric matrix **K**, usually called the stiffness matrix. Similarly, the $\mathbf{M}_{mn}$ and $\mathbf{C}_{mn}$ can be represented in a symmetric 'mass matrix' **M** and a 'damping matrix' **C**, while $p_n$ and $Q_n$ can be considered as components of single-column matrices (i.e. as vectors) **p** and **Q**. Then, Equation 3.40 can be rewritten as a matrix equation:

$$\left(\mathbf{K} - k^2\mathbf{M} + ik\mathbf{C}\right)\cdot \mathrm{p} = ik\rho_0 c\mathbf{Q} \tag{3.41}$$

The matrix components $K_{mn}$, $M_{mn}$, and so on are calculated element by element, that is, the integrations are first performed over the element volumes $V_e$ instead of the total volume $V$ (respectively, over the boundary areas $S_e$ of the considered element). Afterwards, these element matrices are assembled into the global matrices **K**, **M**, and **C**. For this process, a global numbering scheme of all elements and nodes is defined, and all matrix components are renumbered according to this scheme. Next, all $N$ element matrices will be 'expanded', that is, their components are inserted at the proper positions of an $N \times N$ matrix which originally contains only zeros. Finally, all $N$ expanded element matrices are added. The same procedure is applied to the calculation of the vector components $Q_m$ in Equation 3.41. The interested reader will find an illustrative description of the assembly process in the book by Huebner and Thornton (1982).

In practical problems, the total number of elements ($N$) may amount to hundreds of thousands. For the solution of the linear equations (3.41), it is a great advantage that most matrix components are zero and that the non-zero components are concentrated about the main diagonal of the matrices. This is a consequence of the limited 'reach' of the shape functions which connect just a few elements within each node.

It is obvious that the accuracy of the final result depends on the spatial density of nodes. In the literature, the use of at least three to ten nodes per wavelength is recommended. Suppose we want to calculate the sound pressure in a room with a volume of 100 m$^3$ at a frequency of 100 Hz, using five nodes per wavelength. Then, the total number of nodes would be about 320. Increasing the volume or the frequency by a factor of 10 would raise this number to about 320,000. This example shows the limits of FEM when it comes to the calculation of sound pressures in a large room and/or at elevated frequencies. One way to circumvent the difficulties to store large amounts of data and to handle them within acceptable times is to introduce additional nodes per element, which may be placed on the edges and the faces of the elements. Then, however, linear interpolation within the elements must be replaced with a more complicated interpolation scheme. Another possibility is to use more complicated element shapes (e.g. hexahedral ones), which also require higher-order polynomials as interpolation functions.

What has been outlined here is just one example from the numerous applications of FEM, which have been developed and are still being refined. For standard applications, complete computer codes are commercially available, for instance, for the meshing process. It should be mentioned that there are other useful numerical methods for evaluating sound fields based on finite elements, for instance, the FDM (finite difference method). The latter, however, are based on a regular pattern of nodes in contrast to the FEM, where the shapes and the sizes of the elements may be adapted to the geometry of the problem to be studied, for instance, to increase the accuracy of the approximation in critical regions of an enclosure. Another important application of finite elements is the boundary element method (BEM). It can be used to determine the acoustical quantities on the boundary of a domain, for instance, on the interior surface of a loudspeaker horn, from which the acoustic field in the inside can be calculated by well-known integral statements.

## 3.6 STEADY-STATE SOUND FIELD

As derived in Section 3.1, the steady-state acoustical behaviour of a room, when it is excited by a sinusoidal signal with angular frequency ω, is described by a series of the form

$$p_\omega = \sum_n \frac{A_n(r)}{\omega^2 - \omega_n^2 - 2i\delta_n\omega_n} \tag{3.42}$$

where we are assuming $\delta_n \ll \omega_n$ according to Section 3.1. By comparing this with Equation 3.12, we see that the coefficients $A_n$ are functions of the source position $r_0$, of the receiving position $r$ and of the angular frequency $\omega$. If both positions are considered as fixed, Equation 3.42 represents the transfer function of the room for this situation.

Since we have supposed that the damping constants $\delta_n$ are small compared with the eigenfrequencies $\omega_n$, the predominant frequency dependence is that of the denominator. Whenever the frequency is close to one of the eigenfrequencies $\omega_n$, the corresponding series term will become very large or, in other words, the system will behave as a resonator. Since the term $\omega^2 - \omega^2_n$ is responsible for the strong frequency dependence, $\omega_n$ can be replaced by $\omega$ in the last term of the denominator without any serious error. Then, the absolute value of the $n$th series term becomes

$$\frac{|A_n|}{\left[\left(\omega^2-\omega_n^2\right)^2+4\omega^2\delta_n^2\right]^{1/2}}$$

and thus agrees with the frequency characteristics of a resonance system, according to Equation 2.36. Therefore, the stationary sound pressure in a room and at one single exciting frequency proves to be the combined effect of numerous resonances. The relative half-power bandwidths of the resonance curves according to Equation 2.38 are

$$\frac{(\Delta f)_n}{f_n}=\frac{(\Delta\omega)_n}{\omega_n}=\frac{2\delta_n}{\omega_n}$$

or

$$(\Delta f)_n=\frac{\delta_n}{\pi} \tag{3.43}$$

In most full-size rooms, the damping constants lie between 1 and 20 $s^{-1}$. Therefore, our earlier assumption concerning the relative magnitude of the damping constants seems to be justified. Furthermore, the half-power bandwidths of the resonances turn out to be of the order 1 Hz, according to Equation 3.43. This figure is to be compared with the average spacing of eigenfrequencies on the frequency axis after Equation 3.28. If the latter is substantially larger than the average half-width $\langle\delta_n\rangle/\pi$, we expect that most of the room resonances are well separated, and that each of them can be individually excited and detected. In a tiled bathroom, for example, the resonances are usually weakly damped, and thus, one can often detect one or several of them by singing or humming. If, on the contrary, the average half-width of the resonances is much larger than the average spacing of the eigenfrequencies, there will be strong overlap of resonances, and the latter cannot be experimentally separated. Instead, at any frequency, several or many terms of the sum in Equation 3.41 will have significant values; hence, several or many normal modes will simultaneously contribute to the total sound pressure. According to Schroeder (1954) and Schroeder and Kuttruff (1962), a limiting frequency separating both cases can be defined by the requirement that, on average, three eigenfrequencies fall into one resonance half-width:

$$\langle\Delta f_n\rangle = 3\cdot\langle\delta f_n\rangle$$

Or with Equation 3.28:

$$\frac{\langle\delta\rangle}{\pi}=3\cdot\frac{c^3}{4\pi V f^2}$$

Solving for $f$ and introducing $c$ = 343 m/s yield the limiting frequency, the 'Schroeder frequency $f_s$':

$$f_s \approx \frac{5500}{\sqrt{V\langle\delta\rangle}} Hz \approx 2000\sqrt{\frac{T}{V}} Hz \tag{3.44}$$

In this expression, we introduced the 'reverberation time' $T = 6.91/\langle\delta\rangle$ (see Section 3.8). The room volume $V$ has to be expressed in cubic metres.

In large halls, the Schroeder frequency is typically below 50 Hz; hence, there is strong modal overlap in the whole frequency range of interest, and there is no point in evaluating single eigenfrequencies. It is only in small rooms that a part of the important frequency range lies below $f_s$, and in this range, the acoustic properties are determined largely by the values and half-widths of individual eigenfrequencies. To calculate the expected number, $N_{fs}$, of eigenfrequencies in the range from 0 Hz to $f_s$, Equation 3.44 is inserted into Equation 3.26, with the result:

$$N_{f_s} \approx 800\sqrt{T^3/V} \tag{3.45}$$

Thus, in a classroom with a volume of 200 m$^3$ and a reverberation time of 1 s, some 60–70 eigenfrequencies dominate the acoustical behaviour below the Schroeder frequency, which is about 140 Hz. This example illustrates the somewhat surprising fact that the acoustics of small rooms are, in a way, more complicated than those of large ones, where the transfer function is governed by statistics.

It should be noted that Equation 3.44 can also be read as a criterion for the acoustical size of a room, again on the grounds of its modal structure. A given room can be considered as 'acoustically large' if

$$V > \left(\frac{2000}{f}\right)^2 T \tag{3.46}$$

($f$ in Hz, $T$ in seconds, and $V$ in m$^3$). After the preceding discussion, it is not surprising that this limit depends on the frequency.

Here, we restrict the discussion to the frequency range above the Schroeder limit, $f > f_s$. Moreover, the observation point is supposed to be far enough from the sound source to make the direct sound component negligibly small. Hence, if the room under consideration is excited with a pure tone, its steady-state response is made up by contributions of several or even many normal modes with randomly distributed phases. The situation may be elucidated by the vector diagram in Figure 3.7. Each vector or 'phasor' represents the contribution of one term in Equation 3.42 (nine significant terms in this example). The resulting sound pressure is obtained as the vector sum of all components. For a different frequency or at a different point in the room, this diagram has the same general character, but it looks quite different in detail, provided that the change in frequency or location is sufficiently great.

Since the different components can be considered as mutually independent, the central limit theorem of probability theory can be applied to the real part as well as to the imaginary part of the resulting sound pressure $p_\omega$. According to this theorem, both quantities are random variables obeying a Gaussian distribution. This statement implies that the squared absolute value of the sound pressure $p$, divided by the mean of this quantity, $y = |p|^2/\langle|p|^2\rangle$,

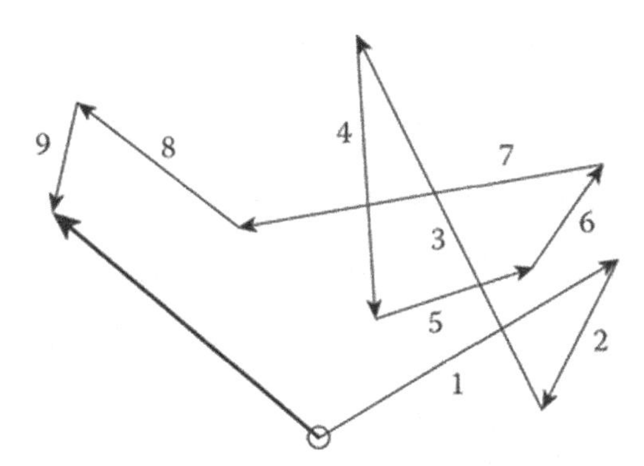


*Figure 3.7* Phasor diagram showing the components of the steady-state sound pressure in a room and their resultant.

which is proportional to the energy density, is distributed according to an exponential law, or more precisely, the probability of finding this quantity between $y$ and $y + \mathrm{d}y$ is given by

$$P(y)\mathrm{d}y = \exp(-y)\mathrm{d}y \tag{3.47}$$

The mean value $\langle y \rangle$, and also the variance $\langle y^2 \rangle - \langle y \rangle^2$ of this distribution, is 1, as is easily checked. The probability that a particular value of $y$ exceeds a given limit $y_0$ is

$$P(y > y_0) = \int_{y_0}^{\infty} \exp(-y)\mathrm{d}y = \exp(-y_0) \tag{3.48}$$

It is very remarkable that the distribution of the energy density is completely independent of the type of the room, that is, on its volume, its shape, or the treatment of its walls.

Figure 3.8a presents a typical 'space curve', that is, the sound pressure level recorded with a microphone along a straight line in a room while the driving frequency is kept constant. Such curves express the space dependence of $p_\omega$ in Equation 3.42 or, more explicitly, in Equation 3.12. Their counterparts are the 'frequency curves', that is, representations of the sound pressure level observed at a fixed microphone position when the excitation frequency is slowly varied. They are based on the frequency dependence of Equation 3.42. A section of such a frequency curve is shown in Figure 3.8b. It would look quite different in detail if recorded at another microphone position or in another room; its general character, however, would be similar to the one shown. The same statement holds for space curves.

Both curves in Figure 3.8 look quite similar: they are highly irregular and show flat peaks and deep valleys. A maximum of the pressure level occurs if in Figure 3.7 many or all arrows happen to point in about the same direction, indicating similar phases of most contributions. Similarly, a minimum appears if these contributions more or less cancel each other. Therefore, the maxima of frequency curves are not related to particular room resonances or eigenfrequencies but are the result of accidental phase coincidences. The general similarity of space and frequency curves is not too surprising: both curves sample the same distribution of squared sound pressure amplitudes, namely, that given by Equation 3.42, but they do it in a different way.

We may ask how much acoustic energy is delivered by a point source operating in a room. The final fate of this energy is its absorption at the boundary of the room (we neglect here the losses occurring in the air). One might suppose that the wall losses depend

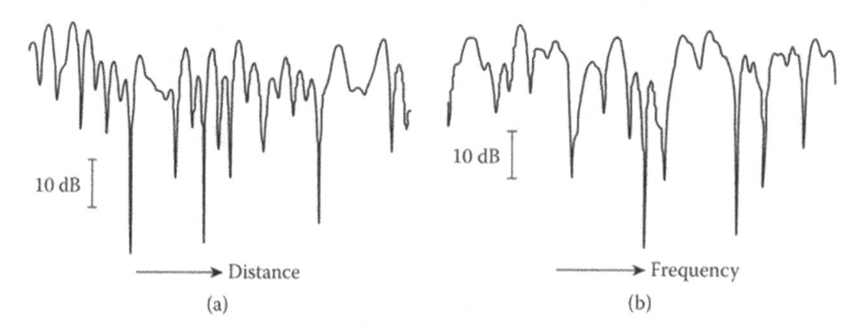


*Figure 3.8* Steady-state sound pressure amplitude (logarithmic representation): (a) along a straight line at constant frequency ('space curve') and (b) at a fixed position with slowly varying driving frequency ('frequency curve').

not only on the physical properties of the boundaries but also on the structure of the wave field, the spatial distribution of maxima and minima of the acoustic pressure, and so on. It was Maling (1967) who studied this problem starting from the Green's function of rectangular rooms after Equation 3.10. He found that the sound power emitted by a point source in a room is the same as that radiated in the free space, provided there is significant overlapping of room resonances, that is, the frequency is well above the Schroeder limit $f_s$ after Equation 3.44. This result, which can easily be generalized to a room of arbitrary shape, is easy to understand: primarily, the point source produces a spherical wave in which the sound pressure is inversely proportional to the distance from the source. Suppose we approach the source from some remote position where the sound pressure of the primary wave is negligibly small; with decreasing distance, the primary sound pressure will grow monotonically. Eventually, it will become dominant so that the 'distortions' of the sound field caused by the room boundaries can be neglected. Accordingly, the power the source emits into the room is the same as it would produce in the free space (see Equation 1.39, second part).

## 3.7 SOME PROPERTIES OF ROOM TRANSFER FUNCTIONS

In this section, some properties of 'frequency curves' are presented without giving their derivations, which the interested reader may look up in the cited references. All these properties are valid for the frequency range of overlapping modes, that is, for $f > f_s$. Furthermore, it is assumed that the distance of the receiving point and the exciting point source is so large that the primary spherical wave mentioned at the end of the preceding section can be neglected.

As explained in Section 1.4, the degree of causality within a time function can be quantified by its autocorrelation function. In the present case, we can apply this concept to the energetic transfer function, either considered in dependence of a space coordinate $y(x)$ or to its frequency dependence $y(f)$. In both cases, we consider the normalized quantity $y = \frac{|p|^2}{\langle |p|^2 \rangle}$ as the significant variable; its distribution function is given by Equation 3.47. For the space dependence in rooms with an isotropic (diffuse) sound field, the autocorrelation function reads (Pierce 1981)

$$\phi_{yy}(\Delta x) = \langle y(x) \cdot y(x + \Delta x) \rangle = 1 + \left( \frac{\sin(k\Delta x)}{k\Delta x} \right)^2 \tag{3.49}$$

with $x$ denoting the coordinate along the straight line where the pressure level is recorded. The acute brackets are to indicate ensemble averages.

The autocorrelation function of frequency curves is given by (Schroeder 1962)

$$\phi_{yy}(\Delta f) = \langle y(f) \cdot y(f + \Delta f) \rangle = 1 + \frac{1}{1 + (\pi \Delta f / \langle \delta_n \rangle)^2} \tag{3.50}$$

The numeral 1 in Equations 3.49 and 3.50 is due to the constant component of the function $y(x,f)$, that is, to the mean value $\langle y \rangle$. Both autocorrelation functions are plotted in Figure 3.9. As long as the autocorrelation function $\phi_{yy}(\Delta x)$ is noticeably different from unity, it indicates that there is still some causal relationship between any two samples of $y$ taken at two points $\Delta x$ apart. A similar consideration applies to $\phi_{yy}(\Delta f)$.

The average distance of adjacent maxima of space curves in a diffuse sound field is

$$\langle \Delta x_{\max} \rangle \approx 0.79\lambda \tag{3.51}$$

For the corresponding quantity of frequency curves, the mean spacing of maxima is

$$\langle \Delta f_{\max} \rangle \approx \frac{\langle \delta_n \rangle}{\sqrt{3}} \approx \frac{4}{T} \tag{3.52}$$

Again, $T = 6.91/\langle \delta \rangle$ denotes the reverberation time, as in Equation 3.44.

A quantity which is especially important with regard to the performance of sound reinforcement systems in rooms is the absolute maximum $y_{max}$ of a frequency curve to be expected within a given frequency bandwidth $B$. In order to estimate it, we assume that the frequency curve can be represented by $N$ equidistant and statistically independent samples. We define the absolute maximum $y_{\max}$ by requiring that there is just one sample which equals or exceeds $y_{\max}$, while the remaining $N - 1$ samples are smaller. The probability of this condition to happen is

$$P_N(y_{\max}) = C \cdot \exp(-y_{\max}) \cdot [1 - \exp(-y_{\max})]^{N-1}$$

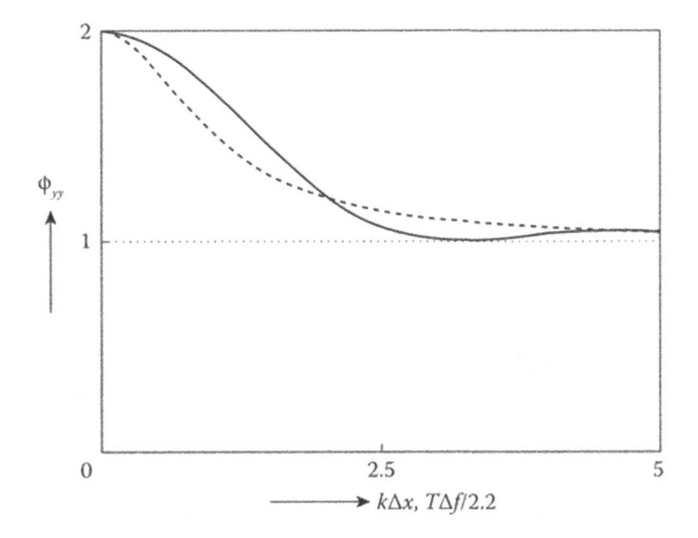


*Figure 3.9* Autocorrelation functions $\varphi_{yy}$ of room responses: solid line represents $\phi_{yy}(\Delta x)$ after Equation 3.49 (the abscissa is $k\Delta x$); dashed line represents $\phi_{yy}(\Delta f)$ after Equation 3.50 (the abscissa is $\pi\Delta f/\langle\delta\rangle = T\Delta f/2.2$).

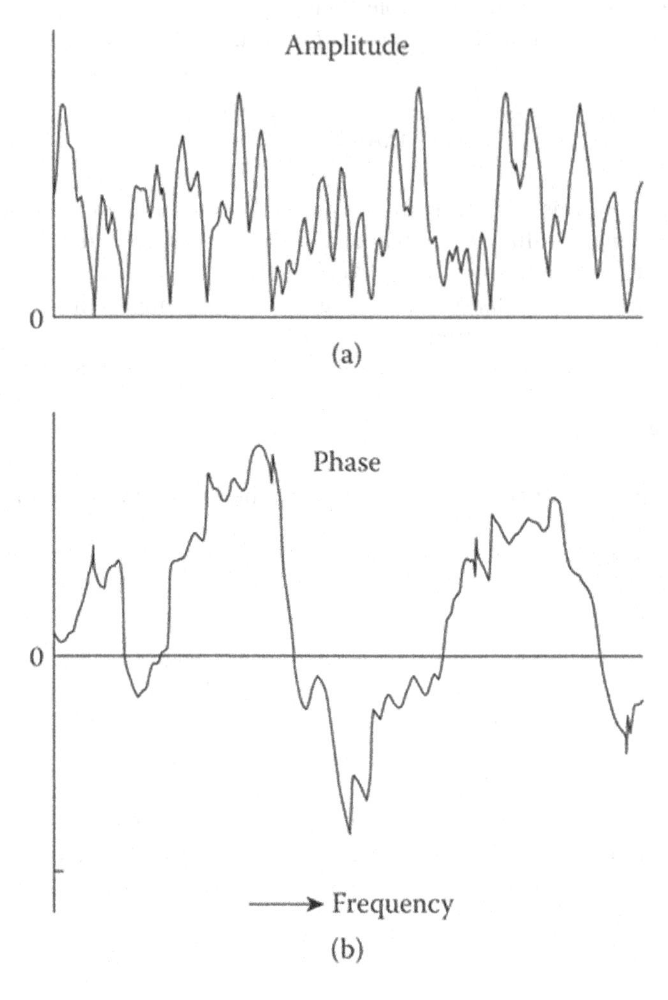


*Figure 3.10* (a) Magnitude and (b) phase of a typical room transfer function above the Schroeder limit $f_s$.

where we have used Equation 3.47. $C$ is some normalization constant. To find the maximum of $P_N$ $(y_{max})$ derivative, this function with respect to $y_{max}$ is set equal to zero, which yields $y_{max} = \ln N$. The level difference $\Delta L_{max} = 10 \cdot \log_{10}(y_{max})$ between $y_{max}$ and the mean value of $y$ is

$$\Delta L_{max} = 10 \cdot \log_{10}(\ln N) = 4.34 \ln(\ln N) \text{ dB}$$

To estimate the number $N$ of samples, one should make sure that their density is high enough to represent the frequency curve. On the other hand, their distance to the frequency axis should be wide enough to assure their independence. A fair compromise between these conflicting requirements is achieved if we take four samples per average spacing $\langle \Delta f_{max} \rangle$. With Equation 3.52, this leads to $N \neq BT$ and, finally, to

$$\Delta L_{max} = 10 \cdot \log_{10}\left[\log_{10}(BT)\right] + 3.6 \text{ dB} = 4.34 \ln[\ln(BT)] \text{ dB} \tag{3.53}$$

This interesting formula is based on the formula by Schroeder (1961a), who derived it in a somewhat different manner. It is certainly not free of some arbitrariness. On the other hand, the number $N$ is not very critical, since the double logarithm depends only slightly on its argument. Later, the same author has given a more accurate formula (Schroeder 1964) for $\Delta L_{max}$ which, however, does not significantly differ from Equation 3.53. In most practical situations, $\Delta L_{max}$ is about 10 dB.

If the driving frequency of a sound source is slowly varied, both the amplitude of the sound pressure in any room and its phase fluctuate in an irregular manner. Apart from these quasi-statistical fluctuations, there is a monotonic variation of the phase angle. The average phase shift per hertz is given by (Schroeder 1961b)

$$\left\langle \frac{d\psi}{df} \right\rangle = \frac{\pi}{\langle \delta_n \rangle} \approx 0.455T \tag{3.54}$$

Figure 3.10 shows in its upper part an amplitude–frequency curve. In contrast to Figure 3.8b, the plotted quantity is not the sound pressure level but the squared absolute value of the sound pressure, $|p_\omega|^2$. The lower part plots the corresponding phase variation obtained after subtracting the monotonic increase according to Equation 3.56. It consists of quasi-stochastic phase fluctuations. Accordingly, the phase spectrum of any signal transmitted in the room will be randomized by these fluctuations.

This can be demonstrated in the following way: A loudspeaker placed in a reverberant room is alternatively fed with two signals which have equal amplitude spectra but different phase spectra. The first signal, for instance, may be a periodic sequence of rectangular impulses (see Figure 3.11a), whereas the second one is a maximum length sequence (see Section 8.2) made up of rectangular impulses with quasi-randomly changing signs (see Figure 3.11b). If the listener is close to the loudspeaker, he can clearly hear that both signals sound quite different, provided the repetition rate $1/T$ is not too high. However, when the listener slowly steps away from the loudspeaker, the room field will prevail over the direct sound signal; the perceived difference of both signals becomes smaller and smaller and finally disappears.

When a room is acoustically excited not by just one harmonic sound signal but by a mixture of several or many sine signals with different frequencies, there will occur some frequency averaging by which the frequency response is smoothed. A similar effect is brought about by spatial averaging of the response of a room to a sinusoidal excitation signal.

The simplest case is that of a room simultaneously excited with $M$ discrete sine signals of different frequencies but of equal strengths. As in the preceding section, we denote the energetic room response to monofrequent excitation with $y = \frac{|p|^2}{\langle |p|^2 \rangle}$. Then, the result of applying $M$ sine signals is $z = y_1 + y_2 + \ldots + y_M$, provided the samples $y_n$ are uncorrelated.

This is the case when the excitation frequencies are sufficiently well separated — a conclusion which can be drawn from the shape of the autocorrelation function $\Phi_{yy}$ (see Figure 3.9, broken line). Under this condition, the average of $z$ is $M$, according to probability theory. Furthermore, the variance of the sum, $\mathrm{Var}(z) = \langle z^2 \rangle - \langle z \rangle^2$ is $M$ times the variance of $y$, which is 1, as can be concluded from Equation 3.47. Hence, the relative variance of the sum $z$ is

$$\frac{\mathrm{Var}(z)}{\langle z \rangle^2} = \frac{1}{M} \tag{3.55}$$

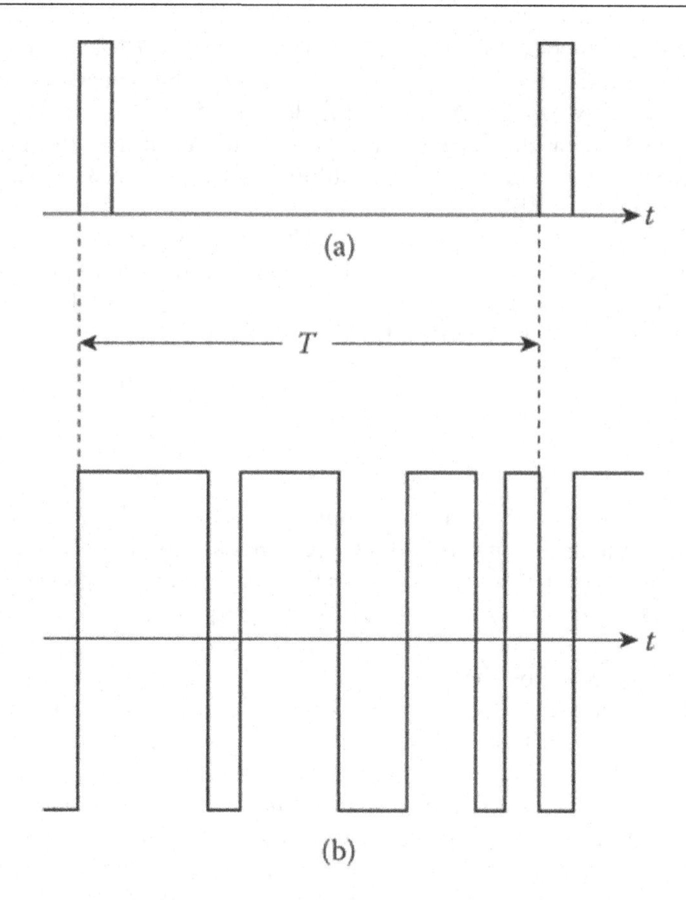


*Figure 3.11* Two periodic signals with equal amplitude spectrum but different phase spectra: (a) periodically repeated rectangular pulses and (b) periodic maximum length sequence of rectangular impulses.

The same result is obtained for the sum of $M$ samples $y_n$ taken at well-separated points of the sound field, while the source is emitting a monofrequent signal.

Matters are more complicated if the enclosure is excited with a signal having a constant frequency spectrum within the bandwidth $B = f_2 - f_1$. This is because then we do not average uncorrelated samples (Schroeder 1969). Then, we consider the average

$$z = \frac{1}{B}\int_{f_1}^{f_2} y(f)\mathrm{d}f \quad \text{with } y = \frac{|p(f)|^2}{\langle |p(f)| \rangle^2} \tag{3.56}$$

Again, we have $\langle y \rangle = 1$ and $\langle z \rangle = 1$ because of the exponential distribution (3.47). The average of $z^2$ is

$$\langle z^2 \rangle = \frac{1}{B^2}\left\langle \left[ \int_{f_1}^{f_2} y(f)\mathrm{d}f \right]^2 \right\rangle = \frac{1}{B^2}\int_{f_1}^{f_2} \mathrm{d}f \int_{f_1}^{f_2} \mathrm{d}f' \langle y(f) y(f') \rangle$$

Evaluation of this double integral leads to the following expression for the 'relative variance' of $z$:

$$\frac{Var(z)}{\langle z \rangle^2} = \frac{2\langle \delta \rangle}{\pi B} \arctan\left(\frac{\pi B}{\langle \delta \rangle}\right) - \frac{\langle \delta \rangle^2}{\pi^2 B^2} \ln\left(1 + \frac{\pi^2 B^2}{\langle \delta \rangle^2}\right) \tag{3.57}$$

where $\delta_n$ has been replaced with the mean damping constant $\langle\delta\rangle$.

This quantity is plotted in Figure 3.12 (dashed line) as a function of the variable

$$\frac{\pi B}{\langle\delta\rangle} = \frac{BT}{2.2}$$

If the power spectrum $w(f)$ of the exciting signal is not constant within the exciting frequency band, the bandwidth $B$ is replaced by an 'equivalent bandwidth':

$$B_{eq} = \left[\int_0^\infty w(f)\mathrm{d}f\right]^2 / \int_0^\infty \left[w(f)\right]^2 \mathrm{d}f$$

Basically, the same procedure as was used in deriving Equation 3.57 can be applied to calculate the space average over a straight path of length $L$. In this case, we consider the average

$$z = \frac{1}{L}\int_0^L y(f)\,\mathrm{d}f \tag{3.58}$$

The result is a somewhat-lengthy formula, the content of which is plotted in Figure 3.12 as a function of $kL$. Both curves in this diagram are quite similar. Therefore, we can use Equation 3.56, too, as an approximate expression for the relative variance of the average (3.58) by replacing the variable $\pi B/\langle\delta\rangle$ with $kL$:

$$\frac{Var(z)}{\langle z\rangle^2} \approx \frac{2}{kL}\arctan(kL) - \frac{1}{(kL)^2}\ln\left[1+(kL)^2\right] \tag{3.59}$$

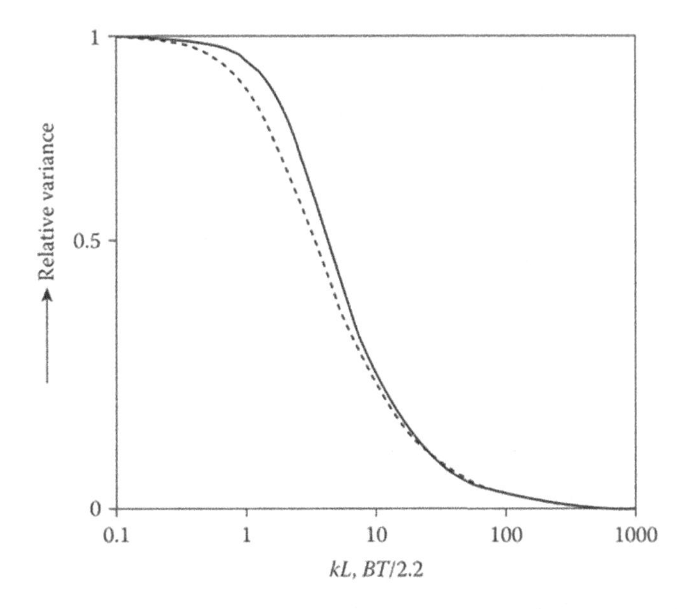


*Figure 3.12* Relative variance of space and frequency averaged room responses (squared magnitude of sound pressure). Solid line represents space averaging over a distance $L$ (the abscissa is $kL$); dashed line represents frequency averaging over a frequency interval $B$ (the abscissa is $\pi B/\langle\delta\rangle = BT/2.2$).

## 3.8 DECAYING MODES, REVERBERATION

If a room is excited not by a stationary signal, as in the preceding sections, but instead by a very short sound impulse emitted at time $t = 0$, we obtain, in the limit of vanishing pulse duration, the impulse response $g(t)$ valid for a particular receiving point of the room. According to the discussion in Section 1.4, this is the Fourier transform of the transfer function:

$$g(t) = \int_{-\infty}^{\infty} p_\omega \exp(i\omega t)\mathrm{d}\omega$$

Generally, the evaluation of this Fourier integral with $p_\omega$ after Equation 3.42 is rather complicated. At any rate, the solution has the form

$$g(t) = \sum_n A_n' \exp(-\delta_n' t)\cos(\omega_n' t + \phi_n') \qquad \text{for } t \geq 0 \tag{3.60}$$

It is composed of sinusoidal oscillations with different frequencies, each dying out with its individual damping constant. This is plausible, since each term of Equation 3.42 corresponds to a resonator whose reaction to an excitation impulse is a damped oscillation. If the wall losses in the room are not too large, the frequencies $p_\omega$ and damping constants $\delta_n'$ differ only slightly from $\omega_n$ and $\delta_n$, as occur in Equation 3.42. As is seen from the more explicit representation (3.12), the coefficients $A_n'$ depend on the location of both the source and the receiving point.

By squaring the expression (3.60), we obtain a quantity which is proportional to the energy density of the decaying sound field:

$$\left[g(t)\right]^2 = \sum_n \sum_m A_n' A_m' \exp\left[-(\delta_n' + \delta_m')t\right] \cdot \cos(\omega_n' t - \phi_n)\cos(\omega_m' t - \phi_m) \tag{3.61}$$

This expression can be considerably simplified by short-time averaging, that is, averaging only the product of cosines but not the slowly varying exponential. Hence, the products with $n \neq m$ will cancel, whereas each term $n = m$ yields a value 1/2. Thus, we obtain for the energy density

$$w(t) = \mathrm{ave}\left\{\left[g(t)\right]^2\right\} = \sum_n A_n'^2 \exp(-2\delta_n' t) \qquad \text{for } t \geq 0 \tag{3.62}$$

where all irrelevant constants have been omitted.

If the energy support to the room is effected not by a very short impulse but by a stationary signal which is switched off at $t = 0$, the energetic room response is given as the sum of all previously generated decays according to Equation 3.62:

$$w_r(t) = \int_{-\infty}^{0} w(t-\tau)\mathrm{d}\tau = \sum_n \frac{A_n'^2}{2\delta_n} \exp(-2\delta_n t) \qquad \text{for } t \geq 0 \tag{3.63}$$

This version of the decay process is called the 'reverberation' of the room. It is one of the most important and striking acoustical phenomena of a room, familiar also to every layman.

Now, we imagine that this sum in Equation 3.63 is rearranged in order of increasing damping constants $\delta_n'$. We suppose the number of significant terms in Equation 3.63 as very large. Then, we can replace the summation by an integration. This is done by collecting the contributions of all terms with damping constants between $\delta$ and $\delta + \mathrm{d}\delta$ in a continuous function called the 'damping density' $H(\delta)\mathrm{d}\delta$, which is normalized by requiring

$$\int_0^\infty H(\delta)\,\mathrm{d}\delta = 1$$

Then, the integral envisaged is

$$w_r(t) = \int_0^\infty H(\delta)\exp(-2\delta t)\,\mathrm{d}\delta \qquad \text{for } t \geq 0 \tag{3.64}$$

Just as the coefficients $A_n$ or $A'_n$, the distribution of damping constants $H(\delta)$ depend on the sound signal and on the locations of the sound source and of the observation point.

From this representation, we can derive some interesting general properties of reverberation. Usually, measurements of the reverberation are based rather on the sound pressure level of the decaying sound field than on the sound pressure itself or, since the energy density $w$ is proportional to the square of the sound pressure (see Equation 1.64):

$$L_\mathrm{r} = 10 \cdot \log_{10}\left(\frac{w_\mathrm{r}(t)}{w_\mathrm{r}(0)}\right) = 4.34 \cdot \ln\left(\frac{w_\mathrm{r}(t)}{w_\mathrm{r}(0)}\right)\mathrm{dB} \tag{3.65}$$

In Figure 3.13, some examples of damping densities are presented, along with the corresponding logarithmic reverberation curves according to Equation 3.64. The distributions are normalized such that they have the same mean value. Only when all the damping constants are equal (Case d) are the decay curves straight.

The decay rate of the sound pressure level is

$$\dot{L}_\mathrm{r} = 4.34\frac{\dot{w}_\mathrm{r}}{w_\mathrm{r}}\,\mathrm{dB/s} \tag{3.66}$$

while the second derivative of the decay level is

$$\ddot{L}_\mathrm{r} = 4.34\frac{w_\mathrm{r}\ddot{w}_\mathrm{r} - \dot{w}_\mathrm{r}^2}{w_\mathrm{r}^2}\,\mathrm{dB/s^2} \tag{3.67}$$

(Each overdot in these formulae means one differentiation with respect to time.) It can be shown that this derivative — along with Equation 3.64 — cannot be negative, which means that the decay curves are curved upwards or are straight lines. For $t = 0$, a logarithmic decay curve has its maximum steepness; the initial slope, as obtained from Equation 3.66, is proportional to the mean value of the distribution $H(\delta)$:

$$\left(\dot{L}_r\right)_{t=0} = -8.69 \cdot \int_0^\infty H(\delta)\delta\,\mathrm{d}\delta = -8.69 \cdot \langle\delta\rangle \tag{3.68}$$

Furthermore, Equation 3.67 leads to

$$\left(\ddot{L}_\mathrm{r}\right)_{t=0} = -17.37 \cdot \left(\langle\delta^2\rangle - \langle\delta\rangle^2\right) \tag{3.69}$$

This means that the second derivative of the level at $t = 0$, which is roughly the initial curvature of the decay curve, is proportional to the variance of the damping density $H(\delta)$.

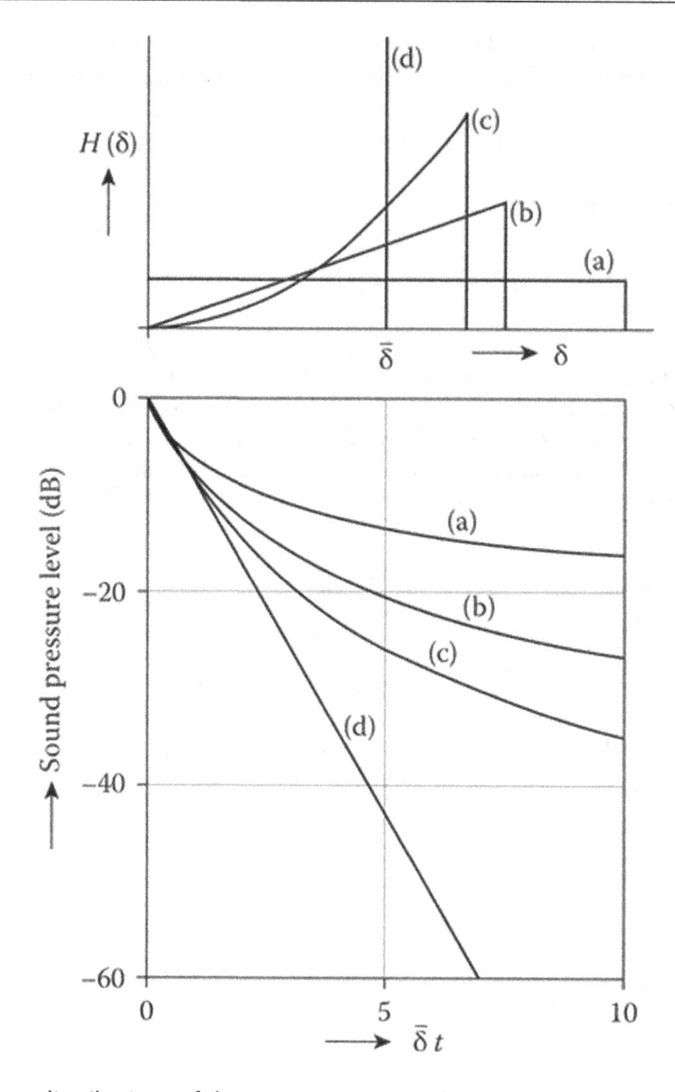


*Figure 3.13* Various distributions of damping constants and corresponding decay curves.

Measured reverberation curves are often straight or nearly straight, apart from some random or quasi-random fluctuations, as shown in Figure 3.14. (The latter are caused by beats between the decaying modes, that is, by incomplete cancellation of the terms with $n \neq m$ in Equation 3.61.) Then, all decay constants can be replaced without much error by their average $\langle\delta\rangle$.

It is usual in room acoustics to characterize the duration of sound decay the 'reverberation time' or 'decay time' $T$, introduced by Sabine. It is defined as the time interval in which the decay level drops down by 60 dB. From

$$-60 = 10 \cdot \log_{10}\left[\exp(-2\langle\delta\rangle)T\right]$$

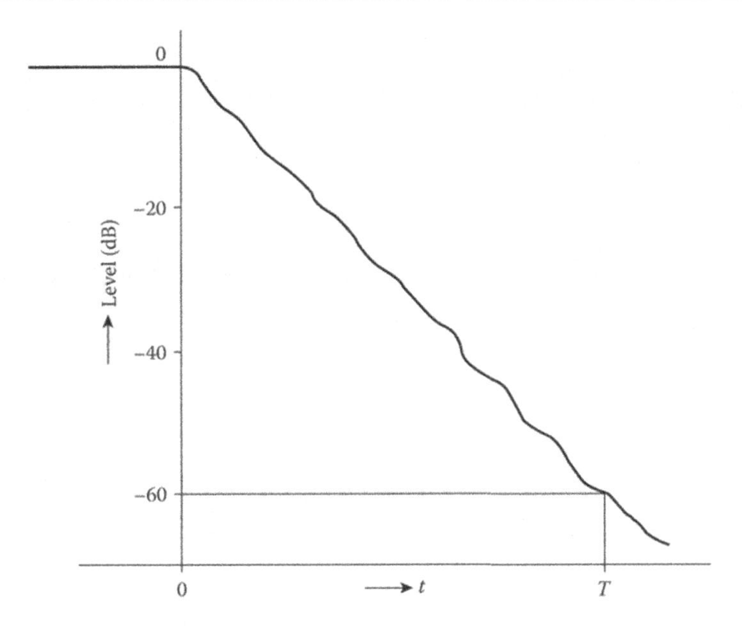


*Figure 3.14* Definition of the reverberation time.

it follows that the reverberation time is

$$T = \frac{3 \cdot \ln(10)}{\langle \delta \rangle} \approx \frac{6.91}{\langle \delta \rangle} \tag{3.70}$$

a relation which has already been used in preceding sections. Typical values of reverberation times run from about 0.3 s (living rooms) up to 10 s (large churches, empty reverberation chambers). Most large halls have reverberation times between 0.7 and 2 s. Thus, the average damping constants encountered in practice are in the range 1 to 20 $s^{-1}$.

## REFERENCES

Astley R.J. Numerical acoustical modelling (finite element modelling). In: Crocker M.J. (ed.), *Handbook of Noise and Vibration Control*. Hoboken: Wiley, 2007, pp. 101–115.

Huebner K.H., Thornton E.A. *The Finite Element Method for Engineers*, 2nd edn. New York: Wiley, 1982.

Maling Jr. G.C., Calculation of the acoustic power radiated by a monopole in a reverberation chamber. *J Acoust Soc Am* 1967; 42: 859.

Morse P.M., Feshbach H. *Methods of Theoretical Physics*. New York: McGraw-Hill, 1953. Chapter 6.

Morse P.M., Ingard K.U. *Theoretical Acoustics*. New York: McGraw-Hill, 1968. Chapter 9.

Pierce A.D. *Acoustics-An Introduction to Its Physical Principles and Applications*. New York: McGraw-Hill, 1981.

Schroeder M.R. The statistical parameters of frequency curves in large rooms (in German). *Acustica* 1954; 4: 594.

Schroeder M.R. Improvement of acoustic feedback stability in public address systems. Proceedings of the Third International Congresses on Acoustics, Stuttgart, 1959. Amsterdam: Elsevier, 1961a, p. 771.

Schroeder M.R. Measurement of reverberation time by counting phase coincidences. Proceedings of the Fifth International Congresses on Acoustics, Stuttgart, 1959. Amsterdam: Elsevier, 1961b, p. 897.
Schroeder M.R. Frequency-correlation functions of frequency responses in rooms. *J Acoust Soc Am* 1962; *34*: 1819.
Schroeder M.R. Improvement of acoustic feedback stability by frequency shifting. *J Acoust Soc Am* 1964; *36*: 1718.
Schroeder M.R. Effect of frequency and space averaging on the transmission responses of multimode media. *J Acoust Soc Am* 1969; *46*: 277.
Schroeder M.R., Kuttruff K.H. On frequency response curves in rooms. Comparison of experimental, theoretical and Monte Carlos results for the average frequency spacing of maxima. *J Acoust Soc Am* 1962; *34*: 76.

Chapter 4

# Geometrical room acoustics

In Section 3.5, we described one possible way to overcome the difficulties in computing sound fields in a realistic room: the application of numerical procedures, in particular the method of finite elements. In this chapter, quite a different approach will be chosen, namely, a substantial simplification of the propagation laws. This is achieved by considering the limiting case of vanishingly small wavelengths, that is, of very high frequencies. This approach is the basis of what is called geometrical room acoustics; it is justified whenever the dimensions of the room and its boundary, including all its details, are large compared with the acoustical wavelength. This condition is not unrealistic in room acoustics; at a medium frequency of 1,000 Hz corresponding to a wavelength of 34 cm, the linear dimensions of the walls and the ceiling, and also the distances travelled by the sound waves, are usually much larger than the wavelength. Even if we regard the reflection of sound from a balcony face, for instance, a geometrical description is applicable, at least qualitatively, keeping possible diffraction effects (see Section 2.7) in the back of the mind.

In geometrical room acoustics, the concept of sound waves is replaced by the concept of sound rays. The latter is an idealization just as much as the plane wave. As in geometrical optics, we mean by a *sound ray* a small sector of a spherical wave with vanishingly small aperture which originates from a certain point. Provided the medium is homogeneous, the sound in a ray travels along a straight line, and its energy remains constant during propagation as long the medium itself does not cause any energy losses. Another fundamental fact is the manner in which a sound ray is reflected from a wall. Much has been said about this in Chapter 2. However, the finite velocity of propagation must be taken into consideration since it is responsible for many typical effects, such as reverberation, echoes, and so on.

Any typical wave effects such as diffraction are neglected in geometrical room acoustics, since propagation in straight lines is its main postulate. Likewise, interference is usually not considered, that is, if several sound field components are superimposed, their mutual phase relations are not taken into account; instead, simply their energy densities or their intensities are added. As explained in Section 2.5, this is permissible if the different components of the sound field are mutually incoherent.

It is self-evident that geometrical room acoustics can reflect only a partial aspect of the acoustical phenomena occurring in a room. This aspect is, however, of great importance because of its conceptual simplicity and the ease of practical sound field computations.

## 4.1 ENCLOSURES WITH PLANE WALLS, IMAGE SOURCES

If a sound ray strikes a solid surface, it is usually reflected from it. If the surface is sufficiently smooth, this process follows the same reflection law as known from in optics. It states that the ray during reflection remains in the plane defined by the incident ray and normal to the

DOI: 10.1201/9781003389873-4

surface, and that the angle between the incident ray and reflected ray is halved by normal to the wall. In vector notation, this law, which is illustrated in Figure 4.1, reads

$$\mathbf{u}' = \mathbf{u} - 2(\mathbf{un}) \cdot \mathbf{n} \tag{4.1}$$

Here, **u** and **u′** are unit vectors pointing in the direction of the incident and the reflected sound ray, respectively, and **n** is the normal unit vector at the point where the arriving ray intersects the surface. The bracket is to indicate the scalar product of the vectors **u** and **n**.

One simple consequence of this law is that any sound ray which undergoes a double reflection in an edge formed by two perpendicular surfaces will travel back in the same direction, as shown in Figure 4.2a, no matter from which direction the incident ray arrives. If the angle of the edge deviates from a right angle by $\delta$, the direction of the reflected ray will differ by $2\delta$ from that of the incident ray (Figure 4.2b). A similar effect occurs in the three-dimensional case of a ray running into the corner formed of three perpendicular planes.

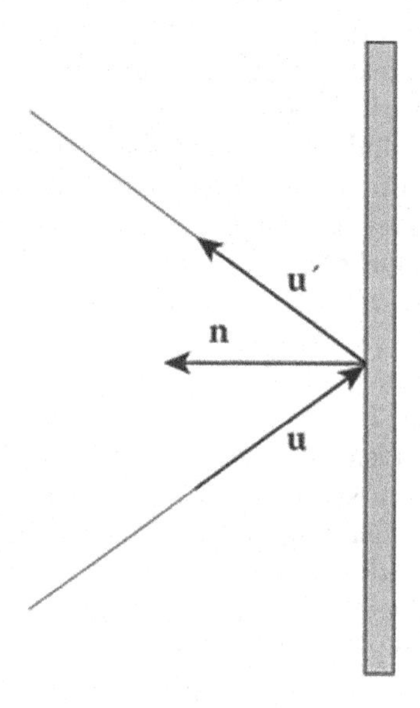


*Figure 4.1* The law of specular sound reflection from a plane.

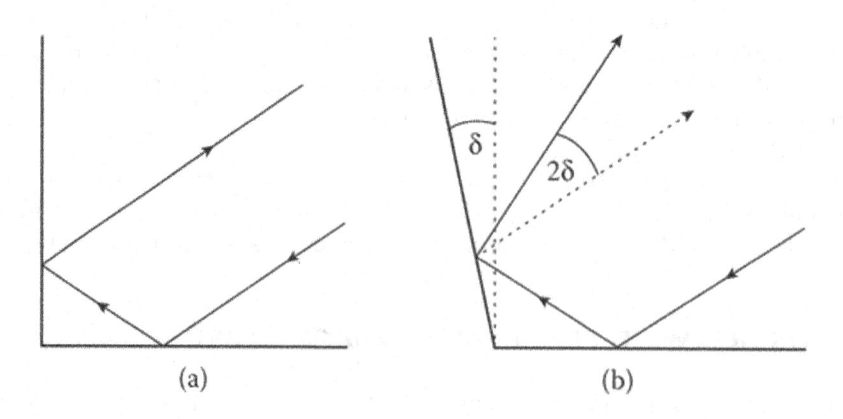


*Figure 4.2* Reflection of a sound ray from a corner (two-dimensional): (a) rectangular corner, (b) non-rectangular corner.

In this section and in the next two sections, the law of specular reflection will be applied to enclosures, the boundaries of which are composed of plane and smooth walls. In this case, one can benefit from the notion of image sources, which greatly facilitates the construction of sound paths within the enclosure. This is explained in Figure 4.3. Suppose that there is a point source A in front of a plane wall or wall section. Then, each ray reflected from this wall can be thought of as originating from a virtual sound source $A'$ which is located behind the wall, on the line perpendicular to the wall and at the same distance from it as the original source $A$. Without the image source, the reflection path connecting the sound source $A$ with a given point $B$ could only be found by trial and error.

Once we have constructed the image source $A'$ associated with a given original source $A$, we can disregard the wall altogether, the effect of which is now replaced by that of the image source. Of course, the image emits exactly the same sound signal as the original source, and its directional characteristics are symmetrical to those of $A$. If the extension of the reflecting wall is finite, then we must restrict the directions of emission of $A'$ accordingly or, put in a different way, for certain positions of the observation point B, the image source may become 'invisible'. This is the case if the line connecting B with the image source does not intersect the actual wall.

Usually, not all the energy striking a wall is reflected from it; part of the energy is absorbed by the wall (or it is transmitted to the other side, which has the same effect as far as the reflected fraction is concerned). The fraction of sound energy (or intensity) which is not reflected is characterized by the absorption coefficient $\alpha$ of the wall, which has been defined in Section 2.1 as the ratio of the non-reflected to the incident intensity. It depends generally, as we have seen, on the angle of incidence and, of course, on the frequency spectrum of the incident sound. Thus, the reflected ray has generally a different power spectrum and a lower intensity than the incident one. Using the picture of image sources, these circumstances can be

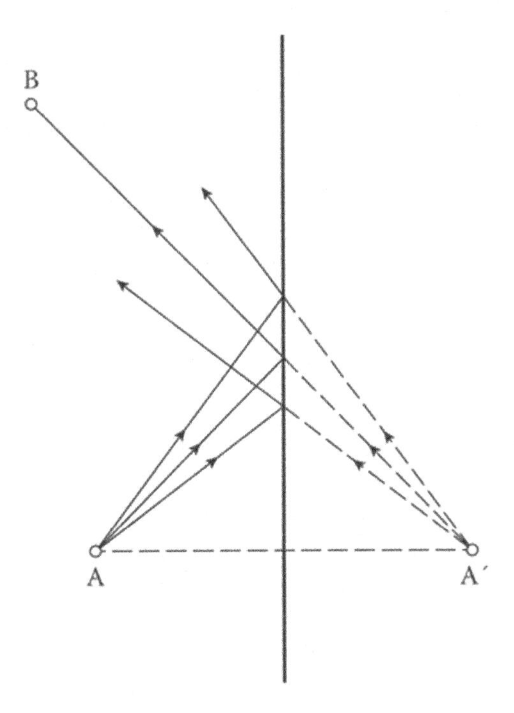


*Figure 4.3* Construction of an image source.

taken into account by modifying the spectrum and the directional distribution of the sound emitted by $A'$. With such refinements, however, the usefulness of the concept of image sources is degraded considerably. It is more convenient to employ some mean value $\alpha$ of the absorption coefficient, for instance, $\alpha_{uni}$ of Equation 2.53, valid for random sound incidence, and accordingly to reduce the intensity of the reflected ray by a factor $1 - \alpha$. As an alternative, the image source can be thought to emit a sound power reduced by this factor.

If a reflected sound ray strikes a second wall, the continuation of the sound path can be found by repeating the mirroring process, as shown in Figure 4.4. Accordingly, a second-order image $A''$ is constructed, which is the mirror image of $A'$ with respect to that wall. Continuing in this way, more and more image sources of increasing order are created. The double reflection reduces the intensity of the contribution of $A''$ by a factor $(1 - \alpha_1) \cdot (1 - \alpha_2)$.

Strictly speaking, the concept of image sources is only exact if the reflecting boundary has the specific impedance +1 or −1. In all other cases, the results obtained with it are not quite correct, since quantities such as the reflection factor or the absorption coefficient are defined for plane waves only, while the waves originating from the real sound source and its images are spherical. This problem has been discussed in detail in Section 2.4. We expect that the errors are tolerable as long as the distance of the source from the reflecting surface is at least a few wavelengths.

For a given enclosure and sound source position, the image sources can be constructed without referring to a particular sound path. Suppose the enclosure is made up of $N$ plane walls. Each of them is associated with one first-order image of the original sound source. Now, each of these image sources is mirrored by all walls except one; hence, there are $N(N - 1)$ images of second order. By repeating this procedure again and again, a rapidly growing number of images is generated with increasing distance from the original source. The number of images of order $i$ is $N(N - 1)^{i-1}$ for $i \geq 1$; the total number of images of order up to $i_0$ is obtained by adding all these expressions:

$$\nu(i_0) = N\frac{(N-1)^{i_0} - 1}{N-2} \tag{4.2}$$

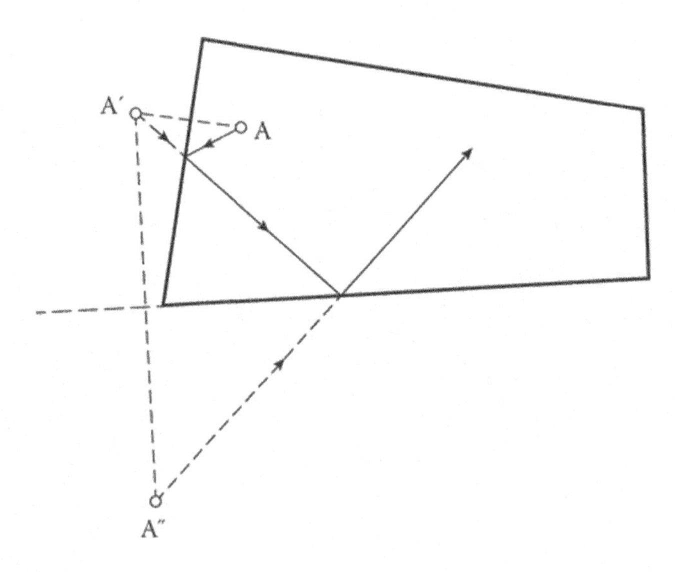


*Figure 4.4* Image sources of first and second order.

For enclosures with high symmetry (see Figure 4.6, for instance), many of the higher-order images coincide. It should be noted, however, that each image source has its own directivity, since it 'illuminates' only a limited solid angle, determined by the limited extension of the walls. Hence, it may well happen that a particular image source is 'invisible' (or, rather, 'inaudible') from a given receiving location. Moreover, only those image sources are valid which have been generated by mirroring at the inside of a wall (i.e. at the side facing the enclosure). These problems have been carefully discussed by Borish (1984). More will be discussed about this subject in Section 9.8.

These complications are not encountered with enclosures of high regularity, the image sources of which form regular patterns. As a simple example, which may also illustrate the usefulness of the image model, let us consider a very flat room, the height of which is small compared with its lateral dimensions. For locations far from the side walls, the effect of the latter may be totally neglected. Then, we arrive at a space which is bounded by two parallel, infinite planes. For the sake of simplicity, we assume that both the sound source *A* and the observation point *B* are located in the middle between both planes, that the source radiates the power *P* uniformly in all directions, and furthermore, that to for both planes the same angle-independent absorption coefficient $\alpha$ is assigned. The corresponding image sources (and image spaces) are depicted in Figure 4.5. The source images form a simple pattern of equidistant points situated on a straight line, and each of them is a valid one, that is, it is 'visible' from any observation point B. According to Equation 1.38, the contribution of an image source of *n*th order to the total energy density in B is $(1-\alpha)^{|n|} \cdot P / 4\pi r_n^2 c$,

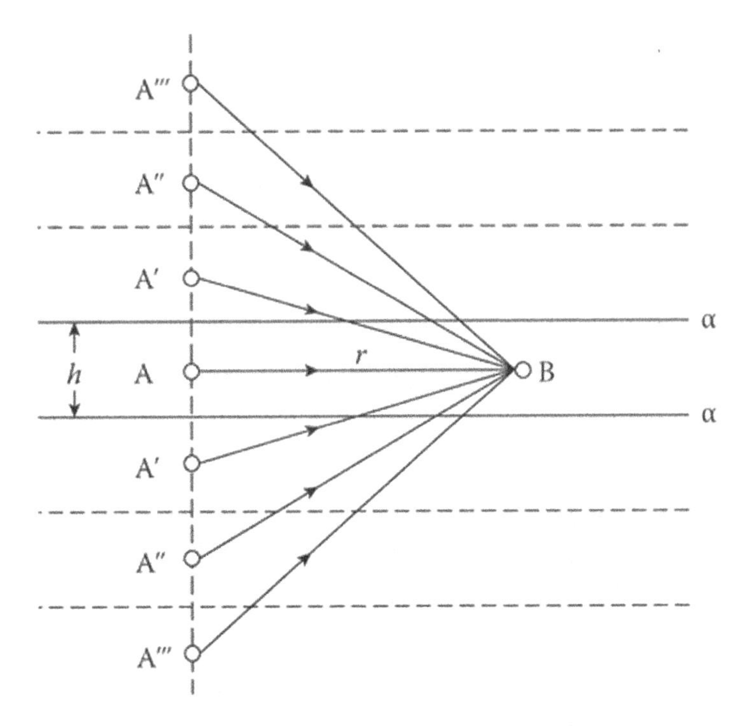


*Figure 4.5* System of image sources of an infinite flat room: *A* is the original sound source; *A′*, *A″*, and so on are image sources; and *B* is the receiving point.

with $r_n = (r^2 + n^2 h^2)^{1/2}$ if $r$ is the horizontal distance of $B$ from the original source $A$ and $h$ is the 'height' of the room. The resulting energy density is

$$w = \frac{P}{4\pi c}\sum_{n=-\infty}^{\infty}\frac{(1-\alpha)^{|n|}}{r^2+n^2h^2} \tag{4.3}$$

which can be evaluated with a pocket calculator. The content of this formula is shown in Figure 9.1a. If desired, one can easily extend this model to angle-dependent absorption coefficients.

Another example is a rectangular room, as depicted in Figure 3.1. For this room shape, certain image sources of the same order coincide but are complementary with respect to their directivity. The result is the regular pattern of image rooms, as shown in Figure 4.6, each of them containing exactly one image source. So the four image rooms adjacent to the sides of the original rectangle contain one first-order image each, whereas those adjacent to its corners contain second-order images, and so on. The lattice depicted in Figure 4.6 has to be completed in the third dimension, that is, we must imagine an infinite number of such patterns one upon the other at equal distances, one of them containing the original room.

In both examples, all image sources are visible. This is because the totality of image rooms, each of them containing one source image, fills the whole space without leaving uncovered regions and without any overlap. Enclosures of less regular shape would produce much more irregular patterns of image sources, and their image spaces would overlap each other in a complicated way. In these cases, the validity or invalidity of each image source with respect

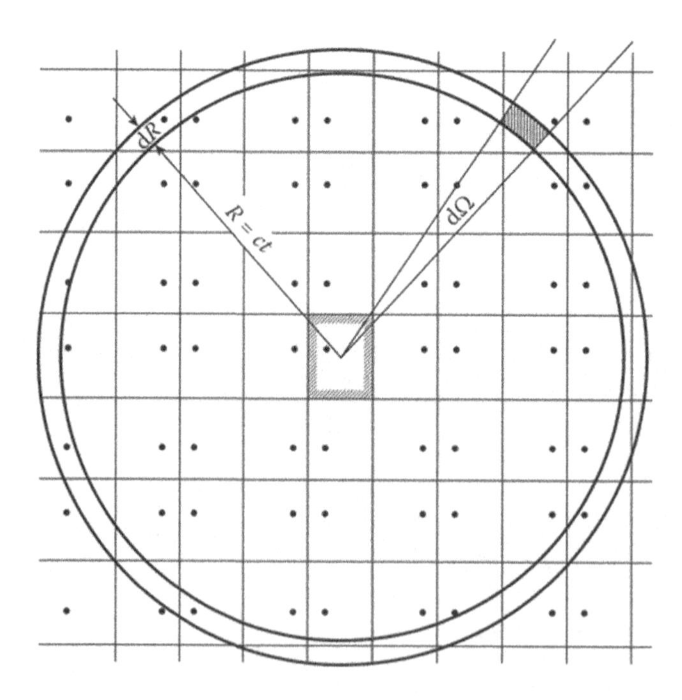


*Figure 4.6* Image sources (marked with dots) in a rectangular room. Each rectangle indicates a mirror image of the original room in the centre and contains an image of the original sound source. The pattern continues in an analogous manner in the direction perpendicular to the drawing plane.

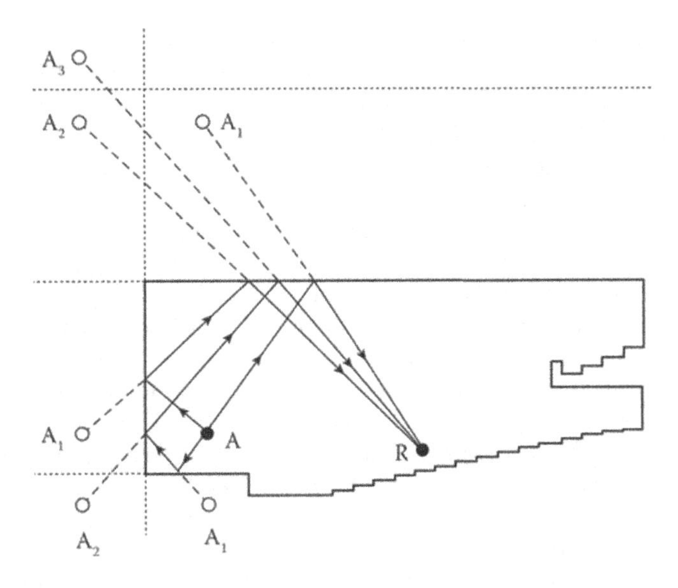


*Figure 4.7* Longitudinal section of an auditorium showing some image sources: $A$ = sound source, $A_I$ = first-order image sources, $A_2$ = second-order image sources, and so on.

to a given receiving point must be carefully examined. A more realistic example is shown in Figure 4.7, along with a few image sources.

When all image sources of a given enclosure have been constructed, we can proceed as before: the total energy density at a given point $R$ is obtained by adding the contributions of all significant image sources under the assumption that all sources, including the original one, emit the same sound signal. Because of the different travelling distances, the waves (or rays) originating from these sources arrive at the receiving point with different delays and strengths. As mentioned before, the strength of a particular contribution must include the absorptivities $\alpha_n$ of all walls, which are crossed by the straight line connecting the image source with the receiving point.

## 4.2 THE TEMPORAL DISTRIBUTION OF REFLECTIONS

In the discussions of Section 4.1, it was tacitly assumed that the output power of the sound source and all its images is constant in time, and that it is correct just to add their intensities in the receiving point. However, if the sound source emits a signal s(t), which contains information as, for instance, speech, this procedure is no longer permissible, since it neglects all time delays caused by the finite sound velocity. If the absorption coefficients of all walls were frequency-independent, the received signal s'(t) would be the superposition of infinitely many replicas of the original signal, each of them with its particular strength $A_n$ and delayed by its particular travelling time $t_n$:

$$s'(t) = \sum_n A_n s(t - t_n) \tag{4.4}$$

Accordingly, the impulse response of the room would be

$$g(t) = \sum_n A_n \delta(t - t_n) \tag{4.5}$$

In reality, a Dirac impulse is deformed each time it is reflected from a wall, that is, the reflected signal is not the exact replica of the original impulse but is changed into a different signal $r(t)$, which we name the 'reflection response' of the surface. Its Fourier transform is the frequency-dependent and complex reflection factor R, as introduced in Section 2.1.

In the following term, 'reflection' will be used with two different meanings: first, it denotes the process which a sound signal undergoes when it hits a wall, and secondly, it means the result of this process. Accordingly, we can say that the impulse response of a room is composed of infinitely many reflections, as shown schematically in Figure 4.8. Each of them is represented by a vertical dash over a horizontal time axis; its arrival time and its relative energy can be seen from its position and length. Such a diagram, which is a highly idealized form of a room impulse response, is also called an 'echogram' or 'reflection diagram'. The third important property of reflections, the direction from which it arrives at the observation point, is not shown in the diagram. The first dash marks the primary sound or the direct sound, arriving at $t = 0$. The subsequent reflections appear, at first, rather sporadically, later at ever increasing density; at the same time, they carry less and less energy. As we shall see later in more detail, the role of the first isolated reflections with respect to our subjective hearing impression is quite different from that of the very numerous weak reflections arriving at later times, which merge into what we perceive subjectively as reverberation. Thus, we can consider the reverberation of a room not only as the common effect of decaying vibrational modes, as we did in Chapter 3, but also as the sum total of all reflections — except the very first ones.

The average rate of reflections received in a point of a real rectangular room can be found by using the system of image rooms and image sound sources, as shown in Figure 4.6. Suppose that at some time $t = 0$ each mirror source emits an impulse of energy $E_0$ in the form of a spherical wave. In the time interval from $t$ to $t + dt$, all those waves will arrive in the centre of the original room which originate from image sources whose distances to the centre are between $ct$ and $c(t + dt)$. These sources are located in a spherical shell with radius $ct$. The thickness of this shell (which is supposed to be very small as compared with $ct$) is $cdt$, and

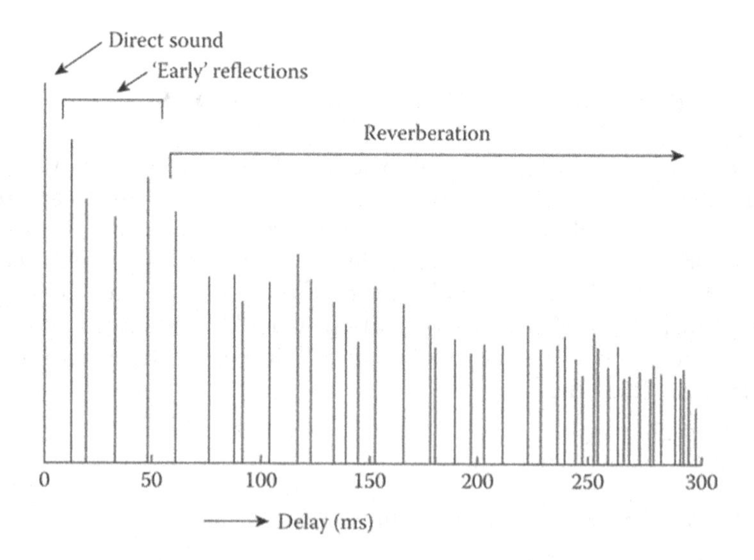


*Figure 4.8* Schematic impulse response of a room.

its volume is $V_s = 4\pi c^3 t^2$ d$t$. Hence, the average number of mirror sources contained in the shell is $V_s/V = 4\pi c^3 t^2$ d$t/V$. This figure is also the average rate of received reflections at time $t$:

$$\frac{\mathrm{d}N_r}{\mathrm{d}t} = 4\pi \frac{c^3 t^2}{V} \tag{4.6}$$

It is interesting to note, by the way, that the preceding approach is the same as that used to estimate the mean density of eigenfrequencies in a rectangular room (Equation 3.26) with formally the same result. In fact, the patterns of mirror sources and the eigenfrequency lattice are in the same relation to each other as are the point lattice and the reciprocal lattice in crystallography: they are spatial Fourier transforms of each other. Moreover, it can be shown that Equation 4.6 applies not only to rectangular rooms but also to rooms with arbitrary shape.

Each reflection corresponds to a bundle of rays originating from the respective image source. In this bundle, the sound intensity decreases proportionally as $(ct)^{-2}$, that is, inversely proportional to the squared distance covered by the rays. Furthermore, the rays are attenuated by absorption in the medium. According to Equation 1.21, this effect can be taken into account by including a factor $\exp(-mx/2)$, which describes the decrement of the pressure amplitude when a plane wave travels a distance $x$ in a lossy medium. Hence, the intensity of a reflection is reduced by the factor $\exp(-mx) = \exp(-mct)$. And finally, the intensity of a ray bundle is reduced by a factor $1 - a$ whenever it crosses a wall of an image room. If this happens $\overline{n}$-times per second on the average, the intensity reduction due to wall absorption after $t$ seconds will be $(1-\alpha)^{\overline{n}t} = \exp\left[\overline{n}t\ln(1-\alpha)\right]$. Therefore, a reflection received at time $t$ has the average intensity

$$\frac{E_0}{4\pi(ct)^2}\exp\left\{\left[-mc + \overline{n}\ln(1-\alpha)\right]t\right\}$$

Dividing this expression by the sound velocity and combining it with Equation 4.6 yields the time-dependent energy density for $t \geq 0$:

$$w(t) = w_0 \exp\left\{\left[-mc + \overline{n}\ln(1-\alpha)\right]t\right\} \quad \text{with } w_0 = E_0 / V \tag{4.7}$$

To complete this formula, we use the average number $\overline{n}$ of wall reflections or wall crossings per second. For this purpose, we refer to Figure 4.9 and consider a sound ray whose angle with respect to the $x$-axis (i.e. to the horizontal axis) is $\beta_x$. It will cross a vertical mirror wall every $L_x/(c \cos \beta_x)$ seconds. ($L_x$, $L_y$, and $L_z$ are the room dimensions, as in Figure 3.1.)

Therefore, the number of such wall crossings per second is

$$n_x(\beta_x) = \left|\frac{c}{L_x}\cos\beta_x\right| \tag{4.8}$$

Similar expressions hold for the crossings of walls perpendicular to the $y$-axis and the $z$-axis. Accordingly, the total number of wall crossings, that is, of reflections which a ray with given direction undergoes per second, is

$$n(\beta_x, \beta_y, \beta_z) = n_x + n_y + n_z$$

with $\cos^2 \beta_x + \cos^2 \beta_y + \cos^2 \beta_z = 1$. This means that each sound ray decays at its own rate, resulting in a bent decay curve according to the discussion in Section 3.8.

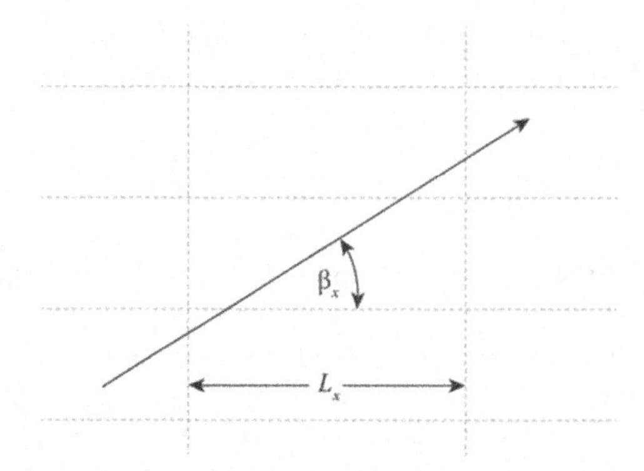


*Figure 4.9* Wall crossings of a sound ray in a rectangular room and its images.

To arrive at a value of $\bar{n}$ which is representative for all sound rays, one has to require that the sound rays change the energy they transport and/or their direction once in a while (Cremer and Müller 1982). Such changes will never happen in a rectangular room with smooth walls. But enclosures with more irregular shape or with diffusely reflecting walls or obstacles which provide for some sound scattering have the tendency to mix the directions of sound propagation. In the ideal case, such randomizing effects will eventually result in what is called a 'diffuse sound field', in which the propagation of sound is completely isotropic.

Under this condition, Equation 4.8 may be averaged over all directions:

$$\frac{c}{L_x}\langle|\cos\beta_x|\rangle = \frac{c}{L_x}\frac{1}{4\pi}2\pi \cdot 2\int_0^{\pi/2}\cos\beta_x \sin\beta_x d\beta_x = \frac{c}{2L_x}$$

The same average is found for $n_y$ and $n_z$. Hence, the total average of reflections per second is

$$\bar{n} = \frac{c}{2}\left(\frac{1}{L_x}+\frac{1}{L_y}+\frac{1}{L_z}\right) = \frac{cS}{4V} \tag{4.9}$$

Here, $S$ is the total area of the room's boundary. By inserting this result into Equation 4.7, we arrive at a fairly general law of sound decay:

$$w(t) = w_0 \exp\left[-ct\frac{4mV - S\ln(1-\alpha)}{4V}\right] \quad \text{for } t \ge 0 \tag{4.10}$$

The reverberation time, that is, the time in which the total energy falls to one millionth of its initial value, is thus

$$T = \frac{1}{c}\cdot\frac{24V\cdot\ln 10}{4mV - S\ln(1-\alpha)} \tag{4.11}$$

or, if we insert 343 m/s for the speed of sound $c$,

$$T = 0.161\frac{V}{4mV - S\ln(1-\alpha)} \tag{4.12}$$

In the preceding text, we have derived by rather simple geometrical considerations the most important formula of room acoustics, which relates the reverberation time of a room to some of its geometrical data and to the absorption coefficient of its walls. We have tacitly assumed that the latter is the same for all wall portions and that it does not depend on the angle at which a wall is struck by the sound rays. In Chapter 5, we shall look more closely into the laws of reverberation using a different approach, but the result will be essentially the same as in Equation 4.12.

The exponential law of Equation 4.10 provides an approximate description of the decaying sound energy in a rectangular room with a diffuse sound field. One must not forget, however, that this formula is the result of two averaging processes. In reality, the reflections arrive at quite irregular times $t_n$, at the observation point — for instance, at a listener's ear — and also their energies vary in quite an irregular way, even if the sound field is diffuse. Accordingly, in exceptional cases, certain details of the decay process may show considerable deviations from Equation 4.10, which may well be relevant for the acoustics of the room. So it may happen, for instance, that one particular reflection with long delay time is much stronger than the majority of its surrounding neighbours and stands out of the general decay process. This can occur when many sound rays are directed to a remote concave wall portion, which concentrates the reflected sound energy in some point or region. Such an isolated component of the echogram is perceived as a distinct echo. Another unfavourable condition is that of many reflections clustered together in a narrow time interval. Since our hearing has a limited time resolution and therefore performs some sort of short-time integration, this lack of uniformity may be audible and may give rise to undesirable effects which are similar to that of a single reflection of exceptional strength.

Particularly disturbing are periodic components hidden in numerous irregularly distributed reflections, since our hearing is very sensitive to periodic repetitions of sound signals. For short periods, that is, for repetition times of a few milliseconds, such periodic components are perceived as a 'colouration' of the sound signal. If the periods are longer and amount to 30, 50, or even 100 ms, distinct repetitions of the original sound signal will be audible. This case, which is frequently referred to as 'flutter echo', is observed if sound is reflected repeatedly to and fro between parallel walls. Flutter echoes are often observed in corridors or other longish rooms where the end walls are rigid while the ceiling, floor, and side walls are absorbent.

## 4.3 THE DIRECTIONAL DISTRIBUTION OF REFLECTIONS, DIFFUSE SOUND FIELD

We shall now consider the third characteristic property of a reflection, namely, the direction from which it reaches an observer. As before, we shall prefer a statistical description. This procedure commends itself not only because of the great number of reflections to be considered but also because no human listener is able to distinguish subjectively the directions of individual reflections. Nevertheless, whether the reflected components arrive uniformly from all directions or whether they all come from one single direction has considerable bearing on what and how we hear in a room, especially in a concert hall. The directional distribution of sound is also important for certain measuring procedures.

First, we assume that the sound source produces a stationary sound signal, and that the same holds for all image sources. Each of them, provided it is visible from the receiving point R, contributes to the received sound energy. Let us consider a narrow cone with the small solid angle $d\Omega$ around a direction characterized by the polar angle $\vartheta$ and the azimuth angle $\varphi$ ($\theta$ and $\phi$ in Figure 2.10). Its tip is located at the receiving point $R$. The contribution of all sources within the 'directional cone' is $I(\varphi,\vartheta)$. This quantity, conceived as a function of the angles φ and ϑ, is called the directional distribution of sound. In general, it depends on the receiver's location. For a rectangular room, the situation is illustrated in Figure 4.6. Obviously, $I(\varphi, \vartheta)$ will only remain finite if there are some losses caused either by air attenuation or by non-vanishing wall absorption. Experimentally, $I(\varphi, \vartheta)$ can be determined by the use of a directional microphone with sufficiently high resolution, at least approximately (see Section 8.5).

If the sound source and its images emit a transient signal, the directional distribution will become time-dependent. Again, the limiting case is that of an impulsive excitation signal with very short duration, idealized as a Dirac impulse. Then, the intensity at time $t$, denoted by $I_t(\varphi, \vartheta)$, is due to those image sources that are lying within the directional cone and, at the same time, within a spherical shell with the radius $ct$ and the thickness $cdt$. For the rectangular room in Figure 4.6, this region corresponds to the cross-hatched area in the upper part. Evidently, the relation between the time-dependent and the steady-state directional distribution is

$$I(\varphi,\vartheta) = \int_0^\infty I_t(\varphi,\vartheta)\,\mathrm{d}t \tag{4.13}$$

If the directional distribution $I(\varphi, \vartheta)$ does not depend on the angles φ and ϑ, the stationary sound field is called 'diffuse' or isotropic. We have already discussed this important condition in Section 2.5. If, in addition to this condition, $I_t(\varphi, \vartheta)$ does not depend on the angles φ and ϑ for all $t$, even the decaying sound field is diffuse.

In a certain sense, the diffuse sound field is the counterpart of a plane wave. Just as certain properties can be attributed to plane waves, so can relationships describing the properties of diffuse sound be established. Some of these have already been discussed in Chapter 2. They are of particular interest to the whole of room acoustics since, although the sound field in a concert hall or theatre is not perfectly diffuse, its directional structure resembles more that of a diffuse field than that of a plane wave. Or put another way, the sound field in a real room, which always contains some shape irregularities, can be approximated fairly well by a sound field with uniform directional distribution. By contrast, a single plane wave is hardly ever encountered in a real situation.

It should be noted that the directional structure of a sound field depends not only on the shape of the room but also on the distribution of absorption over its boundary. As an extreme example, we consider a rectangular room with perfectly rigid walls except for one which absorbs the incident sound energy completely. Its behaviour with respect to the formation and distribution of reflections is elucidated by the image room system depicted in Figure 4.10, consisting of only two 'stores' of height $L$, since the absorbing wall generates no images of the room and the sound source. The elevation angle ε varies in the range $\pm\varepsilon_0$ with

$$\varepsilon_0(t) = \arcsin\left(\frac{L}{ct}\right) \quad \text{for}\, t \geq \frac{L}{c} \tag{4.14}$$

Hence:

$$I_t(\varphi,\vartheta) = \begin{cases} \text{const} & \text{for}\,|\varepsilon| < \varepsilon_0 \\ 0 & \text{for}\,|\varepsilon| < \varepsilon_0 \end{cases} \tag{4.15}$$

It is obvious that the sound field is anything but isotropic in this room. This holds for the steady-state field as well as for the sound decay. With the present meaning of the angle ε, the

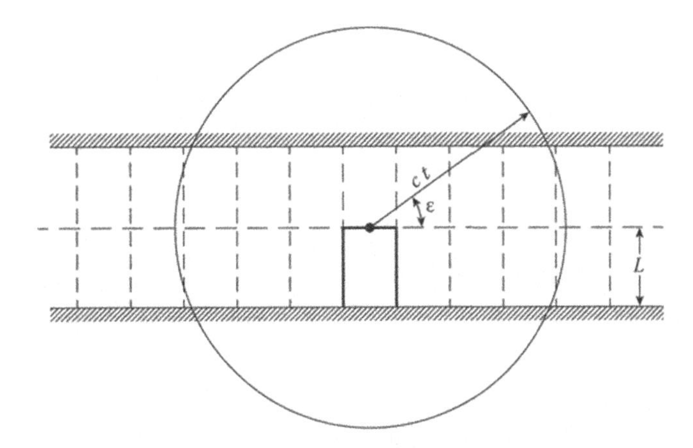


*Figure 4.10* Image rooms of a rectangular room, one side of which is perfectly sound-absorbing. The original room is indicated by solid lines.

element of solid angle becomes cos $\varepsilon d\varepsilon d\varphi$; hence, the integration over all relevant directions yields the following expression for the decaying energy density:

$$w(t) = \frac{2}{c}\int_0^{2\pi} d\varphi \int_0^{\varepsilon_0} I_t(\varphi,\varepsilon)\cdot \cos\varepsilon d\varepsilon = \text{const}\cdot \frac{4\pi L}{ct} \quad \text{for } t \geq \frac{L}{c} \tag{4.16}$$

As a consequence of the non-uniform distribution of the wall absorption, the decay of the reverberant energy does not follow an exponential law but is inversely proportional to the time.

## 4.4 ENCLOSURES WITH CURVED WALLS

In this section, we deal with enclosures the boundaries of which are concavely curved, either completely or partially. Practical examples are rooms with a domed ceiling, as are encountered in many theatres or other performance halls, or the curved rear walls of many lecture theatres. Concavely curved surfaces in rooms are generally considered as critical or even harmful in that they concentrate the sound energy in certain areas and thus impede its uniform distribution throughout the room.

Formally, the law of specular reflection as expressed by Equation 4.1 is valid for curved surfaces as well as for plane ones, since each curved surface can be approximated by many small plane sections. Keeping in mind the wave nature of sound, however, one should not apply this law to a surface the radius of curvature of which is not very large compared to the acoustical wavelength. When the radius of curvature is comparable or even smaller than the wavelength, the surface will scatter an impinging sound wave rather than reflecting it specularly.

Very often, curved walls in rooms or halls are spherical or cylindrical, or parts of them can be approximated by sections of this kind. Then, we can apply the simple laws for reflection of rays at concave or convex mirrors, known from optics. Strictly speaking, those laws are valid only for small segments of curved boundaries, but nevertheless, they are useful for an estimate of the effects to be expected. Figure 4.11a depicts a section of a concave, spherical mirror; its radius of curvature is $R$. We consider a bundle of rays originating from a point $S$ situated on the symmetry axis of the mirror. The mirror collects all reflected rays at some point $S'$; thereafter, they will diverge. This happens when the distance between the source and

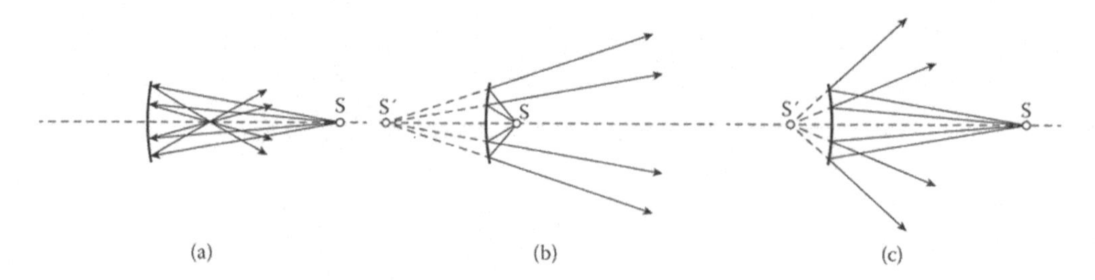


*Figure 4.11* Reflection of a ray bundle from a concave (*a* and *b*) and a convex mirror (*c*).

the mirror is larger than $R/2$. Generally, the source distance $a$, the distance $b$ of the recollection point $S'$ from the mirror, and the radius $R$ are related by

$$\frac{1}{a} + \frac{1}{b} = \frac{2}{R} \tag{4.17}$$

If the source is closer to the mirror than $R/2$ (see Figure 4.11b), the reflected ray bundle is divergent (although less divergent than the incident one) and seems to originate from a point behind the mirror. Equation 4.17 is still valid and leads to a negative value of $b$.

Finally, we consider the reflection at a convex mirror as depicted in Figure 4.11c. In this case, the divergence of any incident ray bundle is increased by the mirror. Again, Equation 4.17 can be applied to find the position of the 'virtual' focus after replacing $R$ with $-R$. As before, the distance $b$ is negative.

In the following, we calculate for a spherical mirror the intensity $I_r$ in the reflected bundle, and comparing it with the intensity $I_0$, we would observe at the same position if the curved reflector would be replaced with a plane one. For this purpose, one should note that the intensity in a divergent ray bundle is generally

$$I = \frac{P_0}{\Omega \cdot r^2} \tag{4.18}$$

where $P_0$ is the power transported by a ray bundle with the aperture $\Omega$ (i.e. the solid angle which defines the width of the bundle); $r$ is the observer's distance from the origin of the bundle, and observer's distance from the surface of the mirror is denoted by $x$. What the reflector does is change the aperture of the bundle while the total power remains unaltered. Application of Equation 4.18 to the incident and the reflected bundle yields

$$I_0 = \frac{P_0}{\Omega_0 (a+x)^2} \quad \text{and} \quad I_r = \frac{P_0}{\Omega_r (b-x)^2}$$

Finally, we invoke the obvious relation $\Omega_r/\Omega_0 = (a/b^2)$ and obtain for the ratio of both expressions:

$$\frac{I_r}{I_0} = \left(\frac{1+x/a}{1-x/b}\right)^2 \tag{4.19}$$

Figure 4.13 plots the level $L_r = 10 \cdot \log_{10} (I_r/I_0)$ derived from this ratio for the three cases depicted in Figure 4.12. The pole in curve $a$ (for $a > R/2$) is due to the concentration of all rays in one point $S'$. Generally, there is a range of increased intensity, that is, of positive $L_r$. From Equations 4.17 and 4.19, it can be concluded that

$$x < 2\left(\frac{1}{b} - \frac{1}{a}\right)^{-1} = \left(\frac{1}{R} - \frac{1}{a}\right)^{-1} \tag{4.20}$$

Outside that range, the level $L_r$ is negative, indicating that the reflected bundle is more divergent than it would be when reflected from a plane mirror. If $a < R/2$ (curve $b$), the intensity is increased at all distances $x$. Finally, the convex mirror (curve $c$) reduces the intensity of the bundle everywhere.

According to Equation 4.19, the limit of $L_r$ for very large distances $x$ is

$$L_r \rightarrow 20 \cdot \log_{10}\left|\frac{b}{a}\right| \quad \text{for}\, x \rightarrow \infty \tag{4.21}$$

Some practical conclusions can be drawn from these findings. A concave mirror may concentrate the impinging sound energy in certain regions, but it may also be an effective scatterer which distributes the reflected energy over a wide angular range. Which of these effects dominates depends on the positions of the source and the observer. Generally, the following rule (Cremer and Müller 1982) can be derived from Equation 4.20: Suppose the mirror in Figure 4.11a is completed to a full circle with radius $R$. Then, if both the sound source and the receiver are outside this volume, the undesirable effects mentioned at the beginning of this section are not to be expected.

Up to now, we discussed the behaviour of spherical reflectors. Cylindrical reflectors can be treated in an analogous way. We just need to replace Equation 4.18 with

$$I = \frac{P_0}{\alpha \cdot r} \tag{4.22}$$

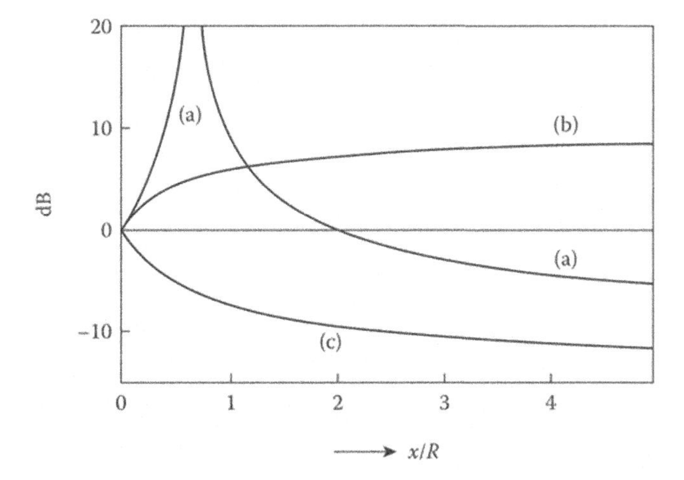


*Figure 4.12* Level difference in ray bundles reflected from a spherical and a plane reflector at distance $x$: (a) concave mirror, $a = 2R$; (b) concave mirror, $a = R/3$; and (c) convex mirror, $a = 2R$.

Here, $\alpha$ is the angle which characterizes the width of a plane ray bundle; $r$ has the same meaning as before. Then, we obtain

$$\frac{I_r}{I_0} = \frac{1 + x/a}{|1 - x/b|} \tag{4.23}$$

As mentioned before, the applied simple laws are valid only for narrow ray bundles, that is, as long as the inclination of the rays against the axis is sufficiently small, or what amounts to the same thing, as long as the aperture of the mirror is not too large. Whenever this condition is not met, the construction of reflected rays becomes more difficult. Either the surface has to be approximated piecewise by circular or spherical sections or the reflected bundle must be constructed ray by ray. As an example, the reflection of a parallel bundle of rays from a concave mirror of large aperture is shown in Figure 4.13. Obviously, the reflected rays are not collected just in one point; instead, they form an envelope which is known as a caustic. Next to the central ray, the caustic reaches the focal point in the distance $b = R/2$ in accordance with Equation 4.17 with $a \to \infty$. Another interesting shape is that of the ellipse or the ellipsoid. Both have two foci, $F_1$ and $F_2$, as shown in Figure 4.14. If a sound source $S$ is situated in one of them, all the rays it emits are collected in the other one. For this reason, enclosures with elliptical floor plan are plagued by quite unequal sound distribution even if neither the sound source nor the listener is in a geometrical focus. The same holds, of course, for halls with a circular floor plan, since the circle is a limiting case of the ellipse.

A striking experience can be made in such halls if a speaker is close to its wall. A listener who is also next to the wall, although distant from the sound source (see Figure 4.15), can hear the speaker quite clearly even if the latter speaks in a very low voice or whispers. An enclosure of this kind is said to form a 'whispering gallery'. The explanation of this phenomenon is simple: many of the sound rays leaving the speaker's mouth hit the wall at grazing incidence and are repeatedly reflected from it. If the wall is smooth and not interrupted by pillars, niches, and so on, the rays remain confined within a narrow band: in other words, the wall conducts the sound along its perimeter. Probably, the best-known example is in St Paul's

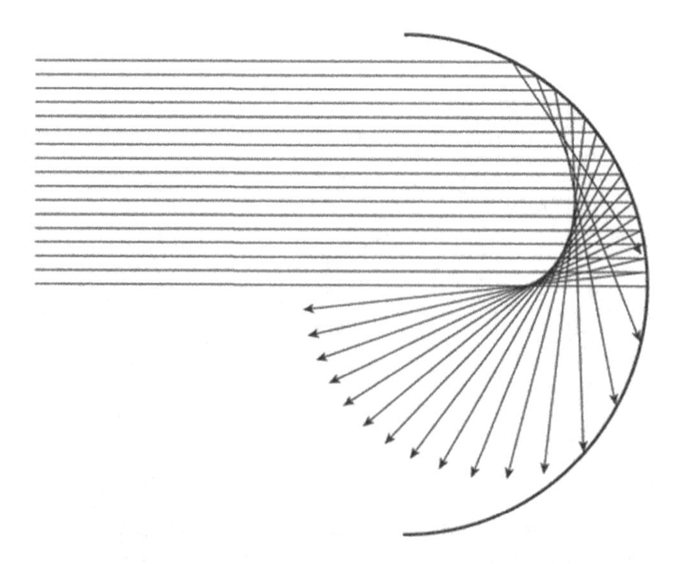

*Figure 4.13* Reflection of a parallel ray bundle from a spherical mirror of large aperture.

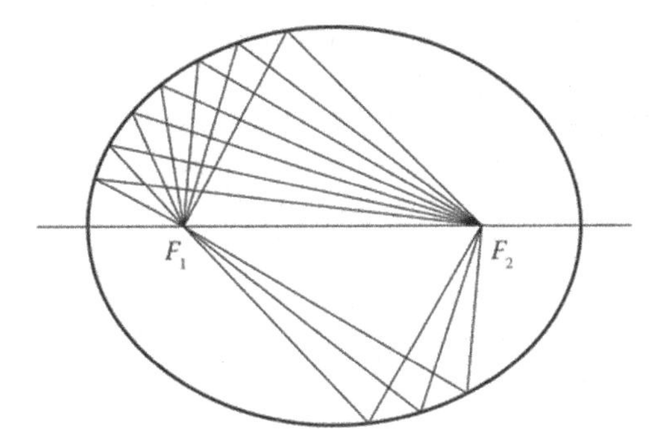


*Figure 4.14* Collection of sound rays in an elliptical enclosure.

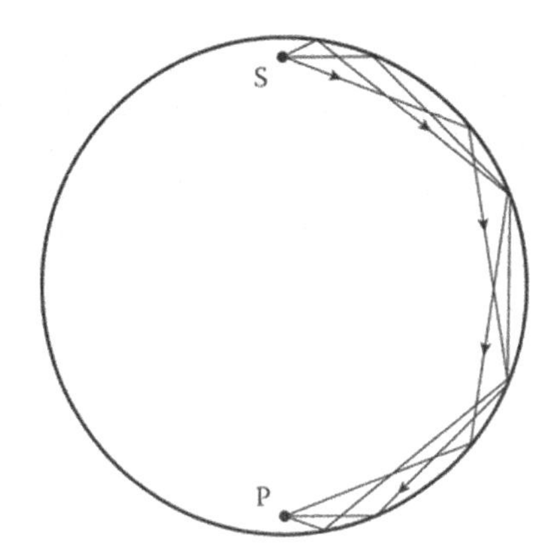


*Figure 4.15* Whispering gallery.

Cathedral in London, which has a gallery at the circular base of the dome. A whispering gallery is an interesting curiosity, but if the hall is used for performances, the acoustical effects encountered in it are rather disturbing.

## 4.5 RADIOSITY INTEGRAL

In Section 2.7, we discussed diffuse sound reflections from acoustically 'rough' boundary portions caused by scattering the incident sound. They occur when the reflecting boundary exhibits irregularities either in geometrical structure or in wall impedance, provided the dimensions of these irregularities are comparable with the acoustical wavelengths (see Figure 2.17). Obviously, diffraction plays the dominant role in the scattering process. Nevertheless, we can treat diffuse wall reflections and their effects within the realm of geometric acoustics. This is achieved by describing just the result of the scattering process, leaving aside the physical details. It is clear that scattering from a rough boundary has a considerable influence on the sound field in the room. Generally, diffuse wall reflections provide for mixing different sound components and thus result in a more uniform distribution of the sound energy throughout the room.

In most cases, the sounds reflected from a rough surface contain a specular component, as shown in the example of Figure 2.18. Completely diffuse reflection is characterized by Lambert's cosine law: Suppose a bundle of parallel or nearly parallel rays with the intensity $I_0$ hits an area element d$S$ under an angle $\vartheta_0$. Then, the intensity of the sound which is scattered in a direction characterized by a radiation angle $\vartheta$, measured at some distance $r$ from d$S$, is given by

$$I(r,\vartheta) = \rho I_0\, \mathrm{d}S \frac{\cos\vartheta \cdot \cos\vartheta_0}{\pi r^2} = \rho B_0 \cos\vartheta \frac{\mathrm{d}\Omega}{\pi} \tag{4.24}$$

The situation is depicted in Figure 4.16. $\mathrm{d}\Omega = \mathrm{d}S \cdot \cos\vartheta/r^2$ is the solid angle under which the element d$S$ appears from a point in distance $r$ and the emission angle $\vartheta$. The circle represents the directional distribution of the scattered sound; the length of the arrow pointing to its periphery is proportional to cos $\vartheta$. The quantity $B_0 = I_0 \cos\vartheta_0$ is the incident energy per second and per unit area. It is called the 'irradiation density', or 'irradiance'; its unit is watts/m$^2$. The factor ρ in Equation 4.24, named as 'reflection coefficient', accounts for the energy loss taking place at dS. Obviously, it is related to the absorption coefficient by $\rho = 1 - \alpha$.

In fact, it does not matter whether the irradiance $B_0$ is brought about by one, two, or a hundred incident rays hitting the element d$S$ at different angles. We can say that a sound ray forgets its history when it is diffusely reflected.

In optics, surfaces with Lambertian scattering are encountered quite frequently. In contrast, in acoustics, and particularly in room acoustics, only partially diffuse reflections can be achieved. Nevertheless, the assumption of totally diffuse reflections comes often closer to the properties of real walls than that of specular reflection, particularly if we are concerned not only with one but, instead, also with many successive reflections of sound from different walls or portions of walls. This is exactly what happens in reverberant enclosures.

According to Equation 4.24, the surface element should be considered as a secondary sound source. Consequently, the distance $r$, which determines the intensity reduction due to the spreading of sound beams, must be measured from the reflecting area element $dS$. This is not so with specular reflection: here, the distance $r$ in the $1/r^2$ law would be the total path length between the sound source and the point of observation, with no regard as to whether this path is straight or contains bents caused by specular reflections.

In the following, we consider an enclosure of arbitrary shape the whole boundary of which is assumed to reflect the impinging sound in a completely diffuse manner, that is, according

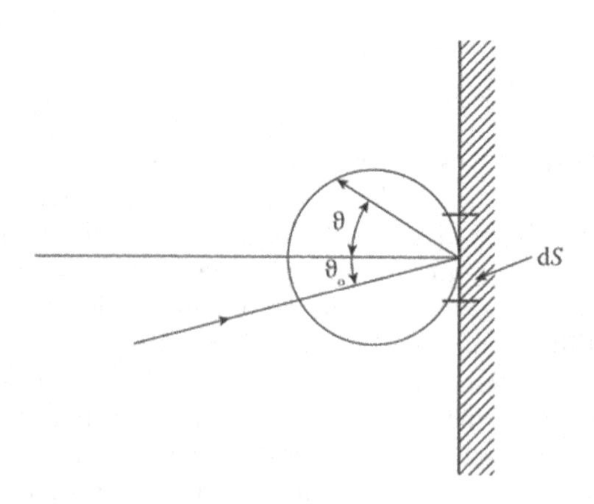


*Figure 4.16* Ideally diffuse sound reflection from an acoustically rough surface.

to Lambert's law. This assumption enables us to describe the sound field within the room in a closed form, namely, by an integral equation. To derive it, consider two wall elements, d$S$ and d$S'$, of a room (see Figure 4.17). Their locations are given by the vectors $\boldsymbol{r}$ and $\boldsymbol{r}'$, respectively, each of them standing for three suitable coordinates. The straight line connecting them has the length $R = |\mathbf{r}' - \mathbf{r}|$, and the angles between this line and the wall normals in d$S$ and d$S'$ are denoted by $\vartheta$ and $\vartheta'$.

Suppose the element $dS'$ is irradiated with the energy B($\mathbf{r}'$)dS′ per second. The fraction it re-radiates into the space is given by the 'reflection coefficient' ρ($\mathbf{r}'$). To avoid unnecessary complications, we assume that ρ($\mathbf{r}'$) is independent of the angles $\vartheta$ and $\vartheta'$. According to Equation 4.24 (second version), the contribution of the element $dS'$ to the intensity received at $\boldsymbol{r}$ is

$$I(\mathrm{r},\vartheta) = B(\mathrm{r}')\frac{\rho(\mathrm{r}')}{\pi}\frac{\cos\vartheta'}{R^2}\mathrm{d}S' \tag{4.25}$$

The total energy per second and per unit area arriving at the element d$S$ is obtained by multiplying this equation with cos $\vartheta$ and integrating it over all wall elements d$S'$:

$$B(\mathrm{r}) = \iint_S \rho(\mathrm{r}')B(\mathrm{r}')\frac{\cos\vartheta\cos\vartheta'}{\pi R^2}\mathrm{d}S' + B_\mathrm{d}(\mathrm{r}) \tag{4.26}$$

$B_\mathrm{d}$ is the direct contribution of some sound source. If this source is a point source located at $\mathrm{r}_\mathrm{s}$ producing the acoustical power $P$, the direct component of the irradiance $B$ is

$$B_\mathrm{d} = \frac{P}{4\pi|\mathrm{r}_\mathrm{s} - \mathrm{r}|^2}\cos\vartheta_d \tag{4.27}$$

$\vartheta_\mathrm{d}$ denotes the angle between the boundary normal at $r$ and the vector $\mathbf{r}_\mathrm{d} - \mathbf{r}$.

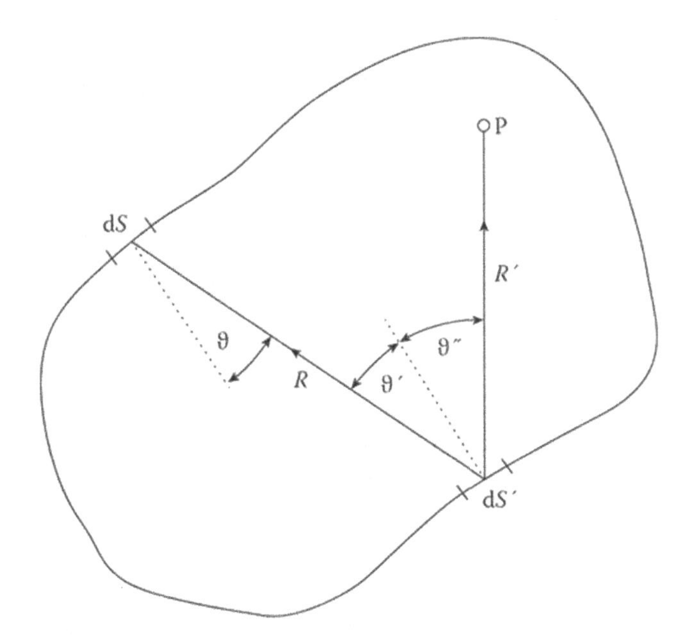


*Figure 4.17* Illustration of Equation 4.26.

The integral equation (4.26) with the irradiance $B$ as the unknown function is known as the radiosity integral. In this form, it applies to steady-state sound fields only. However, when the power output of the sound source varies with the time, the sound field in the enclosure will also be time-dependent. Then, we have to include the finite time which the sound needs for its propagation, for instance, to travel from d$S'$ to d$S$ (Kuttruff 1971, 1976), or from the location of the sound source to the boundary element d$S$:

$$\begin{aligned} B(\mathrm{r},t) &= \iint_S \rho(\mathrm{r}')B(\mathrm{r}',t-R/c)\frac{\cos\vartheta\cos\vartheta'}{\pi R^2}\mathrm{d}S' + B_\mathrm{d}\left(\mathrm{r},t-\frac{\mathrm{r_s}-\mathrm{r}}{c}\right) \\ &= \frac{1}{\pi}\iint_\Omega \rho(\mathrm{r}')B(\mathrm{r}',t-R/c)\cos\vartheta \mathrm{d}\Omega' + B_\mathrm{d}\left(\mathrm{r},t-\frac{\mathrm{r_s}-\mathrm{r}}{c}\right) \end{aligned} \tag{4.28}$$

with $\mathrm{d}\Omega' = \mathrm{d}S' \cdot \cos\vartheta'/R^2$ and $0 \le \vartheta' \le \pi/2$.

Another propagation effect is attenuation by the air. Based on Equation 1.21, the sound pressure amplitude of a wave travelling a distance $x$ in air will be reduced by a factor $\exp(-mx/2)$, and $m$ is the intensity-related attenuation constant of the air. Since the sound intensity is proportional to the square of the sound pressure amplitude, this corresponds to an intensity reduction by $\exp(-mx)$. Accordingly, a factor $\exp(-mR)$ should be inserted in the integrals of Equations 4.26 to 4.28 if attenuation effects are expected. The same holds for all subsequent versions of the radiosity integral, and also for Equation 4.29 (with $x = R'$).

The integral equation (4.28) is fairly general in that it comprehends both the steady-state case (for $B_\mathrm{d}$ and $B$ independent of time $t$) and that of a decaying sound field (if $B_\mathrm{d} = 0$ for $t \ge 0$). Once it is solved, the energy density at a point P inside the room can be obtained from the irradiance $B$:

$$w(\mathrm{r},t) = \iint_S \rho(\mathrm{r}')B(\mathrm{r}',t-R'/c)\frac{\cos\vartheta''}{\pi c R'^2}\mathrm{d}S' + w_\mathrm{d}(\mathrm{r},t-R'/c) \tag{4.29}$$

$R'$ is the distance of the point $P$ from the element d$S'$, while $\vartheta''$ denotes the angle between the wall normal in d$S'$ and the line connecting d$S'$ with $P$ (see Figure 4.17).

Closed solutions of Equation 4.28 are available only for a few simple room shapes. One of them is the spherical enclosure for which

$$\frac{\cos\vartheta\cos\vartheta'}{\pi R^2} = \frac{1}{S} \tag{4.30}$$

with $S$ denoting the surface of the sphere. With this relation, the time-independent integral equation (4.26) simply reads

$$B = \langle \rho B \rangle + B_\mathrm{d}$$

where the brackets indicate averaging over the whole surface. By repeatedly replacing the $B$ in the brackets with $\langle \rho B \rangle + B_\mathrm{d}$, one obtains the solution

$$B = \frac{\langle \rho B_\mathrm{d} \rangle}{1-\langle \rho \rangle} + B_\mathrm{d} \tag{4.31}$$

Another soluble problem is sound propagation between two parallel planes of infinite extent already mentioned in Section 4.1 (see Figure 4.5). We shall come back to this model in Section 9.1.

In general, however, the radiosity integral equation must be numerically solved. According to Miles (1984), this can be achieved by discretization of the boundary, that is, the latter is subdivided into a number of $N$ small patches of finite size, and the integral is approximated by sums. The irradiance $B$ on the $i$th patch is represented by its average $B_i$. For the steady-state condition, one arrives at a system of $N$ linear equations for the unknown $B_i$, which can be solved by standard methods. For time-dependent problems, for instance, for studying sound decay in a room, the time must be discretized too, and the same holds for the distances $R$. Then the equations read:

$$B_{i,n} = \sum_{k=1}^{N} K_{ik}\rho_k B_{k,n-n'} + (B_d)_{i,n} \ (i = 1,2,\ldots,N) \tag{4.32}$$

The first index of the irradiance $B$ denotes the number of the wall element, while the second one is the time index, $n = \text{Int}(t/\Delta t)$, and $n' = \text{Int}(R_{ik}/c\Delta t)$. $R_{ik}$ is the distance between the $i$th and the $k$th wall element. The coefficients $K_{ik}$ are numerically given by

$$K_{ik} = K_{ki} = \frac{1}{\rho}\sum_{\Delta S,}\sum_{\Delta S'}\frac{\cos\vartheta\cos\vartheta'}{R_{ik}}, \tag{4.33}$$

The solution is obtained by a simple recursion, starting with $N$ initial irradiances $B_{i,0}$. Since the distances $R$ have an upper limit, only a limited number of $B_{i,n}$ must be stored in the computer.

## 4.6 MODIFICATIONS AND GENERALIZATIONS, DIFFUSION EQUATION

Equation 4.26 could be called the 'irradiance-based' version of the radiosity integral because the unknown variable is the irradiance B. However, many authors prefer a 'flux-based' formulation (see, for instance, works by Joyce [1978] and Gilbert [1981]). In the introduction of this chapter, a *sound ray* was defined as a narrow sector cut out of a spherical wave; its aperture is given by an infinitesimal solid angle d$\Omega$ (see also Figure 5.1). We call the energy transported per second within this 'cone' as the energy flux $j = I \cdot d\Omega$; it depends on the location $\boldsymbol{r}$, on the time $t$, and on the direction of propagation, indicated by a unit vector $\boldsymbol{u}$. In contrast to the intensity $I$, the energy flux remains constant during propagation, provided dissipation by the air is neglected.

We go back to Equation 4.24 and express $B_0$ by the energy fluxes $j(\mathbf{r}, t, \mathbf{u})$ arriving from all directions at d$S$ and thus 'illuminating' it ($\boldsymbol{u}$ stands for a unit vector pointing in the direction given by angles $\varphi$ and $\vartheta$). As in the derivation of Equation 4.27, the time which the energy needs to traverse the room along one of its chords has to be taken into account. This is done by replacing $t$ with $t - R/c$, where $R$ is the length of the chord; it is determined by backtracking the rays from d$S$ to the boundary (see Figure 4.17). Then the irradiance at d$S$ is

$$B_0 = \iint_{\Omega} j(r, t - R/c, \mathrm{u}')\cos\vartheta' d\Omega'$$

Inserting this expression into the second version of Equation 4.24 yields

$$j(\mathrm{r}, t, \mathrm{u}) = \frac{\rho(\mathrm{r})}{\pi}\iint_{S}(\mathrm{r}, t - R/c, \mathrm{u}')\cos\vartheta' d\Omega' \tag{4.34}$$

This is the alternative we were looking for. Both forms of the radiosity integral as given by Equations 4.28 and 4.34 are equivalent; each of them has its specific advantages. One advantage of the irradiance-based version is that the unknown quantity $B$ depends only on the time and the spatial coordinates.

The radiosity integral has been extended in various ways. So Joyce (1978) has applied it to surfaces with mixed specular-diffuse reflection. For this purpose, he introduced a 'reflection matrix' or 'reflectivity coefficient' $R_j\,(\mathbf{u}', \mathbf{u})$. It characterizes the fraction of the energy which arrives at some boundary element from the direction $\mathbf{u}'$ and is scattered by it into the direction $\mathbf{u}$. With this, the radiosity integral (4.34) reads

$$j(\mathrm{r},t,\mathrm{u}) = \iint_{\Omega} R_j\,(\mathrm{u}',\mathrm{u})\,j(\mathrm{r},t - R/c,\mathrm{u}')\cos\vartheta' \mathrm{d}\Omega' \tag{4.35}$$

Obviously, for a diffusely reflecting boundary, the function $R_j$ is just $\rho/\pi$.

An important case is that of a boundary consisting of one or several portions with specular reflection, while the remaining ones reflect the received energy according to some other law, for example, Lambert's law. Gilbert (1981) solved this problem in the following way: Suppose a ray has left the boundary after a non-specular reflection at $P$. In its further course, it undergoes $N$ specular reflections until it hits another spot $Q$ of non-specular reflection. Then, the delay $R/c$ is given by the total path length $R$ from $P$ to $Q$, including $N$ bents, whereas the intensity of the ray is only reduced according to the reflection coefficients of the $N$ intermediate elements with specular reflection. Quite a different method has been proposed by Fujiwara (1984). It can be applied when the specularly reflecting portion of the boundary is plane. Then, the given room can be mirrored at this portion, resulting in an enclosure with twice its original volume; the sound reflected from the plane portion may be substituted by the sound coming from the mirror image of the original boundary. Fujiwara has applied this method to a two-dimensional room testing various configurations of absorption and sound source location. A rather general approach to combining the radiosity integral with the image source concept is due to Koutsouris et al. (2013). It can be applied for predicting the energy response of general polyhedral enclosures having boundaries with any given absorption and scattering characteristics.

Still, a higher degree of generalization has been achieved by Picaut et al. (1997). These authors considered the distribution function $f(\mathbf{r},\mathbf{v}, t)$, which measures the density of particles in a six-dimensional phase space, comprising three spatial dimensions and three dimensions for the speed vector $\boldsymbol{v}$ of particles. Thus, $f(\mathbf{r},\mathbf{v}, t)\mathrm{d}V_r\,\mathrm{d}V_v$ is the number of particles which are in the element $\mathrm{d}V_r\,\mathrm{d}V_v$ of the phase space at time $t$. Since all sound particles move with the constant sound velocity $c$, the number of dimensions is reduced by 1; the remaining ones are the local coordinates and the two angles $\varphi$ and $\vartheta$ characterizing the direction of particle motion.

Temporal variations of the distribution function may be (1) due to the motion of the particles, (2) due to losses in the medium (air absorption), (3) due to collisions with obstacles embedded in the medium (if there are any; see Section 5.2), (4) due to the action of some sound source, and finally, (5) due to particles scattered from different elements $\mathrm{d}V_v$ of the velocity space elements into the considered one. This happens, for instance, when a particle hits a boundary and is reflected by it — specularly or diffusely. Combining these effects leads to an integro-differential equation from which all interesting parameters (energy density, decay rate, etc.) can be evaluated, at least in principle. Following the procedure which is known from the theory of diffusion (see, for instance, the book of Morse and Feshbach [1953]), Valeau et al. (2006) converted the mentioned equation, which is sometimes referred to as the 'radiative transfer equation', into a more tractable form by the simplifying assumption that the angle dependence of the distribution form is very small or, expressed in acoustical terms, that the sound field is nearly diffuse:

$$j(\mathrm{r},t,\mathrm{u}) \approx \frac{c}{4\pi} w(\mathrm{r},t) + \frac{3}{4\pi} I(\mathrm{r},t)\cos\phi \tag{4.36}$$

As before, $I$ denotes the intensity, while $\varphi$ is the angle between the direction of the energy flux $j$ and that of the intensity. The second term is assumed to be very small compared to the first one. The resulting differential equation for the energy density $w(\mathbf{r},t)$

$$D\Delta w - \frac{\partial w}{\partial t} = -P\delta(\mathbf{r} - \mathbf{r}_s) \tag{4.37}$$

is known as the diffusion equation (it describes heat conduction too, by the way). In this expression, $P$ is the power output of a point source located at $\mathbf{r}_s$, and $\delta$ is Dirac's delta function. $D$ is the so-called diffusion constant:

$$D = \frac{1}{3} c\bar{\ell} \tag{4.38}$$

Here, $\bar{\ell}$ is the mean free path length, that is, the average length of the path which a particle travels between two successive wall reflections. In Section 5.3, it will be shown that this quantity is $4V/S$, with $V$ denoting the volume of the room and $S$ the area of its boundary, provided the sound field in the considered room is diffuse. Finally, the absorptivity of the boundary must be taken into account. Several authors (Jing and Xiang 2007; Navarro et al. 2010) have derived the following boundary condition for the diffusion equation on the basis of Equation 4.36:

$$D\frac{\partial w}{\partial n} + \frac{c}{4} \cdot \frac{\alpha}{1-(\alpha/2)} w = 0 \quad \text{on the boundary} \tag{4.39}$$

The diffusion approach has first been proposed by Ollendorff (1969), who justified it by considering the sound field as a gas of sound particles which perform some kind of Brownian motion due to the numerous irregularities in the shape of real rooms. (Ollendorff neglected the term $\alpha/2$ in the denominator of Equation 4.39. This led him to reverberation times which are longer than the classical Sabine value.)

Solving Equation 4.37 is by no means easier than searching for solutions of the wave equation (3.1); in both cases, we have to determine eigenvalues and eigenfunctions. However, since the diffusion equation deals with the energy neglecting any phase relations, the required number of eigenfunctions is much smaller than that needed for calculating the sound pressure distribution on the basis of Equations 3.1 and 3.2. This holds in particular for evaluating the energy decay. In this case, we can usually content ourselves with the lowest eigenvalue, since the higher ones are associated with components which fade out very rapidly in the decay. Another advantage of the diffusion method is that it requires only moderate computational effort to gain at least some information about the energy distribution and the energy flux within the enclosure. On the other hand, it should be noted that the diffusion equation (4.37) is itself an approximation which is only valid if the sound field in the enclosure can be expected to be nearly diffuse. The same holds for the boundary condition (4.39). Up to now, no clear-cut rules are available to decide whether this assumption is justified or not in a given case.

## REFERENCES

Borish S. Extension of the image model to arbitrary polyhedra. *J Acoust Soc Am* 1984; *75*: 1827.

Cremer L., Müller H.A. *Principles and Applications of Room Acoustics*, Vol. 1, Chapter I.2 and III.2. London: Applied Science, 1982.

Fujiwara K. Steady state sound field in an enclosure with diffusely and specularly reflecting boundaries. *Acustica* 1984; *54*: 266.

Gilbert E.N. An iterative calculation of auditorium reverberation. *J Acoust Soc Am* 1981; *69*: 178.

Jing Y., Xiang N. On boundary conditions for the diffusion equation in room-acoustic prediction: Theory, simulations and experiments. *J Acoust Soc Am* 2007; *123*: 145.

Joyce W.B. Exact effect of surface roughness on the reverberation time of a uniformly absorbing spherical enclosure. *J Acoust Soc Am* 1978; *64*: 1429.

Koutsouris G.I., Brunskog J., Jeong C.H., Jacobsen F. Combination of acoustical radiosity and the image source method. *J Acoust Soc Am* 2013; *133*: 3963.

Kuttruff H. Simulated decay curves in rectangular rooms with diffuse sound field (in German). *Acustica* 1971; *25*: 333.

Kuttruff H. Reverberation and effective absorption in rooms with diffuse wall reflection (in German). *Acustica* 1976; *35*: 141.

Miles R.N. Sound field in a rectangular enclosure with diffusely reflecting boundaries. *J Sound Vibr* 1984; *92*(2): 203.

Morse P.M., Feshbach H. *Methods of Theoretical Physics*, Section 2.4. New York: McGraw-Hill, 1953.

Navarro J.M., Jacobsen F., Escolano J., López J.J. A theoretical approach to room acoustic simulations based on a radiative transfer method. *Acta Acust/Acustica* 2010; *96*: 1078.

Ollendorff F. Statistical room acoustics as a diffusion problem (in German). *Acustica* 1969; *21*: 236.

Picaut J., Simon L., Polack J.D. A mathematical model of diffuse sound field based on a diffusion equation. *Acta Acust/Acustica* 1997; *83*: 614.

Valeau V., Picaut J., Hodgson M. On the use of a diffusion equation for room-acoustic prediction. *J. Acoust Soc Am* 2006; *119*: 1504.

Chapter 5

# Reverberation and steady-state energy density

Probably the most characteristic acoustical phenomenon in a closed room is reverberation: that is, the fact that sound produced in the room will not disappear immediately after the sound source is shut off but remains audible for a certain period of time afterwards, although with steadily decreasing loudness. For this reason, reverberation — or sound decay, as it is also called — as yet yields the least controversial objective criterion for assessing the acoustical qualities of a room. It is this fact which justifies devoting the major part of a chapter to reverberation and to the laws which govern it.

The physical process of sound decay in a room depends critically on the structure of the sound field. Simple laws describing this process can be formulated only when the sound field is isotropic at any point, which means that the sound propagation is uniform in all directions. In this case, we speak of a diffuse sound field (see Section 4.3). Likewise, simple relationships for the steady-state energy density in a room, as will be derived in Section 5.1, are also based on the assumption of a diffuse field. This is the reason that sound field diffuseness plays a key role in this chapter, although this condition is never perfectly fulfilled in practical situations. Even in certain measuring rooms, such as reverberation chambers, where designers take pain to achieve a sound field as diffuse as possible (see Section 8.7), there is always some lack of diffusion. Nevertheless, the assumption of perfect sound field diffuseness is a useful and important approximation of the actual sound field structure. In most instances in this chapter, we shall therefore assume complete uniformity of sound field as far as its directional distribution is concerned.

In Chapter 3, we regarded reverberation as the common decaying of free vibrational modes. In Chapter 4, *reverberation* was understood to be the sum total of all sound reflections arriving at a certain point in the room after the room was excited by an impulsive sound signal. This chapter deals with reverberation from a more elementary point of view in that we consider the energy balance between the energy supplied by the sound source and that absorbed by the boundary. Furthermore, some extensions and generalizations are described, including sound decay in enclosures with imperfect sound field diffuseness. As in Chapter 4, we shall consider the case of relatively high frequencies, that is, we shall neglect interference and diffraction effects which are typical wave phenomena and which only appear in the immediate vicinity of reflecting walls or when the size of obstacles is comparable with the wavelength. We therefore suppose that the applied sound signals are of such a kind that the direct sound and all reflections from the walls are mutually incoherent, that is, they cannot interfere with each other (see Section 2.5). Consequently, their energies or intensities can simply be added together regardless of mutual phase relations. Under these assumptions, sound behaves in much the same way as white light. We shall, however, not consider sound rays so much, but instead, we shall stress the notion of 'sound particles': that is, of small energy packets which travel with constant velocity $c$ along straight lines — except for wall reflections — and are supposed to be present in very large numbers. They cannot interact, in particular; they will

DOI: 10.1201/9781003389873-5

never collide with each other. If they strike a wall with absorption coefficient $\alpha$, only the fraction $1 - \alpha$ of their energy is reflected from the wall. One way to account for absorption is to assign properly reduced energy to the reflected particle. Alternatively, the absorption coefficient can be interpreted as 'absorption probability'.

Of course, the sound particles considered in room acoustics are purely hypothetical and have nothing to do with the sound quanta or phonons known from solid-state physics. To bestow some physical reality upon them, we can consider the sound particles to be short sound pulses with a broad spectrum propagating along sound ray paths. Their exact shape is not important, but they all must have the same power spectrum.

## 5.1 DIFFUSE SOUND FIELDS, ELEMENTARY THEORY OF SOUND DECAY

As mentioned earlier, the uniform and isotropic distribution of sound energy in a room is the crucial condition for the validity of some common and simple formulae which govern the sound decay or the steady-state energy in rooms. Therefore, it is appropriate to deal first with some properties of diffuse sound fields.

Suppose that we select from all sound rays crossing an arbitrary room point $P$ a bundle within a vanishingly small solid angle $d\Omega$. Since the rays of the bundle are nearly parallel, an intensity $I(\varphi, \vartheta)d\Omega$ can be attributed to it, with $\varphi$ and $\vartheta$ characterizing their direction (see Figure 5.1). (The quantity $I(\varphi, \vartheta)d\Omega$ has been called the 'flux density' in Section 4.6.)

Furthermore, we can apply Equation 1.33 to this bundle, according to which the energy density

$$dw = \frac{I(\varphi,\vartheta)}{c} d\Omega \tag{5.1}$$

is associated with it. To obtain the total energy density, we have to integrate this expression over all directions. Since the quantity $I$ is independent of the angles $\varphi$ and $\vartheta$, the integration is achieved just by multiplying Equation 5.1 with $4\pi$. Hence, the total energy density is

$$w = \frac{4\pi I}{c} \tag{5.2}$$

So far, nothing has been said about spatial variations of the energy density. In fact, it is easy to see that it is constant throughout the room, provided the sound field is diffuse. Consider the three circles in *Figure 5.2* representing the directional distribution at the points $P$, $Q$, and $R$. Each pair of points has exactly one sound ray in common. Since the energy propagated along a sound ray does not change with distance (see *Chapter 4*), it follows that they contribute the same amount of energy at both points. Therefore, the three circles must have

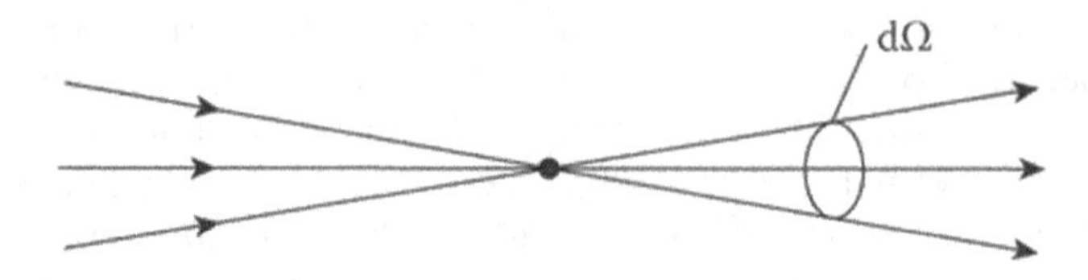


*Figure 5.1* Bundle of nearly parallel sound rays.

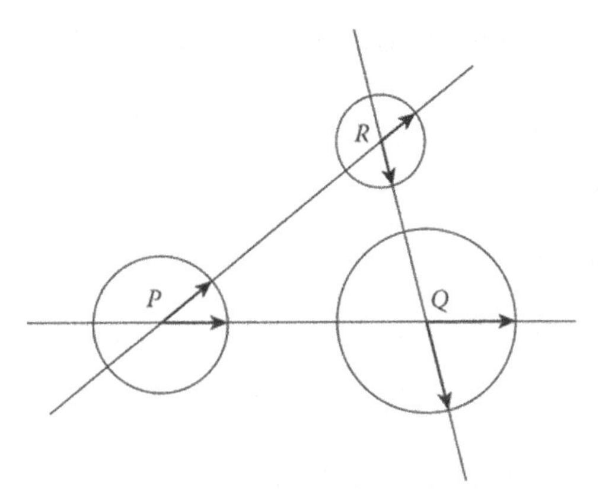


*Figure 5.2* Constancy of energy density in a diffuse sound field.

equal diameters. This argument applies to all points of the space. Thus, we can conclude that in a diffuse sound field, the energy density is everywhere the same, at least under stationary conditions. However, the reverse is generally not true; we can easily imagine sound fields with constant energy density and nonuniform directional distribution.

Another important property of a diffuse sound field has already been derived in Section 2.5. According to Equation 2.51, the energy incident per second on a wall element dS is $\pi I$dS. Hence, the 'irradiance' $B$, already introduced in Section 4.5 as the energy incident per unit time and unit area, is $B = \pi I$, and with Equation 5.2:

$$B = \frac{c}{4} w \tag{5.3}$$

This relation is to be compared with $w = I/c$ valid for a plane progressive wave (see Equation 1.33).

We are now ready to set up an energy balance from which a simple law for the sound decay in a room can be derived. Suppose a sound source feeds the acoustical power $P(t)$ into a room. It is balanced by a temporal increase in the energy content $Vw$ of the room and by the losses due to the absorptivity of its boundary which has the absorption coefficient $\alpha$:

$$P(t) = V\frac{\mathrm{d}w}{\mathrm{d}t} + B\alpha S$$

Or, by using Equation 5.3 and replacing $\alpha S$ with $A$, the 'equivalent absorption area' of the room:

$$P(t) = V\frac{\mathrm{d}w}{\mathrm{d}t} + \frac{cA}{4} w \tag{5.4}$$

When $P$ is constant, the differential quotient is zero, and we obtain the steady-state energy density:

$$w = \frac{4P}{cA} \tag{5.5}$$

If, on the other hand, the sound source is switched off at $t = 0$, that is, if $P(t) = 0$ for $t > 0$, the differential equation (5.4) becomes homogeneous and has the solution

$$w(t) = w_0 \exp(-2\delta t) \quad \text{for } t \geq 0 \tag{5.6}$$

with the initial value $w_0 = w(0)$. The damping constant or decay constant turns out to be

$$\delta = \frac{cA}{8V} \tag{5.7}$$

and is related to the reverberation time $T$ by Equation 3.70. This gives, after inserting the numerical value of the sound speed in air, the following:

$$T = 0.161\frac{V}{A} \tag{5.8}$$

(all lengths expressed in metres and T in seconds). This is probably the best-known formula in room acoustics. It is due to Sabine (1923), who derived it first from the results of numerous ingenious experiments, and later on from considerations similar to the present one. Nowadays, it is still the standard formula for predicting the reverberation time of a room, although it is obvious that it fails for high absorption. In fact, for $\alpha = 1$, it predicts a finite reverberation time, although an enclosure without any reflecting walls cannot reverberate. The reason for the limited validity of Equation 5.9 is that the room is not — as assumed — in steady-state conditions during sound decay and is less so the faster the sound energy decays. In the following sections, more exact decay formulae will be derived which can also be applied to relatively 'dead' enclosures.

## 5.2 FACTORS INFLUENCING SOUND FIELD DIFFUSENESS

In this section, the circumstances will be discussed on which the diffuseness of a sound field depends. Strictly speaking, a real sound field cannot be completely diffuse; otherwise, there would be no net energy flow within the enclosure. In a real room, however, the inevitable wall losses 'attract' a continuous energy flow originating from the sound source. So what we are discussing here is how and to which extent the ideal condition can be approximated.

It is obvious that a diffuse sound field will not be observed in enclosures whose walls have the tendency to concentrate the reflected sound energy in certain regions. In contrast, highly irregular room shapes help establish a diffuse sound field by continuously redistributing the energy in many different directions. Particularly efficient in this respect are rooms with acoustically rough walls, the irregularities of which scatter the incident sound energy in a wide range of directions, as has already been described in Section 2.7. Such walls are referred to as 'diffusely reflecting', either partially or completely. The latter case is characterized by Lambert's law, as expressed in Equation 4.24. Although this ideal behaviour is frequently assumed as a model of diffuse reflection, it will hardly ever be encountered in reality. Any wall or ceiling will, although it may be structured by numerous columns, niches, cofferings, and other 'irregular' decorations, diffuse only a certain fraction of the incident sound, whereas the remaining part of it is reflected into specular directions. The reader will be reminded of Figure 2.18, which shows the scattering characteristics of a particular, irregularly shaped ceiling. On the other hand, in virtually every real hall, a certain part of the sound energy is scattered in non-specular directions, as has been shown by Hodgson (1991), even if the walls and the ceiling are apparently smooth.

Despite the fact that real surfaces produce only partially diffuse reflections, there is a natural tendency towards increasing sound field diffusion. This is due to the fact that the conversion of 'specular sound energy' into non-specular or random energy is an irreversible process; in other words, it will never happen that 'diffuse sound' is re-converted into a sound ray traveling in a well-defined direction. The consequence of this can be demonstrated by splitting up the reflected sound energy into two parts: the portion $s(1–\alpha)$, which is scattered into non-specular directions, and the portion $(1–s)(1–\alpha)$, which is reflected specularly. (The remaining energy fraction $\alpha$ is absorbed.) The parameter $s$ is the 'scattering coefficient'. Then, after $n$ reflections, the specular component will be reduced by a factor $[(1–s)(1–\alpha)]^n$, while the fraction of non-specular energy is $[1–(1–s)^n]\cdot(1–\alpha)^n$. Both components are shown in Figure 5.3 for a particular set of parameters. It is shown that the relative amount of diffuse energy increases monotonically with the order of the reflections; after a few reflections, nearly all energy has been converted into diffuse energy.

This tendency is counteracted by the absorptivity of the boundary and particularly by a non-uniform spatial distribution of the absorption. A typical situation is that of an occupied hall, for example, of a concert hall, since most of the absorption is affected by the audience, which is usually placed on the floor of the hall. It is evident that highly absorptive portions of the boundary extinguish potential ray paths and, hence, impede the formation of a diffuse sound field. According to Equation 4.25, the product of the irradiance $B$ and the reflection coefficient $\rho$ determines the energy which the given wall element contributes to the sound field. Suppose that the reflectivity $\rho$ of a particular wall element is very small. Then, it is not very likely that its irradiance $B$ will completely make up for this deficiency in reflected energy, since $B$ shows — due to the smoothing effect of wall scattering — only small variations along the boundary. On the other hand, even a small diffusely reflecting wall portion is capable of producing a diffuse field, provided the absorption is sufficiently small.

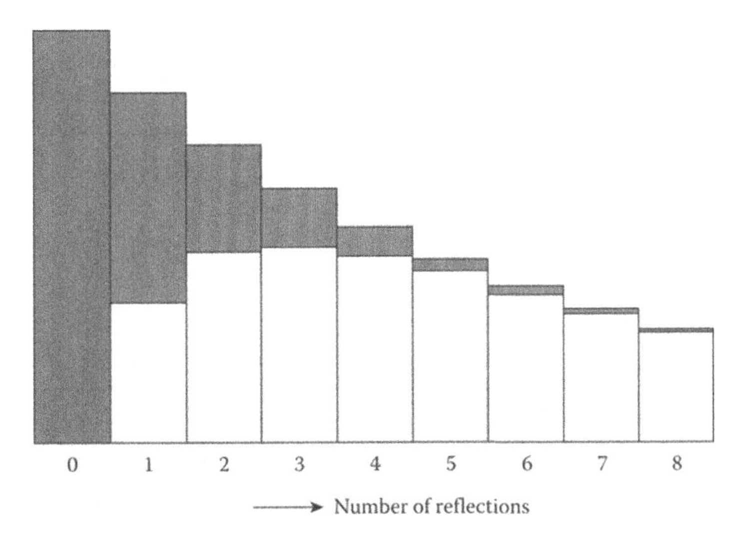


*Figure 5.3* Conversion of specularly reflected energy into diffuse energy by subsequent reflections ($\alpha$ = 0.15, s = 0.4). The total height of the bars corresponds to the totally reflected sound energy; the height of the white bars shows the diffusely reflected energy.

There is still another method of increasing sound field diffuseness, namely, by disturbing the free propagation of sound rays. This is achieved by suitable objects — rigid bodies or shells — which are suspended freely in the room at random positions and orientations and which scatter the arriving sound waves or sound particles more or less in all directions. This method is quite efficient even when it is applied only to parts of the room, or in enclosures with partially absorbing walls. Of course, for the acoustical design of performance halls, this measure is of limited use because most architects would not be keen to have a substantial part of a hall's free space filled with such 'volume diffusers'. For certain measuring rooms, however, the so-called reverberation chambers (see Section 8.7), they are a well-proved way to provide for the necessary sound field diffuseness.

To estimate the efficiency of volume scatterers, we assume $N$ of them to be randomly distributed in a room with volume V, but with constant mean density $\langle n_s \rangle = N/V$. The scattering efficiency of a single obstacle or diffuser is characterized by its 'scattering cross section' $Q_s$, defined by Equation 2.60. If we again stress the notion of sound particles, the probability that a particle will travel at least a distance $r$ without being scattered by a diffuser is $\exp(-\langle n_s \rangle Q_s r)$ or $\exp(-r/\overline{r}_s)$, where

$$\overline{r}_s = \frac{1}{\langle n_s \rangle Q_s} \tag{5.9}$$

is the mean free path length of a sound particle between successive collisions with diffusers. In the next section, we shall introduce the mean free path of sound particles between successive wall reflections $\overline{\ell} = 4V/S$, with $S$ denoting the wall surface of a room. Obviously, the efficiency of volume diffusers depends on the ratio $\overline{r}_s / \ell_s$. If an obstacle is not small compared with the acoustical wavelength, its scattering cross section $Q_s$ is roughly twice its visual cross section. One half of this value represents scattering of energy in different directions; the other half corresponds to the energy needed to form the 'shadow' behind the obstacle by interfering with the incident sound wave. For non-spherical diffusers, the cross section must be averaged over all directions of incidence.

The preceding discussion may be summarized in the following way: Diffusely reflecting room walls help establish isotropy of sound propagation in a room, but they alone do not guarantee high diffuseness of the sound field. At least of equal importance is the amount and distribution of wall absorption. Perfect or nearly perfect sound field diffuseness is only possible if the total absorption in the room is very small, no matter whether it is caused by imperfect wall reflections or by losses within the medium.

## 5.3 MEAN FREE PATH LENGTH AND AVERAGE RATE OF REFLECTIONS

In the following, we imagine that the sound field is composed of numerous sound particles. Our goal is to derive more general laws for the sound decay in a room than that of Equation 5.8. For this purpose, we have, at first, to follow the 'fate' of one sound particle and, subsequently, to average over many of these fates.

In this connection, the notion of the 'mean free path' of a sound particle is frequently encountered in literature on room acoustics. The notion itself appears at first glance to be quite clear, but its use is sometimes misleading, partly because it is not always evident whether it refers to the time average or the particle ('ensemble') average.

We shall start here from the simplest concept: A sound particle is observed during a very long time interval $t$; the total path length $ct$ covered by it during this time is divided by $N$, the number of wall collisions which have occurred in the time $t$. Then, the mean free path length is

$$\overline{\ell} = \frac{ct}{N} = \frac{c}{\overline{n}} \tag{5.10}$$

where $\overline{n} = N/t$ is the average reflection frequency, that is, the average number of wall reflections per second.

Here, the quantities $\overline{\ell}$ and $\overline{n}$ are clearly defined as time averages for a single sound particle; they may differ from one particle to another. In order to obtain averages which are representative of all sound particles, we should average $\overline{\ell}$ and $\overline{n}$ once more, namely, over all possible particle fates. In general, the result of such a procedure would depend on the shape of the room as well as on the directional distribution of sound paths.

Fortunately, we can avoid these complications by assuming the sound field to be diffuse. Then, no additional specification of the directional distribution is required. Furthermore, no other averaging is needed. This is so because — according to the discussion in Section 5.2 — a diffuse sound field is established by nearly non-predictable changes in the particle directions, caused either by a highly complex room shape, by diffuse wall reflections, or by volume scatterers, as described in the preceding section. In any case, the sound particles change their roles and their directions again and again, and during this process, they completely lose their individuality. Thus, the distinction between time averages and particle or ensemble averages is no longer meaningful; it does not matter whether the quantities we are looking for are evaluated by averaging over many free paths traversed by one particle or by averaging, for one instant, over a great number of different particles. Or in short:

time average = ensemble average

Under these conditions, the calculation of the mean free path length of sound particles is quite easy. Let us suppose that a single sound particle carries the energy $e_0$. Its contribution to the energy density of the room is

$$w = \frac{e_0}{V}$$

On the other hand, if the sound particle strikes the wall $\overline{n}$ times per second, the average energy it transports per second and unit area toward the wall is

$$B = \overline{n}\frac{e_0}{S}$$

By inserting these formulae into Equation 5.3, one arrives at the important relation

$$\overline{n} = \frac{cS}{4V} \tag{5.11}$$

This expression, which is the time average as well as the particle average, has already been derived in Section 4.2 for the rectangular room. It is evident that the present derivation is more general and, in a way, more satisfying than the previous one. Inserting its result into Equation 5.10 again leads to the mean free path length:

$$\overline{\ell} = \frac{4V}{S} \tag{5.12}$$

The actual free path lengths are distributed around their mean $\overline{\ell}$ in some way, of course. This distribution depends on the shape of the room, and the same is true for other characteristic values, such as its variance $\overline{\left(\ell - \overline{\ell}\right)^2} = \overline{\ell^2} - \left(\overline{\ell}\right)^2$. As an illustration, Figure 5.4 shows the distributions of free path lengths for three different shapes of rectangular rooms with diffusely reflecting walls; the abscissa is the path length divided by its mean value $\overline{\ell}$. These distributions have been obtained by ray tracing, a method which will be described to some more detail in Section 9.8.

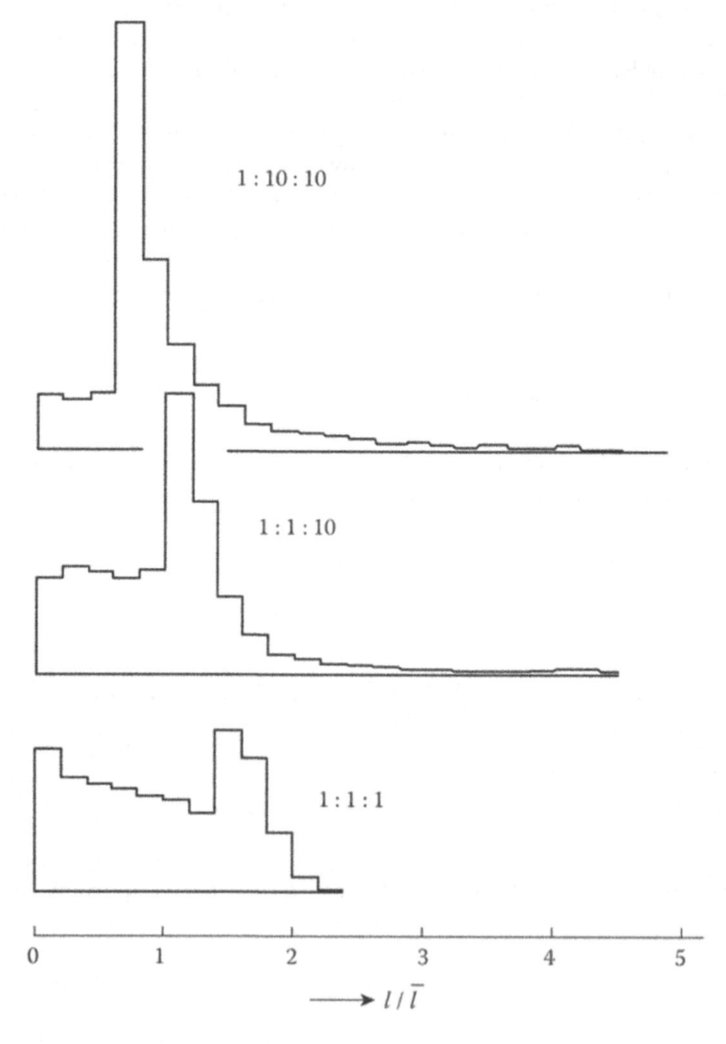


*Figure 5.4* Distribution of free path lengths for rectangular rooms with diffusely reflecting walls. The numbers indicate the relative dimensions of the rooms.

Source: Kuttruff (1970, 1971).

*Table 5.1* Average and the relative variance of the path length distribution in some rectangular rooms with diffusely reflecting walls (ray-tracing results)

| *Relative dimensions* | *Mean value* $\overline{\ell}_{MC} / (\overline{\ell})$ | *Relative variance* ($\gamma^2$) |
|---|---|---|
| 1:1:1 | 1.0090 | 0.342 |
| 1:1:2 | 1.0050 | 0.356 |
| 1:1:5 | 1.0066 | 0.412 |
| 1:1:10 | 1.0042 | 0.415 |
| 1:2:2 | 1.0020 | 0.363 |
| 1:2:5 | 1.0085 | 0.403 |
| 1:2:10 | 0.9928 | 0.465 |
| 1:5:5 | 1.0035 | 0.464 |
| 1:5:10 | 0.99930 | 0.510 |
| 1:10:10 | 1.0024 | 0.613 |

Source: Kuttruff (1970, 1971).

Typical parameters of path length distributions, evaluated in the same way for different rectangular rooms, are listed in Table 5.1. The first column shows the relative dimensions of the various rooms, and the second column lists the mean free path length divided by the 'classical' value of Equation 5.12. These numbers are very close to unity and can be looked upon as an 'experimental' confirmation of Equation 5.12. The remaining deviations from 1 are insignificant and are due to the random errors inherent in the method. It may be added that a similar investigation of rooms with specularly reflecting walls, which are equipped with scattering elements in the interior, yields essentially the same result. Finally, the third column of Table 5.1 contains the 'relative variance' of the path length distributions:

$$\gamma^2 = \frac{\overline{\ell^2} - (\overline{\ell})^2}{(\overline{\ell})^2} \tag{5.13}$$

The significance of this quantity will be discussed in Section 5.5.

## 5.4 NON-UNIFORM BOUNDARY ABSORPTION

As far back as Chapter 4, formulae have been derived for the time dependence of decaying sound energy and for the reverberation time of rectangular rooms (Equations 4.10 to 4.12). In the preceding section, it has been shown that the value cS/4V of the mean reflection frequency, which we have used in Chapter 4, is valid not only for rectangular rooms but also for rooms of arbitrary shape, provided that the sound field in their interior is diffuse. Thus, the general validity of those reverberation formulae has been proven. It may be added that for m → 0 and for small absorption coefficients, Equation 4.12 will become identical with Equation 5.8 since

$$-\ln(1-\alpha) \to \alpha \quad \text{for } \alpha \to 0 \tag{5.14}$$

If the sound absorption coefficient of the walls depends on the direction of sound incidence, which will usually be the case, one should replace $\alpha$ with the average value $\alpha_{uni}$ according to Equation 2.53.

Further consideration is necessary if the absorption coefficient is not constant along the boundary, which is usually the case in real rooms. For the sake of simplicity, we assume that there are only two types of walls in a room with different absorption coefficients, $\alpha_1$ and $\alpha_2$; the areas of these wall portions are $S_1$ and $S_2$, with $S_1 + S_2 = S$ (see Figure 5.5). The subsequent generalization of the results to the case of more than two different sorts of wall will be obvious.

Now, we follow the path of a particular sound particle over $N$ wall reflections, among which there are $N_1$ reflections from $S_1$, and $N_2 = N - N_1$ reflections from $S_2$. These numbers can be assumed to be distributed in some way about their mean values:

$$\bar{N}_1 = N\frac{S_1}{S} \text{ and } \bar{N}_2 = N\frac{S_2}{S} \tag{5.15}$$

$S_1/S$ and $S_2/S$ are the 'a priori probabilities' of a sound particle hitting the wall portion $S_1$ or $S_2$, respectively.

In a diffuse sound field, subsequent wall reflections are stochastically independent from each other, that is, the probability of hitting one or another portion of the wall does not depend on the past history of the particle. In this case, the probability of $N_1$ collisions with wall portion $S_1$ among a total number of reflections $N$ is given by the binomial distribution (Papoulis 1985).

$$P_N(N_1) = \binom{N}{N_1} \cdot \left(\frac{S_1}{S}\right)^{N_1} \cdot \left(\frac{S_2}{S}\right)^{N-N_1} \tag{5.16}$$

With the binomial coefficient:

$$\binom{N}{N_1} \equiv \frac{N!}{N_1!(N-N_1)!}$$

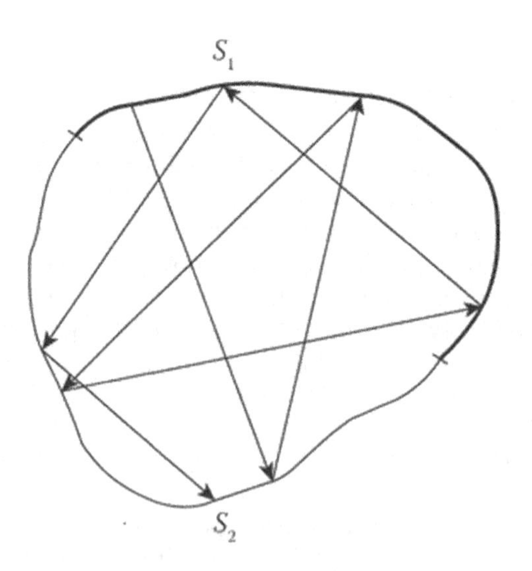


*Figure 5.5* Enclosure with two different types of boundary.

After $N_1$ collisions with $S_1$ and $N - N_1$ collisions with $S_2$, the remaining energy of a sound particle is

$$E_N(N_1) = E_0(1-\alpha_1)^{N_1}(1-\alpha_2)^{N-N_1} \tag{5.17}$$

when $E_0$ is its initial energy. The expectation value of this expression with respect to the distribution (5.16) is

$$\langle E_N \rangle \equiv \sum_{N_1=0}^{N} E_N(N_1) \cdot P_N(N_1) = E_0\left[\frac{S_1}{S}(1-\alpha_1) + \frac{S_2}{S}(1-\alpha_2)\right]^N \tag{5.18}$$

where we have used the binomial theorem. Since $S_1 + S_2 = S$, this result can be written as

$$\langle E_N \rangle = E_0\left[1 - \frac{\alpha_1 S_1 + \alpha_2 S_2}{S}\right]^N \tag{5.19}$$

and with

$$\bar{\alpha} = \frac{1}{S}(S_1\alpha_1 + S_2\alpha_2) \tag{5.20}$$

$$\langle E_N \rangle = E_0(1-\bar{\alpha})^N = E_0 \exp\left[N \cdot \ln(1-\bar{\alpha})\right] \tag{5.21}$$

This latter formula tells us that in case of non-uniform boundary absorption, the sound decay is governed by the arithmetic mean $\bar{\alpha}$ of the absorption coefficients $\alpha_1$ and $\alpha_2$ with the respective areas as weighting factors. Finally, we replace the total number $N$ of wall reflections in time $t$ by its expectation value or mean value $\bar{n}t$ with $\bar{n} = cS/4V$ and obtain for the energy of the 'average' sound particle and, hence, for the total energy in the room:

$$E(t) = E_0 \ \exp\left[\frac{cS}{4V} t \cdot \ln(1-\bar{\alpha})\right] \quad \text{for } t \geq 0 \tag{5.22}$$

From this, we can easily evaluate the reverberation time, that is, the time interval $T$ in which the reverberating sound energy reaches one millionth of its initial value:

$$T = -\frac{24V \cdot \ln 10}{cS \cdot \ln(1-\bar{\alpha})} \tag{5.23}$$

Or after inserting $c$ = 343 m/s and accounting for the air attenuation by an additional term $4mV$ in the denominator as in Equations 4.7 to 4.12:

$$T = 0.161 \frac{V}{4mV - S \cdot \ln(1-\bar{\alpha})} \tag{5.24}$$

with

$$\bar{\alpha} = \frac{1}{S}\sum_i S_i \alpha_i = \frac{A}{S} \tag{5.25}$$

The latter is the generalization of Equation 5.20 for any number of different wall portions $S_i$. In all these formulae, all lengths have to be expressed in metres.

Equation 5.23 or 5.24, together with 5.25, is known as Eyring's reverberation formula, although there are independent derivations due to Norris as well as to Schuster and Waetzmann. Neglecting air attenuation and approximating the logarithm by $-\bar{\alpha}$ as before lead us again to Sabine's famous decay formula:

$$T = 0.161\frac{V}{A} \tag{5.26}$$

In deriving Equation 5.24 together with Equation 5.25, the quantities $N_1$ and $N_2 = N - N_1$ have been treated as statistical variables based on the probability distribution (5.16). One could be tempted to shorten the preceding treatment by replacing these quantities in Equation 5.18 with their mean values or expectation values $NS_1/S$ and $NS_2/S$. Then, one would arrive at

$$E_N = E_0\left[(1-\alpha_1)^{S_1}(1-\alpha_2)^{S_2}\right]^{N/S} = E_0 \exp\left[\frac{S_1}{S}\cdot\ln(1-\alpha_1)+\frac{S_2}{S}\cdot\ln(1-\alpha_2)\right]\cdot\frac{cS}{4V}t$$

using the relation $N = \bar{n}t = (cS/4V)\,t$. In this case, it is not the absorption coefficient α which is averaged but the quantity ln(1 – α), which is often referred to as the 'absorption exponent':

$$\bar{a}' = -\frac{1}{S}\sum_i S_i \ln(1-\alpha_i) \tag{5.27}$$

Formally, the decay law would be the same as before:

$$T = 0.161\frac{V}{S\bar{a}'} \tag{5.28}$$

And again, it could be completed by adding a term $4mV$ to the denominator in order to account for the air attenuation. This expression is known as Millington-Sette's formula.

Equation 5.27 has a strange consequence: Suppose one portion $S_i$ of the boundary has the absorption coefficient $\alpha_i = 1$. This would make the average (5.27) infinitely large, and thus, the reverberation time evaluated with Equation 5.28 would be zero no matter how small $S_i$ is. This is obviously an unreasonable result.

## 5.5 THE INFLUENCE OF UNEQUAL PATH LENGTHS

The averaging rule of Equation 5.27 was the result of replacing a probability distribution by its mean value. However, throughout the preceding section, we have tacitly practised a similar simplification in that we replaced the actual number of reflections in a given time $t$ by its average $\bar{n}t$ For a more correct treatment, one should introduce the probability $P_t(N)$ of exactly $N$ wall reflections occurring in a time $t$ and to calculate E(t) as the expectation value of Equation 5.21 with respect to this probability distribution:

$$E(t) = E_0\sum_{N=0}^{\infty} P_t(N)\cdot(1-\bar{\alpha})^N = E_0\sum_{N=0}^{\infty} P_t(N)\cdot\exp(-Na) \tag{5.29}$$

In the latter expression, the absorption exponent $a = -\ln(1 - \alpha)$ has been introduced.

If, for the moment, $N$ is considered as a continuous variable, the function $\exp(-Na)$ in Equation 5.29 can be expanded in a series for ascending powers of $(N - \bar{n}t)a$ a by setting $N = \bar{n}t + (N - \bar{n}t)$. Truncating this series after its third term yields

$$\exp(-Na) \approx \exp(-\bar{n}ta)\left[1 - \frac{N - \bar{n}t}{1!}a + \frac{(N - \bar{n}t)^2}{2!}a^2\right]$$

Before inserting this expression into Equation 5.29, it should be kept in mind that

$$\sum_N P_t(N) = 1 \quad \text{and} \quad \sum_N (N - \bar{n}t) P_t(N) = 0$$

whereas

$$\sum_N (N - \bar{n}t)^2 P_t(N) = \sigma_N^2$$

is the variance of the distribution $P_t(N)$. Hence,

$$E(t) = E_0 \exp(-\bar{n}ta) \cdot \left(1 + \frac{1}{2}\sigma_N^2 a^2\right) \approx E_0 \exp\left(-\bar{n}ta + \frac{1}{2}\sigma_N^2 a^2\right) \tag{5.30}$$

The latter approximation is permissible if $\sigma_N^2 a^2/2$ is small compared with unity.

Obviously, the probability $P_t(N)$ of $N$ reflections occurring during the time $t$ corresponds to the probability $P_N(x)$ of having travelled the total distance $x$ after $N$ reflections. To illustrate this, Figure 5.6 plots the individual 'fates' of three particles, that is, the number $N$ of their wall collisions as a function of the distance $x$ they travelled. Now, suppose a particle

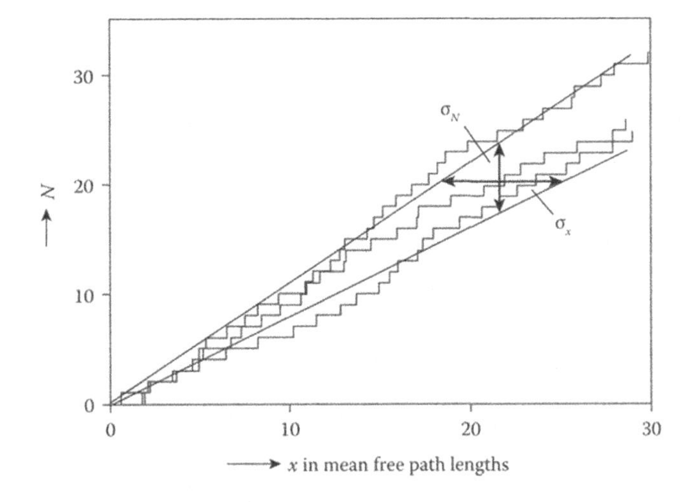


*Figure 5.6* Time histories of three sound particles—relation between $\sigma_x$ and $\sigma_N$.

travels $N$ successive and independent free paths $\ell_1$, $\ell_2$, $\ell_N$. According to the central limit theorem of probability, the variance of the sum $x$ is

$$\sigma_x^2 = N\left(\overline{\ell^2} - \left(\overline{\ell}\right)^2\right) = N\overline{\ell}^2\gamma^2$$

The difference in the bracket is the variance of the path length distribution, and $\gamma^2$ is its relative variance as defined in Equation 5.13. As may be seen from Figure 5.6, the variance $\sigma_N$ is proportional to $\sigma_x$, or more precisely:

$$\frac{\sigma_N}{\sigma_x} \approx \frac{\overline{n}t}{ct} = \frac{1}{\overline{\ell}}$$

From both relations, we conclude that

$$\sigma_N^2 \approx \overline{n}t\gamma^2 \tag{5.31}$$

With this result, we obtain from Equation 5.30

$$E(t) = E_0 \exp\left[-\overline{n}ta\left(1 - \frac{1}{2}\gamma^2 a\right)\right] \tag{5.32}$$

Accordingly, the reverberation time is

$$T = 0.161\frac{V}{Sa''} \tag{5.33}$$

with the modified absorption exponent

$$a'' = -\ln\left(1 - \overline{\alpha}\right) \cdot \left[1 + \frac{\gamma^2}{2}\ln\left(1 - \overline{\alpha}\right)\right] \tag{5.34}$$

In Figure 5.7, the corrected absorption exponent $a''$ is compared with the absorption coefficient $\overline{\alpha}$ as is used in Sabine's formula (Equation 5.26). It plots the relative difference between both quantities for various parameters $\gamma^2$. The curve $\gamma^2 = 0$ corresponds to Eyring's formula (5.24) with $m = 0$. Accordingly, the latter is strictly valid for one-dimensional enclosures only where all paths have exactly the same length. For $\gamma^2 > 0$, $a''$ is smaller than $-\ln\left(1 - \overline{\alpha}\right)$. Hence, the reverberation time is generally longer than that obtained from the Eyring formula.

As long as the mean absorption coefficient is smaller than about 0.3, which is true for most rooms, the Eyring absorption exponent may be approximated by

$$-\ln\left(1 - \overline{\alpha}\right) \approx \overline{\alpha} + \frac{\overline{\alpha}^2}{2}$$

which, after omitting all terms of higher than second order, yields

$$a'' \approx \overline{\alpha} + \frac{\overline{\alpha}^2}{2}\left(1 - \gamma^2\right) \tag{5.35}$$

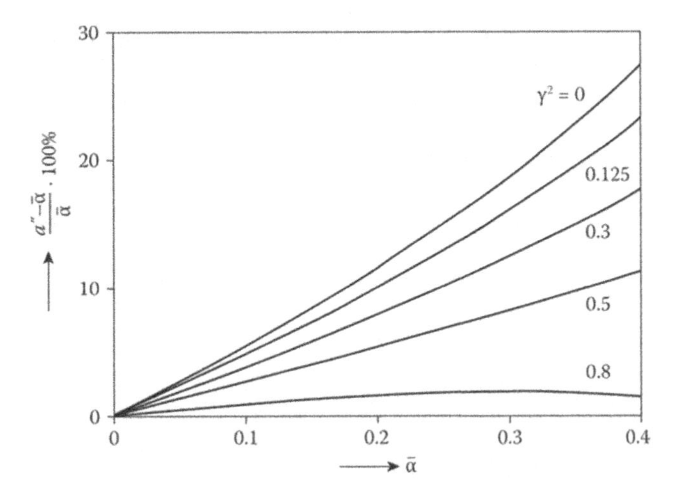


*Figure 5.7* Relative difference between $a''$ and $\alpha$ after Equation 5.34 (in per cent).

With Equations 5.33 and 5.34, we have arrived at a reverberation formula in which the shape of the room is accounted for by the quantity $\gamma^2$. Unfortunately, the latter can only be calculated for a limited number of room shapes. For a sphere, for instance, it turns out to be 1/8. For other shapes, the relative variance $\gamma^2$ can be determined by computer simulation. Results obtained in this way for rectangular rooms have already been presented in Table 5.1. It is seen that for most shapes, $\gamma^2$ is close to 0.4, and it is likely that this value can also be applied to other enclosures, provided that their shapes do not deviate too much from that of a cube. It should be noted that for higher values of the relative variance $\gamma^2$, Equation 5.34 is not very accurate because of the approximations made in its derivation.

For rooms with suspended 'volume diffusers' (see Section 5.1), the distribution of free path lengths is greatly modified by the scattering obstacles. The same applies to the relative variance $\gamma^2$, but not to the mean free path length (Kuttruff 1970, 1971).

## 5.6 ENCLOSURE DRIVEN BY A SOUND SOURCE

In Section 5.1, a differential equation (5.4) was derived for the energy density $w$ in a room in which a sound source with time-dependent power output $P$(t) is operated. In the preceding sections, we used this equation to discuss the sound decay in a room, which can be observed after the sound source has been abruptly switched off at some instant. The present section deals with the acoustic reaction of the room when the sound source produces a time-dependent output power $P(t)$, including the case of a constant power.

The general solution of Equation 5.4 reads

$$w(t) = \frac{1}{V}\int_{-\infty}^{t} P(\tau)\exp\left[-2\delta(t-\tau)\right]d\tau = \frac{1}{V}\int_{0}^{\infty} P(t-\tau)\exp(-2\delta\tau)d\tau \tag{5.36}$$

with $\delta = cA/8V = 6.91/T$. This means the energy density is calculated by convolving the power output $P(t)$ with the 'energetic impulse response' of the enclosure:

$$w(t) = P(t) * \frac{1}{V}\exp(-2\delta t) \tag{5.37}$$

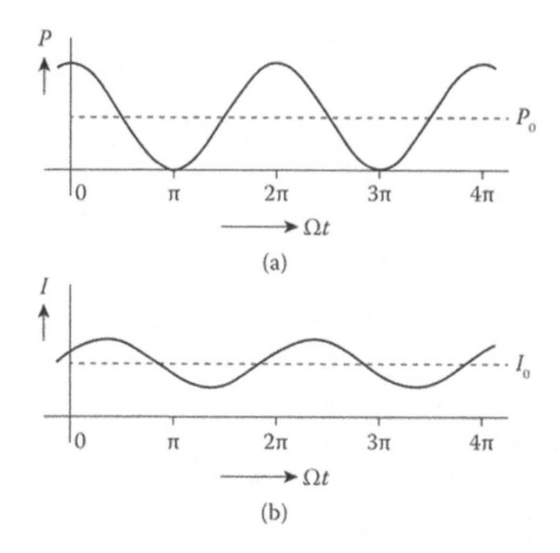


*Figure 5.8* Flattening of the modulation of the energy density affected by transmission in a room: (a) power of the emitted sound signal and (b) energy density in a room point.

As an application of this formula, we consider, as shown in Figure 5.8, a sound source the power output of which varies sinusoidally with the angular frequency Ω. Using complex notation, we write

$$P(t) = P_0[1 + \exp(i\Omega t)] \tag{5.38}$$

In this case, the power delivered by the sound source varies periodically between $2P_0$ and zero. The reaction of the room is obtained by inserting this expression into Equation 5.37 and performing the integration. The result reads

$$w(t) = \frac{P_0}{2\delta V}\left[1 + m \cdot \exp(i\Omega t)\right] = \frac{P_0}{2\delta V}\left\{1 + |m| \cdot \exp\left[i\Omega\left(t - t_0\right)\right]\right\} \tag{5.39}$$

with

$$m(\Omega) = \frac{2\delta}{2\delta + i\Omega} \tag{5.40}$$

and

$$t_0 = \frac{1}{\Omega}\arctan\left(\frac{\Omega}{2\delta}\right) = \frac{1}{\Omega}\operatorname{arc\,cos}|m(\Omega)| \tag{5.41}$$

Since the magnitude of $m(\Omega)$ is always smaller than unity, the reverberation of the room has the effect of smoothing the fluctuations of the energy density imposed by the variable input power; at the same time, it causes a time delay of the fluctuations. For this reason, the function $m(\Omega)$ is called the 'modulation transfer function' (MTF). Of course, the preceding formulae for the MTF are only valid for enclosures the sound decay of which follows the idealized exponential law of Equation 5.6. In real rooms, the decay constant $\delta$ depends on the sound frequency $\omega$. Hence, the MTF is a function of both the modulation frequency $\Omega$

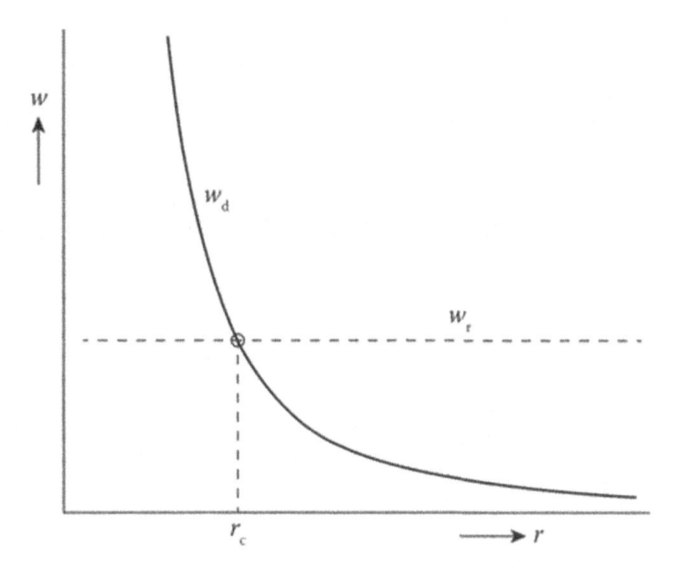


*Figure 5.9* Direct and reverberant energy density $w_d$ and $w_r$ as a function of distance *r*.

of the sound source and the frequency or, more generally, on the frequency spectrum of the exciting sound.

A much simpler case is that of a source with constant power output *P*. Here, Equation 5.36 yields immediately:

$$w_r = \frac{P}{2\delta V} = \frac{4P}{cA} \tag{5.42}$$

with $\delta = cA/8V$ (see Equation 5.7). This formula agrees with Equation 5.5, which was obtained directly from Equation 5.4 by setting the time derivative zero. The subscript 'r' is used to indicate that $w_r$ is the energy density of the 'reverberant field', excluding the contribution of the direct sound. *A* is the 'equivalent absorption area', or shorter: the 'absorption area' which, for the case of *n* wall portions with areas $S_i$ and absorption coefficients α, is given by Equation 5.25.

For a point source with omnidirectional sound radiation, the direct sound intensity in distance *r* is $I_d = cw_d = P/4\pi r^2$; hence, the energy density of the direct component is

$$w_d = \frac{P}{4\pi c r^2} \tag{5.43}$$

In Figure 5.9, $w_r$ and $w_d$, according to Equations 5.42 and 5.43, are presented as functions of distance *r* from the sound source. For a certain distance $r = r_c$, both energy densities are equal. This distance $r_c$ is called the 'critical distance' or 'diffuse-field distance' and is given by

$$r_c = \left(\frac{A}{16\pi}\right)^{1/2} \approx 0.1 \cdot \left(\frac{V}{\pi T}\right)^{1/2} \tag{5.44}$$

In the latter expression, we have introduced the reverberation time *T* from Sabine's formula (5.26); *V* is to be expressed in $m^3$.

Using the critical distance, the total energy density can be expressed as

$$w = w_d + w_r = \frac{4P}{cA}\left(1 + \frac{r_c^2}{r^2}\right) \tag{5.45}$$

Many sound sources have a certain directivity, which can be characterized by their 'gain' or 'directivity' $g$. This is defined as the ratio of the maximum intensity and the average intensity, both at the same distance $r$:

$$g = \frac{I_{\max}}{(P/4\pi r^2)} = 4\pi c r^2 \frac{w_{\max}}{P} \tag{5.46}$$

(In the latter expression, $I_{max}$ has been replaced with $w_{\max} = I_{max}/c$.) Then, Equation 5.44 must be replaced with

$$w_d = \frac{gP}{4\pi c r^2} \tag{5.47}$$

and the critical distance is

$$r_c = \left(\frac{gA}{16\pi}\right)^{1/2} \approx 0.1 \cdot \left(\frac{gV}{\pi T}\right)^{1/2} \tag{5.48}$$

In both equations, it is assumed that the observation point lies in the direction of maximum radiation.

The practical application of Equation 5.43 and the related equations becomes questionable if the average absorption coefficient $\overline{\alpha} = A/S$ of the room's boundary is not small compared with unity. In this case, the contribution of the very first reflections, which are not randomly distributed, of course, is predominant in the total energy density. Therefore, the scope of the preceding formulae with respect to $\overline{\alpha}$ is much smaller than that of the reverberation formulae developed in the preceding sections. This is one of the reasons that the absorption of a room and its reverberation time are usually determined by decay measurements and not by measuring the steady-state energy density and application of Equation 5.42, although this would be possible in principle.

On the other hand, Equation 5.43 can be used to determine the total power $P$ of a sound source from the steady-state energy density $w_r$ in the room. The equivalent absorption area $A$ is obtained from a decay measurement. This procedure is carried out in a 'reverberation chamber' with a long reverberation time and, hence, with low absorption. To obtain reliable results, it is necessary to measure the stationary sound pressure at several positions in the room and to carry out some space averaging, as described in Section 3.7. We should realize that in the measurement of sound power, a relative error of 10% (corresponding to 0.4 dB) can usually be tolerated. However, when it comes to the determination of wall absorption or reverberation time, a higher precision would be required.

If the early part of Equation 5.36 is to be considered in more detail, the integration may not start at zero but at the estimated travel time between sound emission and the first wall reflection. Before that, there cannot be an energy contribution to the integral. If we assume an average first-order reflection delay by $1/\overline{n}$ and set the start of the integration there, the calculation of the energy density based on the sound power yields

$$w'_r = \frac{4P}{cA} e^{-A/S}. \tag{5.49}$$

In level notation, the latter term is represented by a subtractive term of 4.34 $A/S$, thus introducing the equivalent absorption area and the room surface area for correction of some missing energy density with a given sound power which was observed in sound power measurements in reverberation rooms (Vorländer 1995).

## 5.7 APPLICATIONS OF THE RADIOSITY INTEGRAL: THE INFLUENCE OF IMPERFECT DIFFUSENESS

In a way, the theory of reverberation as outlined in the preceding sections is inconsistent in that it is based upon the condition of perfect sound field diffuseness. As has been discussed in Section 5.2, such a sound field would require the absence of any boundary absorption and, hence, cannot decay, strictly speaking. It is noteworthy that — despite the practical impossibility of creating a diffuse field — the decay formulae derived on this basis have proved to be sufficiently precise for most practical purposes.

Nevertheless, we try now to avoid this contradiction by replacing the condition of perfect sound diffusion by a much less stringent one: namely, that the boundary reflects the impinging sound energy in a diffuse manner according to Lambert's cosine law. This condition is also an ideal one, but it is not too far from reality since it can be fulfilled in several successive steps, as discussed in Section 5.2.

Under this premise, the propagation of sound energy within the enclosure can be studied with the integral equation (4.28) as derived in Section 4.5. Since we are interested in the decaying sound field, we can omit the term due to direct irradiation by a sound source. Then, the integral equation reads

$$B(r,t) = \iint_s K(r,r')\rho(r)B(r',t-R/c)\mathrm{d}S' \tag{5.50}$$

with

$$K(r,r') = \frac{\cos\vartheta\cos\vartheta'}{\pi R^2}\mathrm{d}S' \quad \text{and} \quad R = |r-r'| \tag{5.51}$$

Usually, this equation must be solved numerically. For this purpose, it is discretized in space and in time, that is, the boundary area is subdivided into a number of equal elements. Within each of them, the irradiation density $B$ and the reflection coefficient $\rho = 1 - \alpha$ are assumed as constant. At the same time, the continuous time variable is replaced with an index marking the number of a finite time interval $\Delta t$. In this way, the integral equation is converted into a system of linear equations

$$B_{k,n} = \sum_{l=1}^{N} K_{k,l}\rho_l B_{l,n-m} \tag{5.52}$$

The first subscript of $B$ denotes the position of an element, and the second one is the time index. Furthermore,

$$K_{k,n} = \frac{1}{\Delta S}\iint_{\Delta S} K(r,r')\mathrm{d}S\,\mathrm{d}S' \tag{5.53}$$

and $m = \mathrm{Int}(R/c\Delta t)$. This system can be solved by a simple recursion. It was Miles (1984) who developed this method and applied it to rectangular rooms.

Figure 5.10 presents three decay curves obtained in this way for a rectangular room with relative dimensions 1:2:3. For discretizing the integral equation, the boundary of the room has been divided up into 88 quadratic area elements. The patches on the 'floor' (i.e. on one of the walls with dimensions 2:3) are assumed as totally absorbent ($\rho = 0$), whereas the remaining walls are free of absorption ($\rho = 1$). It is evident that this distribution of wall absorption must cause severe deviations from diffuse field conditions.

The curves represent the decay of the irradiance $B$ at three different locations of the enclosure. The unit of the abscissa is the mean free path length $\overline{\ell}$. The quantity plotted on the ordinate is the relative sound pressure level $10 \cdot \log_{10}(B/B_0)$, with $B_0$ denoting an arbitrary normalization constant. All curves show some irregularities at their beginning which fade out very soon, leaving straight lines with equal steepness. (Generally, the independence of

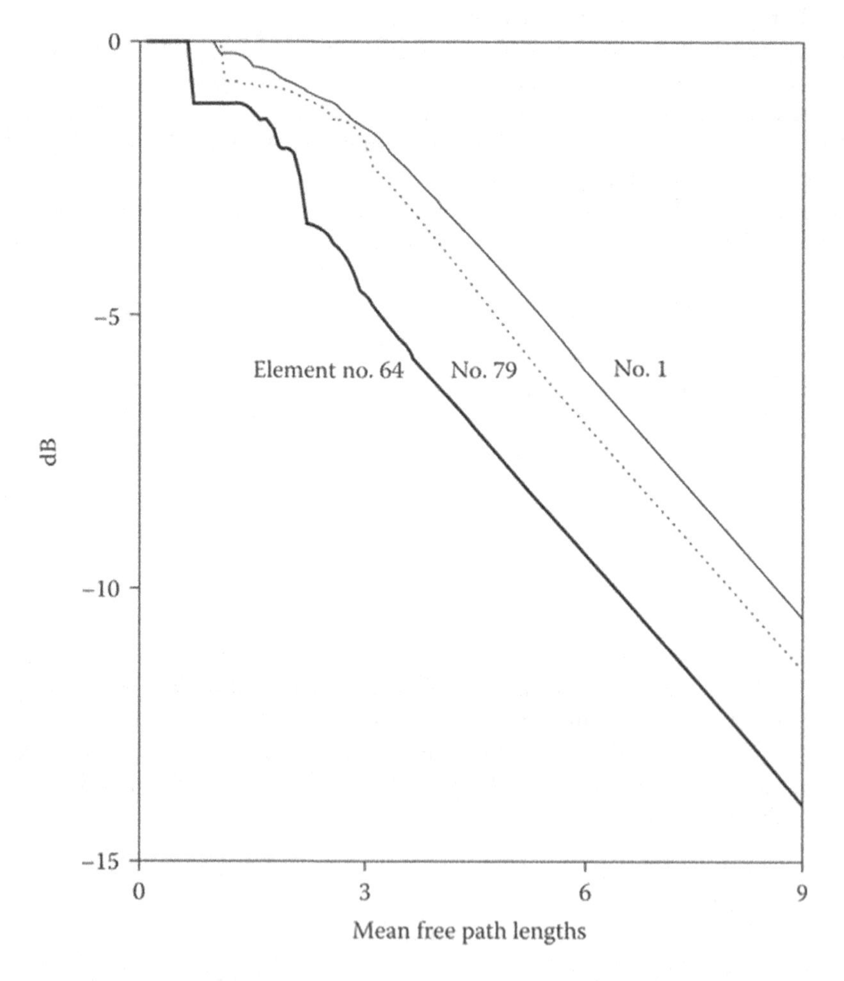


*Figure 5.10* Logarithmic decay curves in a rectangular room with diffusely reflecting boundary, calculated for three different receiver positions by solving Equation 5.50. The relative room dimensions are 1:2:3; the 'floor' with dimensions 2:3 has the absorption coefficient 1, while the remaining walls are free of absorption. The abscissa is $ct/\overline{\ell}$, and the number of mean free path lengths £ which a sound ray travels within a given time $t$.

the final slope on the receiver position has been proven by Miles [1984]) This demonstrates that imperfect diffuseness does not necessarily lead to non-exponential sound decay, nor is it indicated by different decay rates within a room. However, the decay constant itself, that is, the reverberation time of the room, may significantly deviate from that predicted by Eyring's formula. In the example of Figure 5.10, the slope of the straight portions is about -1.57 dB MFP, whereas Eyring's formula (5.24) (with $m$ = 0) would predict a slope of only $10 \cdot \log_{10}(1 - 3/11) = -1.38$ dB/MFP.

The essentially straight decay curves in Figure 5.10 show that the underlying law can be written as

$$E(t) = E_0 \exp\left(-\frac{cS}{4V}a^* t\right) = E_0 \exp\left(-a^* ct / \overline{\ell}\right) \tag{5.54}$$

This expression is equal to Equation 5.22 after replacing the Eyring absorption exponent $-\ln(1-\overline{\alpha})$ with an 'effective absorption exponent' $\alpha^*$. The latter is related to the reverberation time of the room by

$$T = 0.161 \cdot \frac{V}{Sa^*}$$

In the following text, we discuss the possibility of calculating the effective absorption exponent $\alpha^*$ and, hence, the reverberation time on the basis of the integral equation (5.50) without really solving it. Several numerical iteration schemes have been developed (Gilbert 1981; Kuttruff 1995) for this purpose. Although such methods may be too complicated for the everyday use, one of them will be briefly described in the following.

From Equation 5.54, it follows that

$$B(r, t - R/c) = B(r,t) \cdot \exp\left(a^* R / \overline{\ell}\right) \tag{5.55}$$

Furthermore, it is useful to symmetrize the kernel of the integral equation (5.50) for the iteration process by substituting

$$\beta(r) = B(r)\sqrt{\rho(r)} \text{ and } \kappa(r,r') = \sqrt{\rho(r)}K(r,r')\sqrt{\rho(r')}$$

With these steps, the integral equation reads

$$\beta(r) = \iint_{S'} \kappa(r,r')\beta(r')\exp\left(a^* R/\overline{\ell}\right) dS' \tag{5.56}$$

Because of its homogeneity, we may suppose without loss of generality that the function $\beta$ is normalized, that is, $\int \beta^2 ds = 1$. The unknown constant $a^*$ can be considered as a sort of eigenvalue.

Suppose the iteration process has already supplied an $(n - 1)$th approximation $\beta_{n-1}$ and the eigenvalue $a^*_{n-1}$. Then, both these quantities are inserted into the right-hand side of Equation 5.56, from which a new function, $\beta'^2_n$, can be obtained, which, however, will usually not be normalized, instead $\int \beta'^2_n dS = \mu^2 \neq 1$. Suppose from Equation 5.56 we obtain $\mu < 1$. This means that the eigenvalue $a^*_{n-1}$ is too small and thus must be augmented by a factor $1/\mu$, which is of the form $\exp(\Delta a^* R/l) \approx \exp(\Delta a^*)$. Then, we obtain immediately a corrected eigenvalue:

$$a_n^* \approx a_{n-1}^* + \Delta a^* = a_{n-1}^* - \ln \mu \tag{5.57}$$

At the same time, the $n$th approximation must be corrected according to $\beta_n = \beta_n'/\mu$. Now, the next iteration step is carried out with the function $\beta_n$ and the eigenvalue $a_n^*$.

The iteration is started by inserting $\beta_0 = 1/\sqrt{S}$ into the integral at the right-hand side of Equation 5.56. As an initial value of $a^*$, we could use the Eyring absorption coefficient $a_0^* = -\ln(\rho)$ or even $a_0^* = 0$. The process converges relatively fast, and a sufficiently accurate result is arrived at after a few iterations. The irradiance $B$ is obtained as a by-product, so to speak.

In Table 5.2, the reverberation times of a few rectangular rooms with different distributions of absorption are listed, calculated with the described iteration method (last column). For comparison, the results obtained with the Eyring absorption exponent $-\ln(1-\bar{\alpha})$and with the Sabine value $\bar{\alpha}$ are also shown.

Evidently, the results obtained from iteration deviate more or less from the Eyring reverberation time. The difference is least for relatively 'proportionate' rooms with uniform distribution of boundary absorption (case 1). This behaviour is not very surprising, since a symmetric room shape and absorption distribution favour the formation of a diffuse sound field. Of particular interest are the differences between the reverberation times and their opposite signs, as observed for the rectangular room 10 × 20 × 30 m³. If the absorption is concentrated on one of the largest walls with the area 20 × 30 m² (case 4), the iteration method yields a significantly smaller reverberation time than the Eyring formula. If the absorbing wall is one of the smallest walls (10 × 20 m², case 4), the result is reversed: the Eyring reverberation time is the longer one. Similar results have been obtained for many configurations with quite a different method, namely, by ray-tracing, described in Section 9.8.

How can we understand such deviations? To find an explanation, let us have a look at Figure 5.11, which shows a section through a relatively flat or long rectangular room. In Figure 5.11a, it is assumed that the 'floor' is highly absorbing, while the remaining walls are rigid or nearly rigid. This corresponds to the configuration of case 4 in Table 5.2. As indicated, a floor point $P$ is 'irradiated' from all directions of the half-space, while a point $Q$ at the ceiling does not receive energy reflected from the floor — neither directly nor indirectly — since the

*Table 5.2* Comparison of reverberation times, obtained for several arrangements of surface absorption in enclosures with diffusely reflecting walls

| Case no. | Shape and size of the enclosure | Distribution of absorption | Reverberation time (in seconds) after: Sabine | Eyring | Iteration |
|---|---|---|---|---|---|
| 1 | Cube: 10 × 10 × 10 m³ | Uniform, α = 1/6 | 1.61 | 1.47 | 1.55 |
| 2 | Cube: 10 × 10 × 10 m³ | Floor: α = 1 | 1.61 | 1.47 | 1.35 |
| 3 | Rectangular: 10 × 20 × 30 m³ | Uniform, α = 3/11 | 1.61 | 1.38 | 1.52 |
| 4 | Rectangular: 10 × 20 × 30 m³ | Floor (20 × 30 m²): α = 1 | 1.61 | 1.38 | 1.21 |
| 5 | Rectangular: 10 × 20 × 30 m³ | Long side wall (10 × 30 m²): α = 1 | 3.22 | 2.99 | 3.06 |
| 6 | Rectangular: 10 × 20 × 30 m³ | Short side wall (10 × 20 m²): α = 1 | 4.83 | 4.61 | 5.93 |

*Note:* If some wall has α = 1, the remaining ones of the enclosure have α = 0.

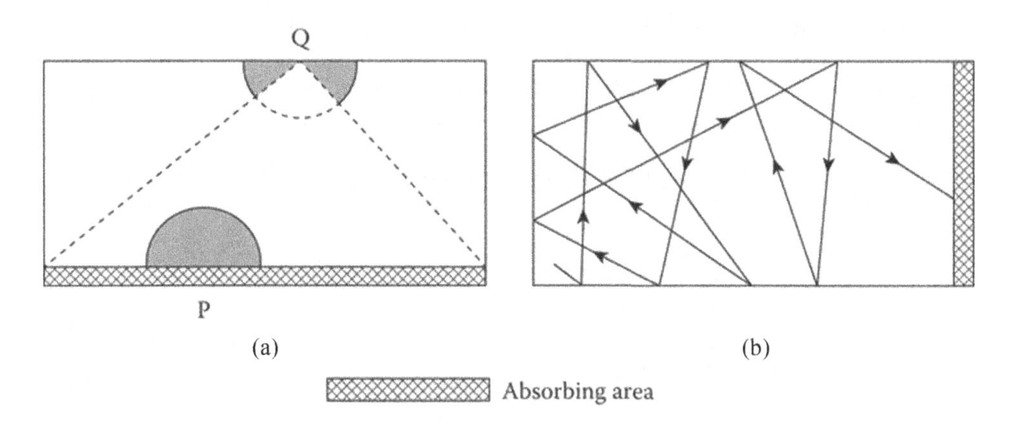


*Figure 5.11* Sound absorption in a flat room: (a) with absorbing 'floor' and (b) with one absorbing side wall.

latter absorbs all the arriving energy. Hence, the floor receives and absorbs more energy than it would under diffuse field conditions.

In Figure 5.11b, the absorbing surface is a short side wall, as assumed in case 6 of the table. In this case, sound particles in the left part of the enclosure are not strongly affected by the absorbing wall. Therefore, we expect that the right half of the enclosure contains less energy than the left one, and hence, the absorbing wall is hit by less sound particles per second than in a diffuse field. Obviously, this effect predominates the one responsible for the increased absorption in the case of Figure 5.11a. The relatively good agreement of reverberation times observed for the case 5 — one long side wall $10 \times 30$ m$^2$ absorbing — is probably due to two opposite effects which compensate each other, at least partially.

From a practical point of view, the most important configuration is probably that of case 4, because this can be conceived as a crude model of most auditoria. This holds in particular when these are occupied with the highly absorbing audience, which represents a lopsided distribution of boundary absorption. For a real hall, the influence of incomplete diffuseness is less dramatic since a certain fraction of the ground area will always remain unoccupied, and since the absorption coefficient of the audience is less than unity, as was assumed in Table 5.2. Nevertheless, the tendency of the non-uniform absorption to reduce the reverberation time below the Eyring value is certainly present in almost every auditorium and should be taken into consideration. Maybe it is the reason that the reverberation time of a completed hall often turns out to be shorter than originally intended.

## REFERENCES

Gilbert E.N. An iterative calculation of reverberation time. *J Acoust Soc Am* 1981; *69*: 178.

Hodgson M.J. Evidence of diffuse surface reflections in rooms. *J Acoust Soc Am* 1991; *89*: 765.

Kuttruff H. Distribution of free path lengths in rooms with diffusely reflecting walls (in German). *Acustica* 1970; *23*: 238.

Kuttruff H. Distribution of free path lengths in rooms with scattering elements (in German). *Acustica* 1971; *24*: 356.

Kuttruff H. A simple iteration scheme for the computation of decay constants in enclosures with diffusely reflecting boundaries. *J Acoust Soc Am* 1995; *98*: 288.

Miles R.N. Sound field in a rectangular enclosure with diffusely reflecting boundaries. *J Sound Vibr* 1984; *92*(2): 203.

Papoulis A. *Probability, Random Variables, and Stochastic Processes*, 2nd edn., Int student edn. New York: McGraw-Hill, 1985.

Sabine W.C. *Collected Papers on Acoustics*. Cambridge: Harvard University Press, 1923.

Vorländer M. Revised relation between the sound power and the average sound pressure level in rooms and consequences for acoustic measurements. *Acustica* 1995; *81*: 332.

Chapter 6

# Sound absorption and sound absorbers

Of considerable importance to the acoustics of a room are the loss mechanisms that reduce the energy of sound waves when they are reflected from its boundary as well as during their free propagation in the air. They influence the strengths of the direct sound and of all reflected components and, therefore, all acoustical properties of the room.

The attenuation of sound waves in the free medium becomes significant only in large rooms and at relatively high frequencies; for scale model experiments, however, it causes serious limitations. We have to consider it inevitable and something which cannot be influenced by the efforts of the acoustician. Nevertheless, in reverberation calculations, it has to be taken into account. Therefore, it may be sufficient to give here a brief description of the causes of air attenuation and to present the relevant experimental data.

The situation is different in the case of the absorption to which sound waves are subjected when they are reflected from the boundary of a room. The magnitude of wall absorption and its frequency dependence vary considerably from one material to another. There is also an unavoidable contribution to the wall absorption which depends on certain physical properties of the medium, but it is so small that, in most cases, it can be neglected.

Since the boundary absorption is of decisive influence on the sound field in a room, the acoustical designer should understand the various absorption mechanisms and know the various types of sound absorbers. In fact, the well-considered use of sound absorbers is one of his most important design tools. In particular, sound absorbers are usually employed to meet one of the following objectives:

- To adapt the reverberation of the room, for instance, a performance hall, to the intended use of it.
- To suppress undesired sound reflections from remote walls which might be heard as echoes.
- To reduce the acoustical energy density, and hence the sound pressure level, in noisy rooms, such as factories, large bureaus, and so on.

This chapter discusses the principles and mechanisms of the most important types of sound absorbers.

## 6.1 ENERGY LOSSES IN THE MEDIUM

In the derivation of the wave equation (1.5), it was tacitly assumed that the changes in the state of the air, caused by the sound waves, occurred without any losses. This is not quite true, however. We shall refrain here from a proper modification of the basic equations and from a quantitative treatment of attenuation. Instead, the most prominent loss mechanisms

DOI: 10.1201/9781003389873-6

are briefly described in the following text. As in Section 1.2 and Chapter 4, the propagation losses will be formally characterized by the attenuation constant $m$ as defined in Equation 1.21. The decrease in the sound intensity $I$, which is proportional to the square of the sound pressure in a plane sound wave, is expressed by

$$I(x) = I_0 \cdot \exp(-mx) \tag{6.1}$$

The attenuation in air is mainly caused by the following effects:

1. Equation 1.5 is based on the assumption that the changes in the state of a volume of gas take place adiabatically, that is, there is no heat exchange between neighbouring volume elements. The equation states that a compressed volume element has a slightly higher temperature than an element which is rarefied by the action of the sound wave. Although the temperature differences occurring at normal sound intensities amount to small fractions of a degree centigrade only, they cause a heat flow because of the thermal conductivity of the air. This flow is directed from the warmer to the cooler volume elements. The changes of state are therefore not taking place entirely adiabatically. According to basic principles of physics, the energy transported by these thermal currents cannot be reconverted completely into mechanical, that is, into acoustical energy; some energy is lost forever. And this happens in every sound period. The corresponding portion of the attenuation constant $m$ increases with the square of the frequency.
2. In a plane sound wave, the length of each volume element undergoes periodical changes in the direction of sound propagation. This deformation of the original element can be considered as a superposition of an omnidirectional compression or rarefaction and of a shear deformation, that is, of a pure change of shape. The former one is counteracted by the elastic reaction of the medium which is proportional to the amount of compression, whereas the shear is controlled by viscous forces which are proportional to the shear velocity. Hence — as with every frictional process — mechanical energy is irreversibly converted into heat. This 'viscous portion' of the attenuation constant $m$ also increases proportionally with the square of the frequency.
3. Under normal conditions, the aforementioned causes of attenuation in air are negligibly small compared with the attenuation caused by what is called 'thermal relaxation'. It can be briefly described as follows. As long as the system is in thermal equilibrium, the total energy contained in a polyatomic gas is distributed among several energy stores (degrees of freedom) of the gas molecules, namely, in the form of translational, vibrational, and rotational energy. If the gas is suddenly compressed, that is, if its energy is suddenly increased, the whole additional energy will be stored at first in the form of translational energy. Afterwards, a gradual redistribution among the other stores will take place. Or in other words, the establishment of a new equilibrium requires a finite time. If compressions and rarefactions change periodically as in a sound wave, a thermal equilibrium can be maintained — if at all — only at very low frequencies; with increasing frequency, the instantaneous contents of the molecular energy stores will lag behind the external changes and will accept or deliver energy at the wrong moments.

The latter process is a sort of 'internal heat conduction', which weakens the sound wave just like the normal heat conduction with which we are more familiar. The characteristic quantity is the energy loss per wavelength, also known as the 'wavelength-related attenuation constant'. It has a maximum when the duration of one sound period is comparable with a specific time interval, called 'relaxation time', being characteristic of the time lag in internal energy distribution.

Figure 6.1 shows, for a single relaxation process, the attenuation constant $m$ multiplied by the wavelength $\lambda$ as a function of the product of frequency and relaxation time. The broad frequency range in which it appears is characteristic of a relaxation process. Moreover, the relaxation of a medium causes not only a substantial increase in absorption but also a slight change in sound velocity, which, however, is not of importance in this connection.

For mixtures of polyatomic gases such as air, which consists mainly of nitrogen and oxygen, matters are more complicated because there are many more possibilities of internal energy exchange which we shall not discuss here.

Because of their importance in room acoustics, particularly for the calculation of reverberation times, a few numerical values of the intensity-related absorption constant $m$ are listed in Table 6.1 (Bass 1995).

Even if the walls, the ceiling, and the floor of a room were completely rigid and smooth, they would cause some sound absorption which, however, is quite small. It becomes noticeable only when there are no other absorbents or absorbing wall portions in the room. This is the case for measuring rooms which are specially built to obtain a high reverberation time (reverberation chambers; see Section 8.7).

Physically, this kind of absorption is again caused by the heat conductivity and viscosity of the air. According to Equation 1.4, the periodic temperature changes caused by a sound wave are in phase with the corresponding pressure changes — apart from the slight deviations discussed before. Therefore, the maximum sound pressure amplitude which is observed

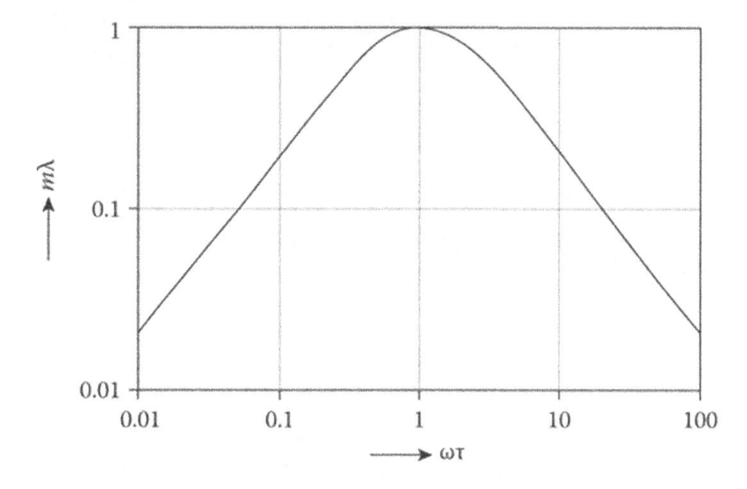


*Figure 6.1* Wavelength-related attenuation constant $m\lambda$ as typical for a relaxation process. The abscissa is the product of angular frequency $\omega$ and the relaxation time $\tau$.

*Table 6.1* Attenuation constant $m$ of air at 20°C and normal atmospheric pressure, in $10^{-3}$ $m^{-1}$

| | *Frequency (kHz)* | | | | | | |
|---|---|---|---|---|---|---|---|
| *Relative humidity (%)* | *0.5* | *1* | *2* | *3* | *4* | *6* | *8* |
| 40 | 0.60 | 1.07 | 2.58 | 5.03 | 8.40 | 17.71 | 30.00 |
| 50 | 0.63 | 1.08 | 2.28 | 4.20 | 6.84 | 14.26 | 24.29 |
| 60 | 0.64 | 1.11 | 2.14 | 3.72 | 5.91 | 12.08 | 20.52 |
| 70 | 0.64 | 1.15 | 2.08 | 3.45 | 5.32 | 10.62 | 17.91 |

Source: Adapted from Bass (1995).

immediately in front of a rigid wall should be associated with a maximum of 'temperature amplitude'. In reality, this is impossible: because of its high thermal capacity, the wall surface remains virtually at a constant temperature. Therefore, there exists some boundary layer next to the wall in which strong temperature gradients will develop and, hence, a periodically alternating heat flow will be directed to and from the wall. This energy transport occurs at the expense of the sound energy, since the heat which was produced by the wave in a compression phase can only be partly reconverted into mechanical energy during the rarefaction phase.

A related effect becomes evident when we consider a sound wave impinging obliquely onto a rigid plane. In this case, the normal component of the particle velocity vanishes at the boundary, but the parallel component does not, at least if calculated without accounting for the viscosity of air. This, however, cannot be true, since in a real medium, the molecular layer immediately on the wall is fixed to the latter, which means that the parallel velocity component vanishes as well. For this reason, our assumption of a perfectly reflecting wall is not correct, in spite of its rigidity. In reality, another boundary layer is formed between the region of unhindered parallel motion in the air and the wall; at oblique incidence, high viscous forces, and hence a substantial conversion of mechanical energy into heat, take place in this layer.

The effect of both loss processes can be accounted for by ascribing an absorption coefficient to the wall. It can be shown that the thickness of both boundary layers is inversely proportional to the square root of the frequency. Since, on the other hand, the gradients of the temperature and of the parallel velocity component increase proportionally with frequency, both contributions to the absorption coefficient are proportional to the square root of the frequency. Their dependence on the angle of incidence, however, is different. The viscous portion is zero for normal sound incidence, whereas the heat flow to and from the wall does not vanish at normal incidence.

Both effects are very small even at the highest frequencies relevant in room acoustics. Those can be safely neglected for practical design purposes.

## 6.2 SOUND ABSORPTION BY MEMBRANES AND PERFORATED SHEETS

For the acoustics of a room, it makes no difference whether the apparent absorption of a wall is brought about by dissipative processes, that is, by conversion of sound energy into heat, or by parts of the energy penetrating through the wall into the outer space. In this respect, an open window is a very effective absorber, since it acts as a sink for all the arriving sound energy.

A less-trivial case is that of a wall or some part of a wall forced by a sound field into vibration. (Strictly speaking, this happens more or less with any wall, since completely rigid walls do not exist in the real world.) Then, a part of the wall's vibrational energy is re-radiated into the outer space. This part is withdrawn from the incident sound energy, viewed from the interior of the room. Thus, the effect is the same as if it were really absorbed. It can therefore also be described by an absorption coefficient. In practice, this sort of 'absorption' occurs with doors, windows, light partition walls, suspended ceilings, circus tents, and similar 'walls'.

This process, which may be quite involved especially for oblique sound incidence, is very important in all problems of sound insulation. From the viewpoint of room acoustics, it is sufficient, however, to restrict discussions to the simplest case of a plane sound wave impinging perpendicularly onto the wall, whose dynamic properties are completely characterized by its mass inertia. Then, the propagation of bending waves on the wall can be left without consideration.

Let us denote the sound pressures of the incident and of the reflected waves on the surface of a wall (see Figure 6.2a) by $p_1$ and $p_2$, and the sound pressure of the transmitted wave by $p_3$. The total pressure acting on the wall is then $p_1 + p_2 - p_3$. It is balanced by the inertial force $i\omega M'v$, where $M'$ denotes the mass per unit area of the wall and $v$ is the velocity of its motion.

This velocity is equal to the particle velocity of the wave radiated from the rear side, for which $p_3 = \rho_0 c v$. Therefore, we have $p_1 + p_2 - \rho_0 c v = i\omega M' v$, from which we obtain the wall impedance:

$$Z = \frac{p_1 + p_2}{v} = i\omega M' + \rho_0 c \tag{6.2}$$

Inserting $\zeta = Z/\rho_0 c$ into Equation 2.11 (with $\theta = 0$) yields

$$\alpha = \left[1 + \left(\frac{\omega M'}{2\rho_0 c}\right)^2\right]^{-1} \approx \left(\frac{2\rho_0 c}{\omega M'}\right)^2 \tag{6.3}$$

The latter approximation is permissible if the mass reactance of the wall is large compared with the characteristic impedance of air. In any case, the 'absorption' of a light wall or partition becomes noticeable only at low frequencies.

At a frequency of 100 Hz, the absorption coefficient of a glass pane with 4 mm thickness is — according to Equation 6.3 — as low as 0.02, approximately. For oblique or random incidence, this value is a bit higher due to the better matching between the air and the glass pane, but it is still very low. The increase in absorption with decreasing frequency has the effect that rooms with many windows sometimes sound 'crisp', as the reverberation at low frequency is not as long as it would be in the same room without windows.

The absorption caused by vibrations of single-leaf walls and ceilings is thus very low. Matters are different for double-leaf or more complex walls, provided that the partition on the side of the room under consideration is mounted in such a way that vibrations are not hindered, and provided that it is not too heavy. Because of the interaction between the leaves and the enclosed volume of air, such a system behaves as a resonance system. This will be discussed in the next section.

It is a fact of great practical interest that a rigid perforated plate or panel has essentially the same acoustical properties as a membrane or a foil. This is explained in Figure 6.2b. Each hole in a plate may be considered as a short tube or channel with length *b;* the mass of air contained in it, divided by the cross section, is $\rho_0 b$. Obviously, the airstream directed towards the panel must contract to pass a hole; the factor of contraction is $S_2/S_1$, where $S_1$ is the area of the hole and $S_2$ is the panel area per hole. After the passage of the hole, the airstream expands to its former cross section. This holds not only for a constant airflow but also for the alternating flow as occurring in sound waves. By the same factor $S_2/S_1$, the flow velocity in the holes will be increased, and also the inertial reaction of the included air. Hence, the 'equivalent mass' of the perforated panel per unit area is

$$M' = \frac{\rho_0 b'}{\sigma} \tag{6.4}$$

with

$$\sigma = \frac{S_1}{S_2} \tag{6.5}$$

$\sigma$ is the perforation ratio of the plate.

In Equation 6.4, the geometrical length of the hole has been replaced with an 'effective length':

$$b' = b + 2\Delta b \tag{6.6}$$

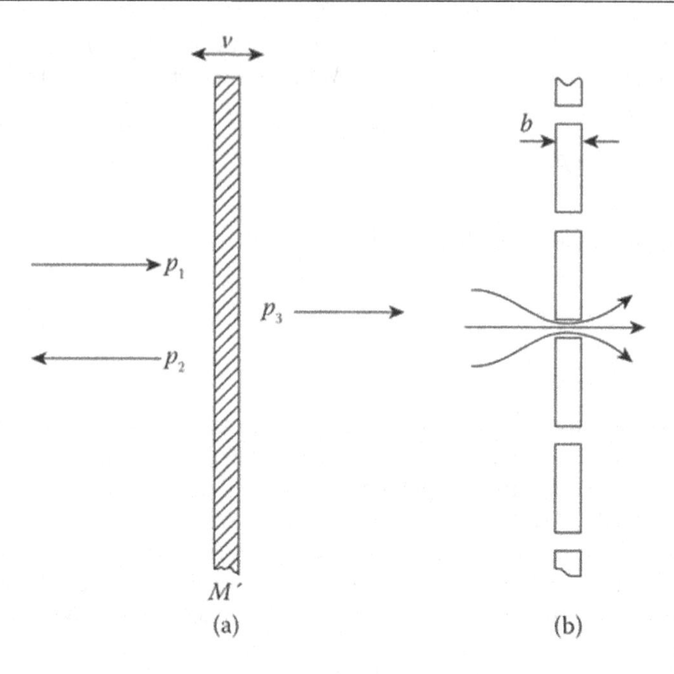


*Figure 6.2* (a) Pressures acting on a layer with mass $M'$ per unit area and (b) perforated panel.

The correction $2\Delta b$, called 'end correction', accounts for the fact that the streamlines (see Figure 6.2b) cannot contract or expand abruptly but only gradually before entering or after leaving an aperture. For a circular aperture with radius $a$, the end correction is

$$\Delta b = \frac{\pi}{4} a \tag{6.7}$$

Finally, the absorption coefficient of a perforated panel is obtained from Equation 6.3 by substituting $M'$ from Equation 6.4.

In deriving Equation 6.4, it has been assumed that the panel itself remains at rest when a sound wave strikes it. However, often, the perforated plate is so light that it will vibrate as a whole. In this case, $M'$ in Equation 6.5 must be replaced by the 'effective mass':

$$M^{?} = \left( \frac{1}{M'} + \frac{1}{M'_0} \right)^{-1} \tag{6.8}$$

where $M'$ is the equivalent mass per square metre of the 'clamped' panel after Equation 6.4, and $M'_0$ is the specific mass of the solid part of the panel.

If frictional forces within the holes and other loss processes are neglected, the absorption coefficient in Equation 6.3 represents the fraction of incident sound energy, which is transmitted by a wall or a perforated panel. It characterizes the sound transparency of the wall — or its insulation against sound waves. Therefore, this equation plays an important role in sound insulation, where it is known as the 'mass law'.

Let us illustrate this by an example: To yield 90% energy transparency ($\alpha = 0.9$), the quantity $\omega M'/2\rho_0 c$ in Equation 6.3 must be 1/3. At 1,000 Hz, this is the case with a mass layer with an (equivalent) mass per unit area of about 45 g/m². If realized by a perforated panel, this can be achieved, for instance, by a 1 mm thick sheet with 7.5% perforation and with holes having a diameter of 2 mm.

## 6.3 RESONANCE ABSORBERS

In this section, we come back to the idealized resonator discussed in Section 2.3. It consists basically of a membrane, which is arranged in front of a rigid wall and parallel to it. A sound wave impinging onto the membrane will excite vibrations of it, which are controlled by its specific mass $M'$ and by the reaction of the air cushion behind. Furthermore, there are vibrational losses that can be represented by some loss resistance $r_s$.

The absorption coefficient of this configuration can be calculated by using Equation 2.34 along with Equation 2.11. The result is plotted in Figure 6.3 for various ratios $r_s/\rho_0 c$ and under the additional assumption that $M'\omega_0 = 10\rho_0 c$; the abscissa is the angular frequency divided by the resonance frequency. The maximum absorption coefficient 1 is only reached with exact matching, that is, for $r_s = \rho_0 c$.

For $r_s > \rho_0 c$, the maximum absorption is less than unity and the curves are broadening. This behaviour is similar to that of a thin porous layer which is arranged at some distance in front of a rigid wall (compare Figure 2.7). The frequency-dependent absorption characteristics make this device a useful tool for the control of reverberation, mainly in the low- and mid-frequency range. In practical applications, the 'membrane' consists usually of a panel of wood, chipboard, or gypsum (see Figure 6.4a). The vibrational losses occurring in this system may have several physical reasons. One of them has to do with the fact that any kind of panel must be fixed at certain points or along certain lines to a supporting construction, which forces the panel to be bent when it vibrates. Now, every elastic deformation of a solid, including those by bending, is associated with internal losses that depend on the material and other circumstances. In metals, for instance, the intrinsic losses of the material are relatively small, but they may be substantial for plates made of wood or of plastic. If desired, the losses may be increased by certain surface layers or by porous materials placed in the space between the panel and the rigid rear wall.

According to the discussion of the preceding section, the mass layer can also be realized as a perforated or slotted sheet (see Figure 6.4b). This device is often called a Helmholtz resonator, although the original Helmholtz resonator consists of a single, rigid-walled cavity with a narrow opening, as described in the next section. If the apertures in the sheet are very narrow,

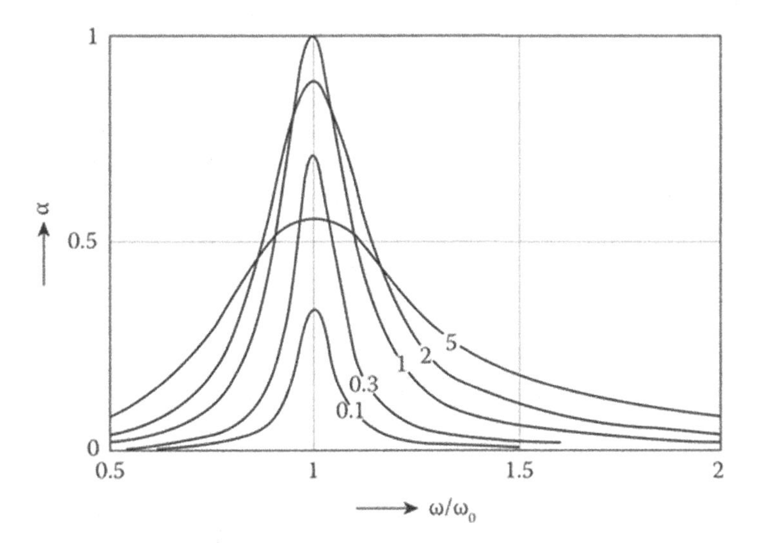


*Figure 6.3* Absorption coefficient (calculated) of resonance absorbers as a function of frequency at normal sound incidence, for $M'\omega = 10\rho_0 c$. Parameter is the ratio $r_s/\rho_0 c$.

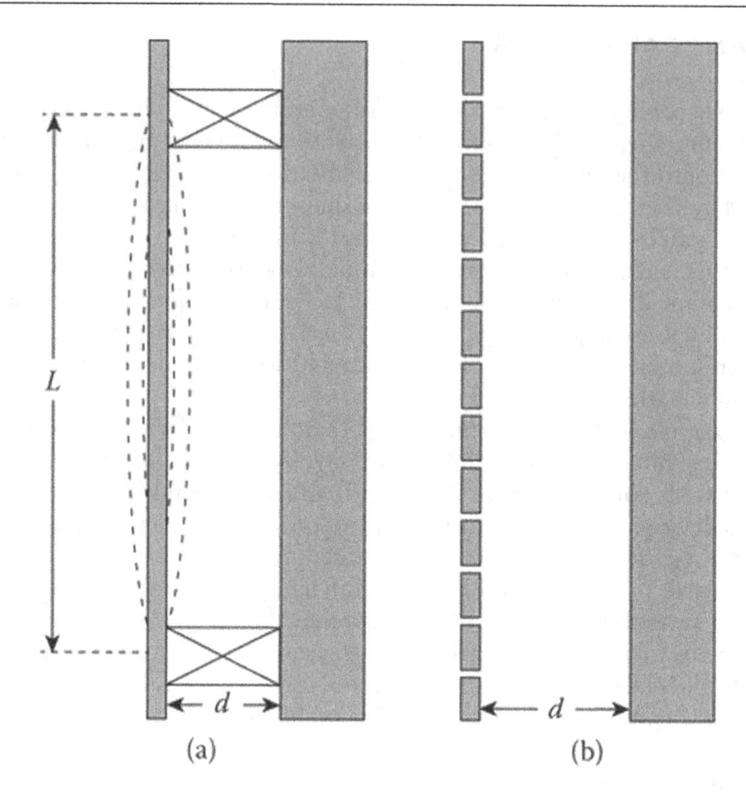


*Figure 6.4* Resonance absorbers: (a) with vibrating panel and (b) with perforated panel.

the frictional losses occurring in them may be sufficient to ensure the low $Q$-factors that are needed for high efficiency of the absorber. Such devices are known as microperforated absorbers and have found considerable attention (Maa 1985), prompted by the technical progress in drilling numerous tiny holes into a panel. When applied to transparent materials such as acrylic glass, they offer the possibility of manufacturing transparent resonance absorbers, although it may prove difficult in practice to keep apertures with diameters in the sub-millimetre range free of dust particles and other obstructions.

For wider holes, it is usually necessary to provide for additional losses. This can be achieved, for instance, by covering the holes with a porous fabric. Another method of adapting the magnitude of $r_s$ to a desired value is to fill the air space behind the panel partially or completely with porous material.

For normal sound incidence, the (angular) resonance frequency of this absorber is given by Equation 2.35, provided the space behind the membrane or perforated panel is empty. For its practical application, the following form may be more useful:

$$f_0 = \frac{600}{\sqrt{(M'd)}} \text{Hz} \tag{6.9}$$

where $M'$ is in kg m$^{-2}$ and $d$ in cm. If the air space behind the panel is filled with porous material, the changes of state of the air will be rather isothermal than adiabatic. This means that Equation 1.5 is no longer valid and has to be replaced with $p/p_0 = \delta\rho/\rho_0$; the sound velocity becomes $c_{\text{iso}} = \sqrt{p_0 / \rho_0}$. Accordingly, the numerical value in Equation 6.9 is reduced by a factor $\sqrt{\kappa}$, that is, by about 20%, and is 500.

This formula is relatively reliable if the mass layer consists of a perforated panel, as in Figure 6.4b, or of a flexible membrane. Then, the way in which the mass layer is fixed has no influence on its acoustical effect, at least as long as the sound waves arrive frontally. Matters are different for non-perforated wall linings made of panels with noticeable bending stiffness. Since such panels must be fixed in some way, for instance, on battens that are mounted on the wall (Figure 6.4a), their vibrations are controlled not only by the air cushion behind but also by their bending stiffness. Accordingly, the resonance frequency will be higher than that given by Equation 6.9, namely,

$$f_0' = \sqrt{f_0^2 + f_1^2} \tag{6.10}$$

with $f_1$ denoting the lowest bending resonance frequency of a panel supported (not clamped) at two opposite sides. The typical range of $f_1$ is 10–30 Hz, whereas $f_0$ is typically 50–100 Hz. This shows that the influence of the bending stiffness on the resonance frequency can be neglected in most practical cases, and Equation 6.9 may be applied to give at least a clue to the actual resonance frequency.

Another consequence of the strong lateral coupling between adjacent elements of an unperforated panel is that the angle dependence of sound absorption is more complicated than that in Equation 2.11 (with frequency-independent $\zeta$). In fact, a sound wave with oblique incidence excites a forced bending wave in the panel. The propagation of this wave is strongly affected by the way the panel is fixed and by the air layer behind it. Since general statements on its influence on the absorption coefficient and its dependence on the incidence angle are not possible, we shall not discuss this point in detail.

For resonators with perforated panels, lateral coupling of surface elements is affected by lateral sound propagation in the air space behind the panels. It can be hindered by lateral partitions made of rigid material or by filling the air space with porous and, hence, sound-absorbent materials, like glass or mineral wool. Neighbouring elements of the panel can then be regarded as independent; the wall impedance and, similarly, the resonance frequency are independent of the direction of sound incidence. In any case, it is difficult to assess correctly the losses of a resonance absorber which determine its absorption coefficient. Therefore, the acoustical consultant must rely on his experience or on a good collection of typical absorption data. In cases of doubt, it may be advisable to measure the absorption coefficient of a sufficiently large sample in a reverberation chamber (see Section 8.7).

Resonance absorbers of the described type are typically mid-frequency or low-frequency absorbers. Their practical importance stems from the possibility of choosing their significant data (dimensions, materials) from a wide range so as to give them the desired absorption characteristics. By a suitable combination of several types of resonance absorbers in a room, the acoustical designer is able to achieve a prescribed frequency dependence of the reverberation time. The most common application of vibrating panels is to compensate for the high absorption of the audience at medium and high frequencies, and thus to equalize the reverberation time to some extent. This is the reason for the generally favourable acoustical conditions which are frequently met in halls whose walls are lined with wooden panels or are equipped with a suspended decoration ceiling made of thin plaster, for instance. Thus, it is not, as is sometimes believed by laymen, some sort of 'amplification' caused by 'resonance', which is responsible for the good acoustics of concert halls lined with wooden panels. Likewise, audible decay processes of the wall linings, which are sometimes also believed to be responsible for good acoustics of such halls, do not occur in practical situations, although they might be possible in principle. If a resonance system with the relative half power bandwidth (reciprocal of the $Q$-factor) $\Delta\omega/\omega_0$ is excited by an impulsive signal, its amplitude will decay with a damping constant $\delta = \Delta\omega/2$ according to Equation 2.38; thus, the 'reverberation time' of the resonator is

$$T' = \frac{13.8}{\Delta\omega} = \frac{2.2}{\Delta f} \tag{6.11}$$

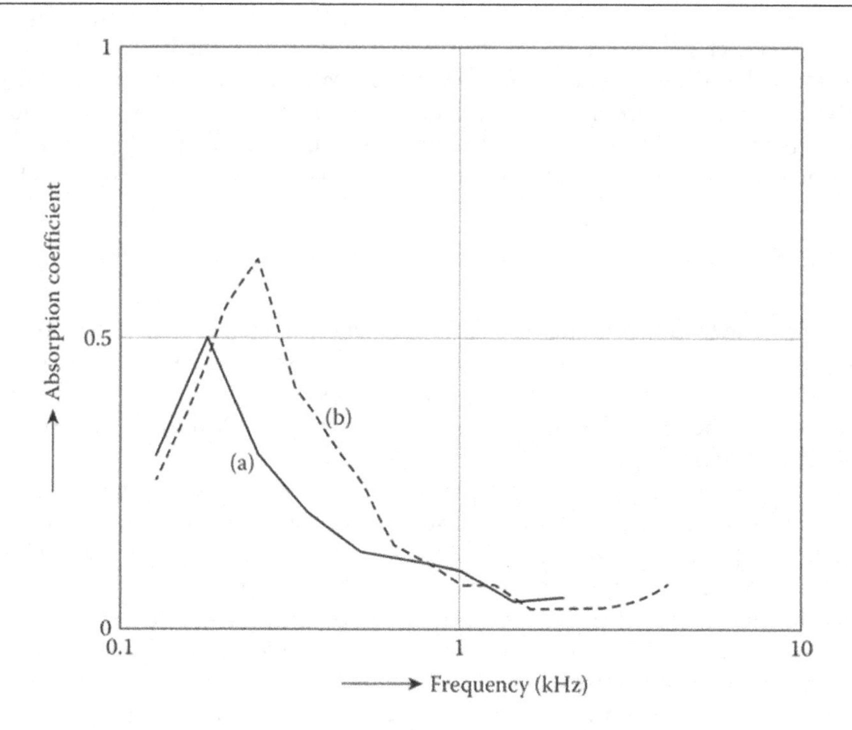


*Figure 6.5* Absorption coefficient of resonance absorbers at random sound incidence (measured in a reverberation chamber; see Section 8.7): (a) wooden panel, 8 mm thick, $M' = 5$ kg/m², 30 mm away from rigid wall, with 20 mm Rockwool in the air gap; (b) panels, 9.5 mm thick, perforated at 1.6% (diameter of holes 6 mm), 50 mm distant from rigid wall, air space filled with glass wool.

To be comparable with the reverberation time of a room, the frequency half-width $\Delta f$ must be about 2 Hz at most. However, the half-width of a resonating wall lining is larger than this by several orders of magnitude.

In Figure 6.5, the absorption coefficients of a wooden wall lining and of a resonance absorber with perforated panels are plotted as functions of the frequency, measured at omnidirectional sound incidence.

## 6.4 HELMHOLTZ RESONATORS

Sometimes, sound-absorbent elements are not distributed so as to cover uniformly the wall or the ceiling of a room, but instead, they are single or separate objects arranged either on a wall or in free space. Examples of this are separately standing or seating persons, empty chairs, small wall openings, or lamps; musical instruments, too, can absorb sound. To sound absorbers of this sort, we cannot attribute an absorption coefficient, since the latter refers to a uniform surface. Instead, their absorbing power is characterized by their 'absorption cross section' or their 'equivalent absorption area', which is defined as the ratio of sound energy $P_{abs}$ they absorb per second and the intensity $I_0$ which the incident sound wave would have at the place of the absorbent object if it were not present:

$$A_a = \frac{P_{abs}}{I_0} \tag{6.12}$$

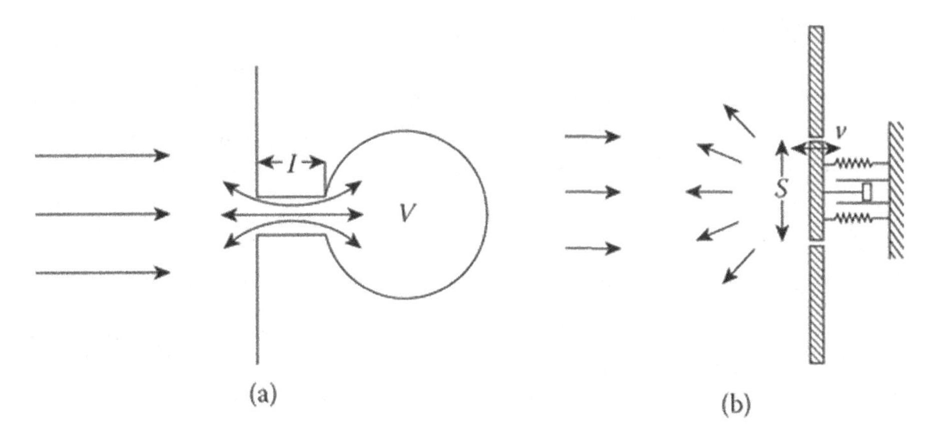


*Figure 6.6* Helmholtz resonator: (a) realization and (b) schematic.

This definition is similar to that of the scattering cross section in Equation 2.60.

When calculating the reverberation time of a room, the absorption of these types of absorbers is taken into account by adding their absorption areas $A_i$ to the sum $A = \Sigma \alpha_i S_i$ (see Equation 5.25). The same holds for all other formulae and calculations in which the total absorption or the mean absorption coefficient $\bar{\alpha}$ of a room appears, as, for instance, for the calculation of the steady-state energy density as described in Equation 5.5.

In this section, we discuss discrete sound absorbers with pronounced resonant behaviour. Their characteristic feature is an air volume which is enclosed in a rigidly walled cavity and is coupled to the surrounding space by an aperture, as shown in Figure 6.6a. The latter may also be a channel or a 'neck'. The whole structure is assumed to be small compared with the wavelength of sound, and thus, it has one single resonance in the interesting frequency range. It is brought about by the interaction of two elements: the air contained in the neck or in the aperture, which acts as a mass load, and the air within the cavity, which can be regarded as a spring counteracting the motion of the air in the neck. Arrangements of this type are called 'Helmholtz resonators'; examples of these are all kinds of bottles, vases, and similar vessels. In ancient times, such resonators, known as Vitruvius sounding vessels, played an unknown, possibly only a surmised, acoustical role in antique theatres and other spaces.

Figure 6.6 depicts a Helmholtz resonator along with its schematic presentation. The basic parameters of the resonator are the length $l$ and the cross-sectional area $S$ of the neck, which determine the mass $M = \rho_0 lS$ of the air enclosed in the neck. A further characteristic parameter is the volume $V$ of the attached cavity. The air contained in it reacts to small changes $p_i$ of the pressure like a spring with the stiffness $s$ (see Figure 6.6b) and thus determines the resonance frequency of the resonator:

$$\omega_0 = \sqrt{s / M} \tag{6.13}$$

The stiffness of the air cushion is given by

$$s = \frac{\rho_0 c^2 S^2}{V} \tag{6.14}$$

Now we have to discuss the losses occurring in the resonator. There are two types of losses: those which are due to the internal friction of the air oscillating in the aperture, represented by some resistance $R_0$. If desired, $R_0$ can be increased, for instance, by introducing some porous material in the aperture. The second sort of loss is caused by the re-radiation of sound into the free space. It is characterized by the radiation resistance $R_r$ of the aperture. We imagine this aperture as an oscillating piston, as in Figure 6.6b, mounted flush in a rigid wall of infinite extension. The lateral dimensions of the aperture are assumed as small compared with the wavelength. Then, the radiation impedance of the aperture is given by

$$R_r \approx \frac{\rho_0 \omega^2 S^2}{2\pi c} = 2\pi\rho_0 c\left(\frac{S}{\lambda}\right)^2 \tag{6.15}$$

Now, suppose the resonator is excited at its resonance frequency $\omega_0$. In this case, the mass reactance $i\rho M$ and the reactance $1/i\omega s$ of the stiffness cancel each other. Hence, the ratio of the total force acting on the 'piston' and its velocity $v$ is real:

$$\frac{F}{v} = R_0 + R_r$$

The energy converted into heat per second by viscous losses is

$$P_{abs} = R_0\overline{v^2} = \frac{R_0}{(R_0 + R_r)^2}\overline{F^2} \tag{6.16}$$

For given radiation resistance $R_r$, $P_{abs}$ assumes its maximum if $R_0 = R_r$. This is the condition of complete matching to the surrounding medium.

The external force $F$ which acts upon the piston is $F = 2pS$, with $p$ denoting the sound pressure in an arriving sound wave; the factor 2 accounts for its reflection from the rigid wall surrounding the piston. By application of the second equation (1.39), we can express the sound pressure $p$, and hence the force $F$, in terms of the intensity $I$ of the incident sound wave:

$$\overline{F^2} = 4S^2\overline{p^2} = 4\rho_0 c S^2 I \tag{6.17}$$

Now we are ready to evaluate Equations 6.16 and 6.12 with $R_0 = R_r$ by substituting from Equations 6.14 and 6.17, and we obtain as a final result

$$P_{abs} = \frac{\lambda_0^2}{2\pi} I \text{ and } A_{max} = \frac{\lambda_0^2}{2\pi} \tag{6.18}$$

where $\lambda_0$ is the wavelength corresponding to the resonance frequency. According to Equation 2.38, the $Q$-factor of the resonator is

$$Q_A = \frac{M\omega_0}{2R_r} \tag{6.19}$$

Now, we can insert $M = s/\omega_0^2 = \rho_0 c^2 S^2/\omega_0^2 V$ into Equation 6.19 and express $R_r$ by Equation 6.15, with the following result:

$$Q_A = \frac{\pi}{V}\left(\frac{c}{\omega_0}\right)^3 = \frac{\pi}{V}\left(\frac{\lambda_0}{2\pi}\right)^3 \tag{6.20}$$

We notice that the maximum absorption area $A_{max}$ of a resonator matched to the sound field is much larger than one might expect. On the other hand, the $Q$-factor is very large too, which means that the relative half-power bandwidth, which is the reciprocal of the Q-factor, is very small, that is, large absorption will occur only in a very narrow frequency range. This is clearly illustrated by the following example: Suppose a resonator is tuned to a frequency of 100 Hz corresponding to a wavelength of 3.43 m. This can be achieved conveniently by a resonator volume of 1 l. If it is matched ($R_0 = R_r$), the resonator's absorption area is at maximum and is as large as 1.87 m$^2$. Its Q-factor is — according to Equation 6.20 — about 500; the relative half-width is thus 0.002. This means it is only in the range from 99.9 to 100.1 Hz that the absorption area of the resonator exceeds half its maximum value. Therefore, the very high absorption area in the resonance is paid for by the exceedingly narrow frequency bandwidth. This is why the application of such weakly damped resonators does not seem very useful. It is more promising to increase the losses, and hence the bandwidth, at the expense of maximum absorption (see also Figure 6.3).

Finally, we investigate the problem of audible decay processes, which we have already touched on in the preceding section. The 'reverberation time' of the resonator can again be calculated by the relation (6.11):

$$T' = \frac{13.8}{\Delta\omega} = 13.8\frac{Q_A}{\omega_0}$$

In many cases, this time cannot be ignored when considering the reverberation time of a room. What about the audibility of this decay process?

It is evident from the derivation of Equation 6.18 that the same amount of energy per second which is converted into heat inside the resonator is re-emitted, since we have assumed $R_0 = R_r$. Its maximum radiation power is thus $P_s = \left(\lambda_0^2 / 2\pi\right) \cdot I$, where $I$ denotes the intensity of the incident sound wave. Thus, the intensity of the re-radiated spherical wave at distance $r$ is

$$I_s = \left(\frac{\lambda_0}{2\pi r}\right)^2 \cdot I = \left(\frac{c}{r\omega_0}\right)^2 \cdot I \tag{6.21}$$

Both intensities $I$ and $I_s$ are equal at a distance

$$r = \frac{c}{\omega_0} = \frac{55}{f_0}\,\text{m} \tag{6.22}$$

The decay process of the resonator is therefore only audible in its immediate vicinity. In the example mentioned earlier, this critical distance would be 0.55 m; at substantially larger distances, the decay cannot be perceived.

## 6.5 SOUND PROPAGATION IN POROUS MATERIALS, THE RAYLEIGH MODEL

Nearly all practically used sound absorbers contain some porous material which is exposed in one or another way to the arriving sound wave. The present section contains a more detailed discussion of the dissipation processes taking place in such materials.

In Section 6.1, inevitable reflection losses of sound waves impinging on smooth surfaces were briefly described. These are caused by viscous and thermal processes and occur within a boundary layer next to the surface, produced by the sound field. The thickness of this layer is typically in the range of 0.01 to 0.2 mm, depending on the sound frequency.

The absorption due to these effects is negligibly small if the surface is smooth (see Figure 6.7a). It is larger, however, at rough surfaces, since the roughness increases the zone in which the losses occur (Figure 6.7b). And it is even more pronounced if the material contains pores, channels, and voids connected with the air outside. A material of this kind is drawn schematically in Figure 6.7c. Then, the pressure fluctuations associated with the external sound field give rise to alternating airflows in the pores and channels, which will be filled more or less by the lossy boundary layer, so to speak. The consequence is that a significant amount of mechanical energy is withdrawn from the external sound field and is converted into heat.

This latter mechanism of sound absorption concerns all porous materials with pores accessible from the outside. Examples are the 'porous layers' which the reader encountered in Section 2.3 and which can be thought of as woven fabrics or thin carpets. The standard materials for practical application in room acoustics are mineral wool — also named Rockwool — and glass wool. They are manufactured from inorganic fibres by pressing them together, often with the addition of suitable binding agents. Materials of this kind are commercially available in the form of plates or loose blankets and found also wide application in sound insulation as well as in heat insulation. Other common materials are porous plaster or foams of certain polymers. The latter should have open cells; otherwise, their absorption is rather low. It should be mentioned that a brick wall may show noticeable absorption as well.

The most important characteristic parameters of a porous material are their porosity $\sigma$ and their specific flow resistance $\Xi$. The porosity is the fraction of volume which is not occupied by the solid structure. It can be determined by immersing a sample in a suitable liquid. A more reliable way is by comparing the isothermal compressibility of air contained in a

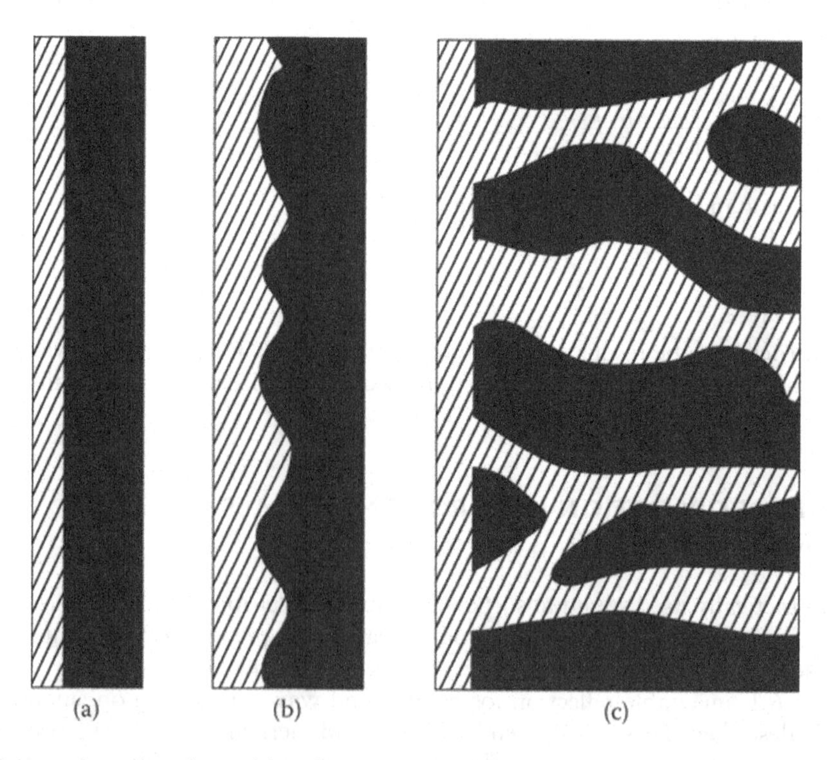


*Figure 6.7* Lossy boundary layer: (a) in front of a smooth surface, (b) in front of a rough surface, and (c) in front of and within a porous material.

solid chamber before and after inserting a sample of the material into it. For most materials of practical interest, the porosity exceeds 0.5 and may come close to unity.

The definition of the specific flow resistance is similar to that of the flow resistance of a porous layer in Equation 2.27. Suppose a constant airflow with velocity $v_s$ is forced through a porous sample with the thickness $d$; the pressure difference between both sides of the sample required to maintain this flow is $p - p'$. Then, the specific flow resistance is

$$\Xi = \frac{p - p'}{v_s d} \tag{6.23}$$

that is, it is the flow resistance per metre. Its unit is 1 kg-m$^{-3}$ · s$^{-1}$ = Pa·s·m$^{-2}$. A somewhat obsolete unit is 1 Rayl/cm = $10^3$ Pa·s·m$^{-2}$.

This is certainly not the place to present a comprehensive and exact description of sound absorption in porous materials which covers the full variety of materials and effects. In order to understand the basic processes, it is sufficient to restrict the discussion to an idealized model of a porous material, the so-called Rayleigh model, which qualitatively exhibits the essential features. Furthermore, we shall see the viscous processes in the foreground, that is, we neglect the less-prominent effects which are due to heat conduction.

We consider now a solid body which is traversed by equal and equidistant thin parallel channels (see Figure 6.8). It is supposed that the surface of that system, being located at $x = 0$, is perpendicular to the axes of the channels; in the positive $x$-direction, the model is assumed unbounded. (A practical realization of the Rayleigh model is the microperforated absorber mentioned in Section 6.3.)

For discussing the sound propagation in a single channel, we suppose that it is so narrow that the profile of the airstream is determined mainly by the viscosity of the air and not by inertial forces, that is, that the flow is laminar. This is always the case at sufficiently low frequencies; a more precise criterion will be presented in Equation 6.26. Then, the lateral distribution of flow velocities inside the channel is nearly the same as that with a stationary airflow, and likewise, the specific flow resistance of the channel has nearly the same value as for constant flow velocity.

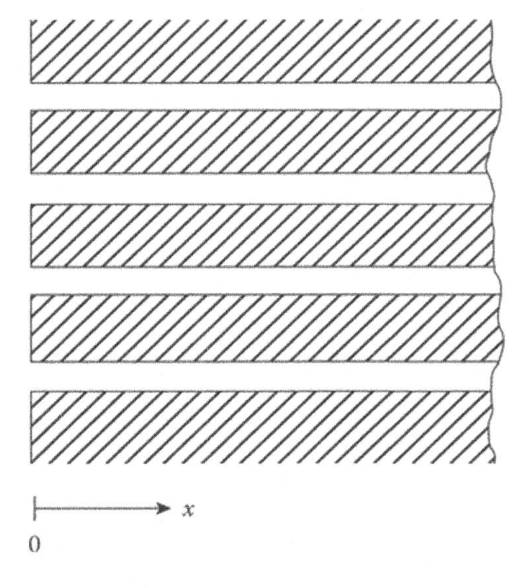


*Figure 6.8* Rayleigh model.

To adapt Equation 6.23 to a single channel, we replace the average flow velocity $v_s$ with the average velocity $\bar{v}$ within one channel. It exceeds $v_s$ by a factor $1/\sigma$ since the flow is confined to the inside of the channels. Furthermore, $(p - p')/d$ is replaced with the negative gradient of the pressure. Then, Equation 6.23 reads

$$-\frac{\partial p}{\partial x} = \sigma \, \Xi \, \bar{v} \tag{6.24}$$

To pass from a stationary airflow to sound, we must include the inertial forces according to Equation 1.2a. This is done by adding the term $\rho_0 \left(\partial \bar{v} / \partial t\right)$ to the right-hand side of Equation 6.24:

$$-\frac{\partial p}{\partial x} = \rho_0 \frac{\partial \bar{v}}{\partial t} + \sigma \, \Xi \, \bar{v} \tag{6.25}$$

It can be shown that the assumptions of the present treatment are fulfilled as long as

$$\frac{\rho_0 \omega}{\sigma \, \Xi} \leq 4 \tag{6.26}$$

The second relation we need is the equation of continuity (1.3), which is not affected by the internal friction of the air; its one-dimensional formulation is

$$\rho_0 \frac{\partial \bar{v}}{\partial x} = -\frac{\partial \rho}{\partial t} = -\frac{1}{c^2} \frac{\partial p}{\partial t} \tag{6.27}$$

Now, the following plane-wave expressions with the unknown propagation constant $k'$

$$p = \hat{p} \, \exp\left[ i\left(\omega t - k'x\right)\right], \quad \bar{v} = \hat{\bar{v}} \, \exp\left[ i\left(\omega t - k'x\right)\right]$$

are inserted into Equations 6.25 and 6.27. This yields two homogeneous equations for $p$ and $\bar{v}$:

$$\begin{aligned} k'p - \left(\omega \rho_0 - i\sigma\Xi\right)\bar{v} = 0 \quad \text{and} \\ \omega p - \rho_0 c^2 k' \bar{v} = 0 \end{aligned} \tag{6.28}$$

They have a non-zero solution if the determinant, formed of the coefficients $p$ and $\bar{v}$, is zero. From this condition, one obtains the complex propagation constant

$$k' = \beta' - i\gamma' = \frac{\omega}{c}\left(1 - i\frac{\sigma\Xi}{\rho_0 \omega}\right)^{1/2} \tag{6.29}$$

with the phase constant $\beta' = \omega/c'$ and the attenuation constant $\gamma' = m'/2$ (compare Equation 1.17).

Using this result, we obtain from one of both parts of Equation 6.28 the ratio of sound pressure to velocity, that is, the characteristic impedance in the channel:

$$Z_c' = \frac{p}{\bar{v}} = \rho_0 c \left(1 - i\frac{\sigma\Xi}{\rho_0 \omega}\right)^{1/2} \tag{6.30}$$

For very high frequencies — or for very wide channels — these expressions approach the values $\omega/c$ and $\rho_0 c$, valid for free sound propagation, because then the viscous boundary layer occupies only a very small fraction of the cross section. In contrast, at very low frequencies Equation 6.29 yields

$$\beta' = \gamma' \approx \left(\frac{\sigma\omega\Xi'}{2\rho_0 c^2}\right)^{1/2} \tag{6.31}$$

since $\sqrt{-i} = (1-i)/\sqrt{2}$. The attenuation in this range is considerable: the reduction of the amplitude per wavelength is as high as $20 \cdot \log_{10} [\exp(2\pi)] = 54.6$ dB.

Since the 'outer' flow velocity $v_s$ in Equation 6.23 is $\sigma\bar{v}$, we obtain from Equation 6.30 the characteristic impedance of the material (channels plus walls):

$$Z_0' = \frac{p}{v_s} = \frac{\rho_0 c}{\sigma}\left(1 - i\frac{\sigma\Xi}{\rho_0\omega}\right)^{1/2} \tag{6.32}$$

It should be noted that the Rayleigh model, even at normal sound incidence, is only useful for a qualitative understanding of the effects in a porous material but not for a quantitative description of real absorbent materials. This has several reasons. At first, the skeleton of the material is not entirely rigid, as was assumed so far. Furthermore, the heat exchange between the air contained in the channels and the solid skeleton causes an additional absorption, which has been neglected. It is due to the fact that the compressions and rarefactions of the air occur neither adiabatically nor according to an isothermal law but somehow in between these limiting cases (see Section 2.1). And finally, the pores within the materials are usually not well-separated and smooth channels but consist of irregularly shaped cavities with many interconnections. For this reason, the surface of a porous layer is not locally reacting to the sound field except at very low frequencies, where the attenuation within the material is high enough to prevent lateral coupling (see Equation 6.31). Several authors have tried to account for this fact by introducing a 'structure factor' which, however, is difficult to determine. From a practical point of view, it therefore seems more advantageous to do without a more rigorous treatment and to determine the absorption coefficient by measurement. This procedure is recommended all the more because the performance of porous absorbers depends only partially on the properties of the material and, to a greater extent, on its arrangement, on the covering, and on other constructional details, which vary substantially from one situation to another.

## 6.6 POROUS ABSORBERS

Now, we apply the preceding relations to a sound-absorbing wall lining consisting of a homogeneous layer of porous material. The simplest way to achieve this is to attach this layer directly on a rigid wall (see Figure 6.9). A sound wave arriving at the surface of the layer will be partially reflected from it; the remaining part of the sound energy will penetrate into the material and again reach the surface after its reflection from the rigid rear wall. Then, it will again split up into one portion penetrating the surface and another one returning to the rear wall, and so on. This qualitative consideration shows that the reflected sound wave can be thought of as being made up of an infinite number of successive contributions, each of them weaker than the preceding one because of the attenuation of the interior wave. Furthermore, it shows that the total reflection factor, and hence the absorption coefficient, may show maxima and minima, depending on whether the various components interfere constructively or destructively at a given frequency.

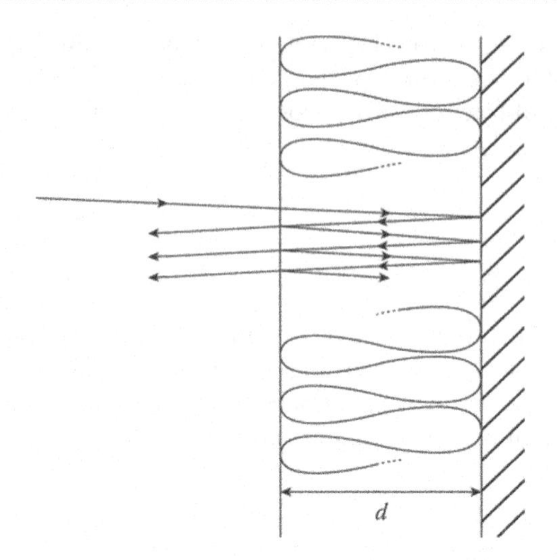


*Figure 6.9* Porous layer in front of a rigid wall.

For a quantitative treatment at normal sound incidence, we refer to Equation 2.25:

$$Z_1 = -iZ_0'\cot(k'd) \tag{6.33}$$

As in the preceding section, $k'$ and $Z_0'$ are the propagation constant and the characteristic impedance of the material, respectively, while $d$ is the thickness of the layer. From this expression, the reflection factor and the absorption coefficient can be calculated using Equations 2.10 and 2.11 with $Z = Z_1$ and $\theta = 0$. For the following discussion, it is useful to separate the real and the imaginary part of the cotangent:

$$\cot(k'd) = \frac{\sin(2\beta'd) + i\cdot\sinh(2\gamma'd)}{\cosh(2\gamma'd) - \cos(2\beta'd)} \tag{6.34}$$

Then, the following qualitative conclusions can be drawn:

1. For a layer which is thin compared with the sound wavelength, that is, for $k'd \ll 1$, cot $(k'd)$ can be replaced with $1/k'd$. Hence, the surface of the layer has a very large impedance and accordingly low absorption. In other words, substantial sound absorption cannot be achieved by just applying some kind of paint or wallpaper to a wall.
2. If the sound wave inside the porous material is strongly attenuated during just one round trip, that is, if $\gamma'd \gg 1$, the cotangent in Equation 6.34 becomes $i$, and the wall impedance is the same as that of an infinitely thick layer (see Equation 6.32):

$$Z_1 = \frac{Z_0'}{\sigma} = \frac{\rho_0 c}{\sigma}\left(1 - i\frac{\sigma\Xi}{\rho_0\omega}\right)^{1/2} \tag{6.35}$$

Figure 6.10a shows the absorption coefficient calculated from Equation 6.35. For high frequencies, it approaches asymptotically the value

$$\alpha_\infty = \frac{4\sigma}{(1+\sigma)^2} \tag{6.36}$$

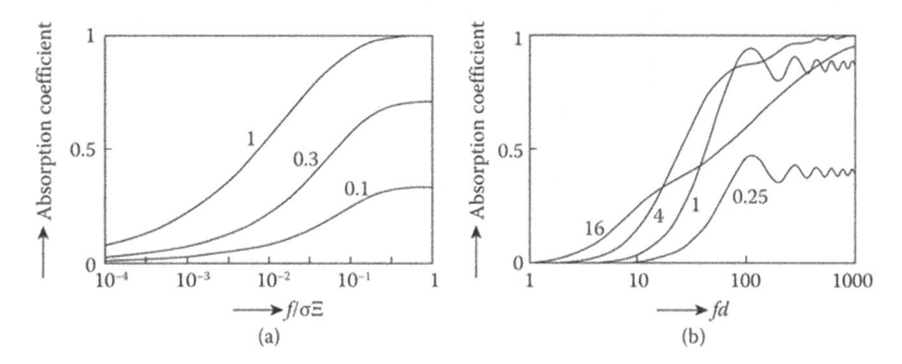


*Figure 6.10* Absorption coefficient of a porous layer (Rayleigh model) in front of a rigid wall, normal sound incidence: (a) infinite thickness and (b) finite thickness, $\sigma = 0.95$. The abscissa is $fd$ in Hz·m; the parameter is $\Xi d/\rho_0 c$ ($\Xi$ in Pa·s·m²). (For $\Xi d/\rho_0 c = 0.25$, the condition (6.26) is only partially fulfilled.)

3. If, on the contrary, $\gamma' d << 1$, that is, if the attenuation inside the material is small, the periodicity of the trigonometric functions in Equation 6.34 will dominate in the frequency dependence of the cotangent, and the same holds for that of the absorption coefficient represented in Figure 6.10b. The latter has a maximum whenever $\beta' d$ equals $\pi/2$ or an odd multiple of it, that is, at all frequencies for which the thickness $d$ of the layer is an odd multiple of $\lambda'/4$, with $\lambda' = 2\pi/\beta'$ denoting the acoustic wavelength inside the material. For higher values of $\Xi d$, the fluctuations fade out. As a figure of merit, we can consider $(fd)_{0.5}$, the value of the product $fd$ for which the absorption coefficient equals 0.5. Its minimum value is $(fd)_{0.5} = 23$, which is achieved for $\Xi d/\rho_0\, c = 6$.

As is evident from Figure 6.10, a porous layer is basically a high-frequency absorber. This holds not only for an absorber based on the Rayleigh model but, more or less, also for all sorts of absorbers employing porous materials arranged, as in Figure 6.9, and is a consequence of the very mechanism of sound dissipation. A rather trivial way to increase the absorption in the low-frequency range is to increase the thickness $d$ of the layer. Another way is to abandon the concept of a homogeneous absorber or, more specifically, to provide for some air space between the rear side of the absorber and the rigid backing (see Figure 6.11). For frontal sound incidence, the absorption coefficient of this configuration can be calculated with Equations 2.22 and 2.23 by including the impedance of the air gap, that is, by setting $Z_r = -i\rho_0 c \cot(kd)$ (according to Equation 2.25):

$$Z_1 = iZ_0' \cdot \frac{Z_0' - \rho_0 c \cdot \cot(kd)\cot(k'd')}{Z_0' \cdot \cot(k'd') + \rho_o c \cdot \cot(kd)} \tag{6.37}$$

A typical result obtained in this way is shown in Figure 6.11, calculated for $d = 2d'$ and $\sigma\Xi/\rho_0 c = 1.5$. The effect of the air gap is obvious; it is remarkable that the increase of low-frequency absorption is achieved without any additional porous material. The pronounced fluctuations of the absorption coefficient are caused by resonances of the air gap. Qualitatively, the effect of the air space can easily be understood by having a look at Figure 6.12. When the porous layer is fixed immediately on the wall (Figure 6.12a), it is close to the point of vanishing particle velocity (dashed line); hence, almost no air is forced through the pores. The air backing (Figure 6.12b) shifts the active layer to a position where the particle velocity is larger, causing higher flow velocity within the pores. (In the limiting case of a very thin porous sheet, we arrive at the stretched fabric, which we have already dealt with

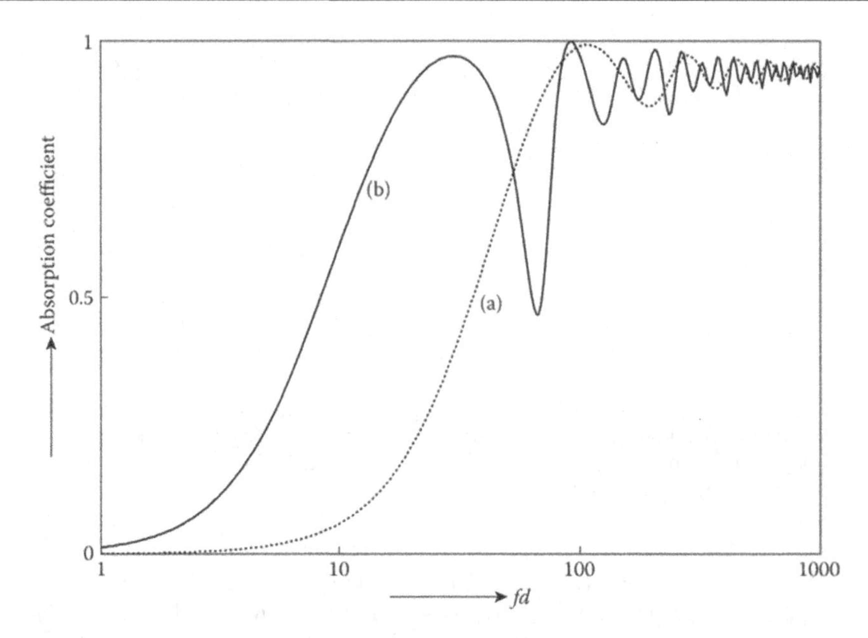


*Figure 6.11* Absorption coefficient of a porous layer in front of a rigid wall, normal sound incidence ($\Xi d'/\rho_0 c$ = 1.5, $d'$ = thickness of the layer): (a) without air gap and (b) with an air gap behind the porous layer. The depth $d$ of the gap is $2d'$, and the porosity $\sigma$ is 0.95.

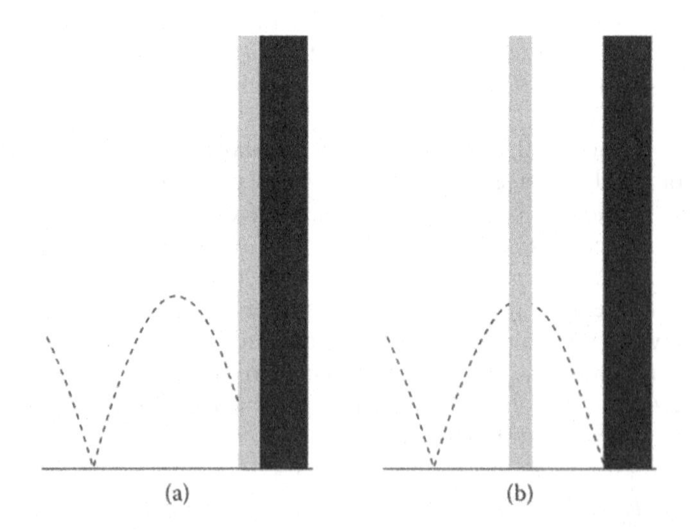


*Figure 6.12* Porous layer in front of a rigid wall: (a) without air backing and (b) with air backing. (Dashed line: particle velocity in the absence of the porous layer.)

in Section 2.3.) With regard to oblique or random sound incidence, it may be advantageous to partition the back space and, thus, to hinder wave transmission parallel to the surface of the layer.

A third possibility to increase the low-frequency absorption of an absorber after Figure 6.9 is to cover its front side with some membrane or perforated panel. This acts as a mass layer of the kind we have discussed in Section 6.2. Accordingly, its acoustic effect is accounted for by an additional term $i\omega M'$ in the expression for the wall impedance, for instance, in Equation 6.33

or 6.37, with $M'$ denoting the specific mass of the covering. Usually, the wall impedance of the uncovered arrangement has a large negative imaginary part at low frequencies. The added mass of the covering reduces this imaginary part and thus increases the absorption coefficient according to Figure 2.2. In other words, the mass layer has the tendency to change the original arrangement into sort of a resonance absorber, as discussed in Section 6.4.

In addition, the increase of low-frequency absorptivity by such a covering offers an important practical advantage. Usually, porous materials as Rockwool or glass wool, for instance, are not very hard wearing, nor do their surfaces look very pleasant. Many of them will, in the course of time, shed small particles, which must be prevented from polluting the air in the room. If the absorbent portions of the wall are within the reach of people, a suitable covering is desirable too, as a protection against unintentional or thoughtless damage of their surface. And finally, the architect usually wishes to hide the Rockwool layer behind a surface which can be painted and cleaned from time to time.

To prevent purling (or to keep water away from the pores, as, for instance, in swimming baths), it is often sufficient to bag the absorbent materials in very thin plastic foils. Furthermore, purling can be avoided by a somewhat-denser porous front layer on the bulk of the material.

Coverings for mechanical protection are usually made of wood, metal, or plastic panels, impervious or perforated, as described in Section 6.2. In the latter case, the apertures may be so small that the holes or slots are only visible at short distances. When such a lining is cleaned or provided with a new paint, care must be taken that the holes are not obstructed, which would make the absorber useless. Often, a combination of perforated panels and a limp porous fleece behind is employed as a protective layer. Thus, the covering of a porous lining, originally introduced for acoustical reasons, has an additional function of practical importance.

We close this section by presenting a few typical results as measured in the reverberation chamber, that is, at random sound incidence. It is a standard procedure since long, although it has the peculiarity that, for highly absorptive test samples, it occasionally yields absorption coefficients in excess of unity — a result which is physically impossible. Reasons for this strange behaviour will be discussed in Section 8.7, where this method is described in detail.

In Figure 6.13, the absorption coefficient of two homogeneous Rockwool layers with different densities and flow resistances is shown as a function of the frequency. Both test samples are 50 mm thick. Obviously, the denser material shows an absorption coefficient close to 1 even at somewhat-lower frequencies.

Figure 6.14 shows the influence of an air gap on the absorption coefficient. The porous sheet is 30 mm thick and has a density of 46.5 kg $m^{-3}$. In one case, it is mounted directly in front of a rigid wall (solid curve); in the other case, there is an air space of 50 mm between the sheet and the wall (dotted curve). The air gap is partitioned off by wooden lattices with a pattern of 50 cm × 50 cm. As expected, the second method of mounting leads to a considerable increase of absorption especially at low frequencies. The strong fluctuations which appeared in the curves of Figures 6.10 and 6.11 have been smoothed out — a consequence of the random sound incidence. It is evident that an air space behind the absorbent material considerably improves its effectiveness (or helps save material and costs). As mentioned before, the partitioning is to prevent lateral sound propagation within the air space at oblique sound incidence.

The influence of a perforated covering is demonstrated in Figure 6.15. In both cases, the porous layer has a thickness of 50 mm and is mounted directly onto the wall. The fraction of hole areas, that is, the perforation, is 14%. The mass load corresponding to it is responsible for an absorption maximum at 800 Hz, which is not present with the bare material. This resonance absorption can sometimes be very desirable. At a higher degree of perforation, this

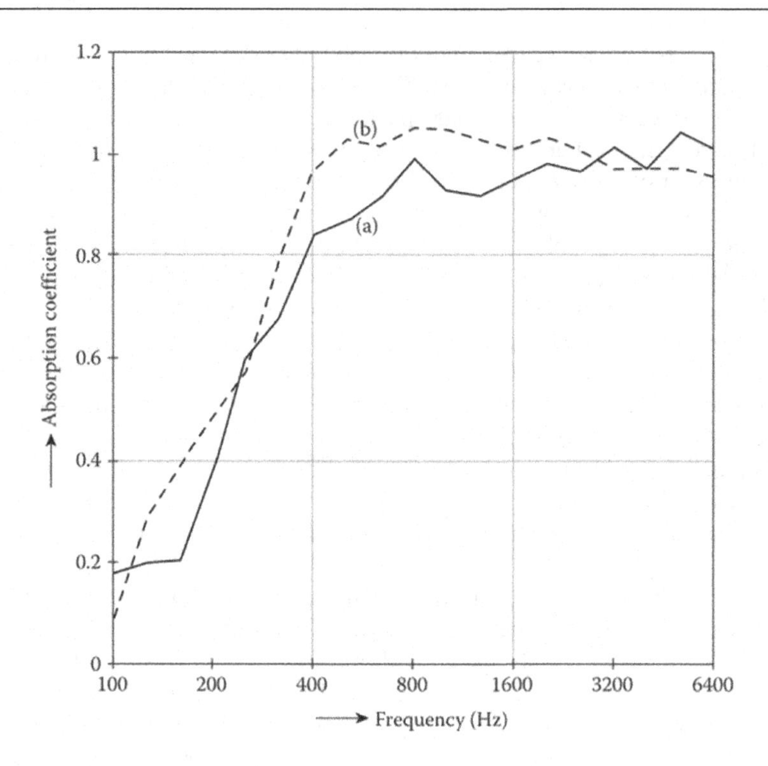


*Figure 6.13* Absorption coefficient (reverberation chamber) of Rockwool layers, 50 mm thick, immediately on concrete: (a) density 40 kg · m$^{-3}$, $\Xi$ = 12.7 × 10$^3$ Pa s m$^{-2}$ and (b) density 100 kg m$^{-3}$, $\Xi$ = 22 × 10$^3$ Pa s m$^{-2}$.

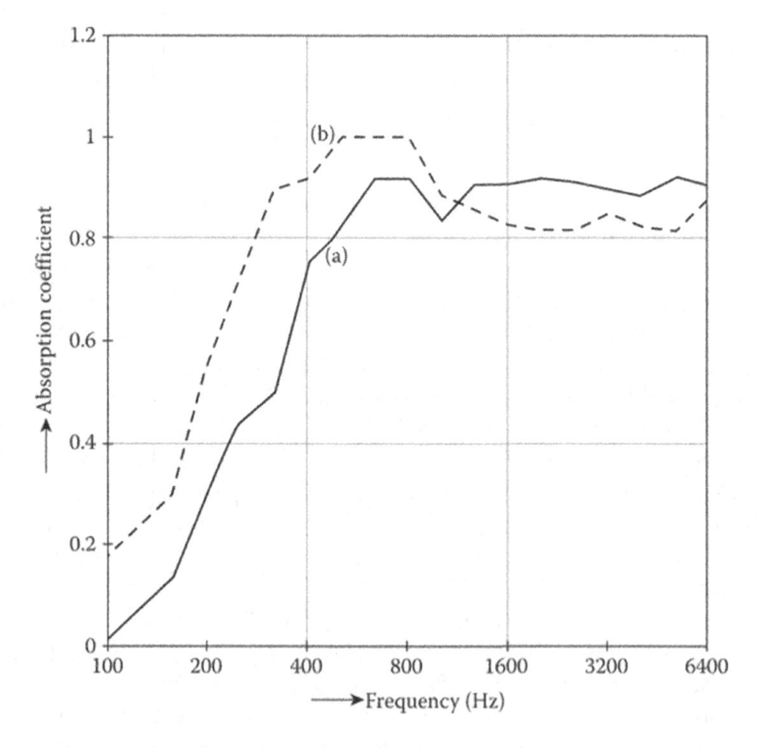


*Figure 6.14* Absorption coefficient (reverberation chamber) of a Rockwool layer 30 mm thick, density 46.5 kg·m$^{-3}$, $\Xi$ = 12 × 10$^3$ Pa·s·m$^{-2}$: (a) mounted immediately on concrete and (b) mounted with 50 mm air backing, laterally partitioned.

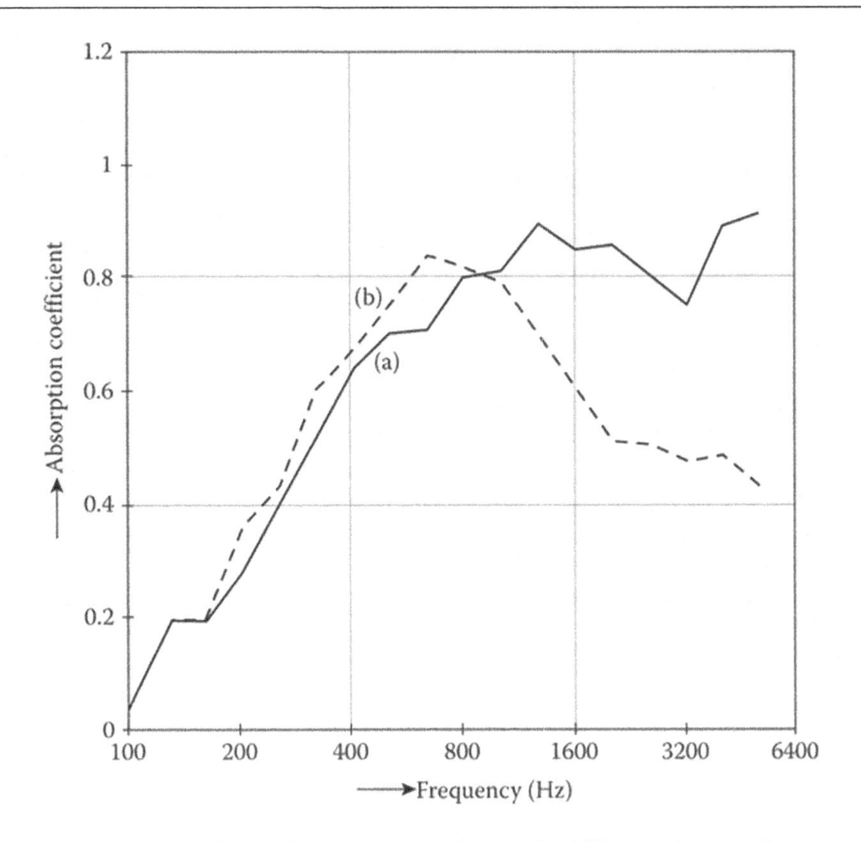


*Figure 6.15* Absorption coefficient (reverberation chamber) of 50 mm glass wool, mounted immediately on concrete: (a) uncovered and (b) covered with a protective panel, 5 mm thick, 14% perforation.

influence is much less pronounced, and with a perforation of 25% or more, the effect of the covering plate can virtually be neglected.

## 6.7 AUDIENCE AND SEAT ABSORPTION

The purpose of most medium- to large-size halls is to accommodate a large number of spectators or listeners and thus enable them to watch events or functions of common interest. This is true for concert halls and lecture rooms, for theatres and opera houses, for churches and sports halls, cinemas, council chambers, and entertainment halls of every kind. The important acoustical properties are therefore those which are present when the rooms are occupied or at least partially occupied. These properties are largely determined by the audience itself, especially by the sound absorption effected by it. The only exceptions are broadcasting and television studios, which are not intended to be used with an audience present.

The sound absorption caused by audience is mainly due to people's clothing and its porosity. Since clothing is not usually very thick, the absorption is considerable only at medium and high frequencies; in the range of low frequencies, it is relatively small. Since people's clothing differs from individual to individual, only average values of the audience absorption are available, and it is quite possible that these values are changing with the passage of time according to changing fashion or season. Furthermore, audience absorption depends on the kind of seats and their arrangement, on the occupancy density, on the way in which the audience is exposed to the incident sound, on the interruption of 'blocks' by aisles, stairs, and so on, and not least on the structure of the sound field. It is quite evident that a person seated at the rear of a narrow box with a small opening, as was typical in 18th- to 19th-century theatres, absorbs much less sound energy than a person sitting among steeply raked rows of seats and who is thus well exposed to the sound. Therefore, it is not surprising that there are considerable differences in the data on audience absorption which have been given by different authors.

There are two ways to determine experimentally the sound absorption of audience and seats. One is to place seats and/or persons into a 'reverberation chamber' (see Section 8.7) and evaluate their absorption from the change in reverberation time they cause. This has the advantage that sound field diffusion, which is a prerequisite for the applicability of the common reverberation formulae, can be established by adequate means. On the other hand, it may be doubtful whether a 'block' consisting 20 or 25 seats is representative for an extended area covered with occupied or unoccupied seats. In the second method, completed concert halls are used as reverberation chambers, so to speak: the absorption data are derived from reverberation times measured in them. The structure of the sound field is unknown in the second method, but provided the shape of the hall is not too exotic, it can at least be considered as typical for such halls. The same holds for the absorption coefficients of the walls and the ceiling.

The absorption of persons standing or seated singly is characterized most appropriately by their absorption cross section or equivalent absorption area $A$, already defined in Equation 6.12. For calculating the reverberation time, the absorption area of each person is added to the sum of Equation 5.25, which in this case reads

$$\bar{\alpha} = \frac{1}{S}\left(\sum_i S_i \alpha_i + N_p A\right) \tag{6.38}$$

$N_p$ being the number of persons. Table 6.2 lists a few of these absorption areas as a function of the frequency measured by Kath and Kuhl (1965) in a reverberation chamber.

When listeners are seated close together as in a fully occupied room, it seems that the absorption effected by the audience is proportional rather to the floor area it occupies than to the number of occupied seats (Beranek 1960). In other words, the absorptivity of a closed audience should be characterized by an apparent 'absorption coefficient', since this figure seems to be almost independent of the density of chairs. (However, according to a more recent publication [Choi et al. 2015], this is not always true; see Table 6.6.) The same holds

*Table 6.2* Equivalent absorption area of persons, in m²

| | *Frequency (Hz)* | | | | | |
|---|---|---|---|---|---|---|
| *Kind of person* | *125* | *250* | *500* | *1,000* | *2,000* | *4,000* |
| Male standing in heavy coat | 0.17 | 0.41 | 0.91 | 1.30 | 1.43 | 1.47 |
| Male standing without coat | 0.12 | 0.24 | 0.59 | 0.98 | 1.13 | 1.12 |
| Musician seated, with instrument | 0.60 | 0.95 | 1.06 | 1.08 | 1.08 | 1.08 |

Source: Adapted from Kath and Kuhl (1965).

for unoccupied seats. In Table 6.3, absorption coefficients of seated audience and of unoccupied seats, as measured in a reverberation chamber, are listed. Although the absorption of an audience in a particular hall may differ from the data shown, the latter at least demonstrates the general features of audience absorption: at increasing frequencies, the absorption coefficients increase at first. For frequencies higher than 2,000 Hz, however, there is a slight fall-off. This decrease is presumably due to mutual shadowing of absorbent surface areas by the backrests or listeners' bodies. At high frequencies, this effect becomes more prominent, whereas at lower frequencies, the sound waves are diffracted around the listeners' heads and shoulders.

These results are interesting in that they show general trends, but it is not clear how representative they are for any type of chairs. The effect of upholstered chairs essentially consists of an increase in absorption at low frequencies, whereas at frequencies of about 1,000 Hz and above, there is no significant difference between the absorption of audiences seated on upholstered or on unupholstered chairs.

Large collections of data on seat and audience absorption have been published by Beranek (1996) and by Beranek and Hidaka (1998). These authors determined absorption coefficients from the reverberation times of completed concert halls using the Sabine formula, Equation 5.26, with 5.25. It should be noted that their calculations are based upon the 'effective seating area' $S_a$, which includes not only the floor area covered by chairs but also a strip of 0.5 m around the actual area of a block of seating, except for the edge of a block when it is adjacent to a wall or a balcony face. This correction is to account for the 'edge effect', that is, diffraction of sound which generally occurs at the edges of an absorbent area (see Section 8.7). Absorption coefficients for closed blocks of seats — unoccupied as well as occupied — are listed in Table 6.4. The data shown are averages over three groups of halls with different types of seat upholstery. They show the same general frequency dependency as the absorption coefficients listed in Table 6.3, apart from the slight decrease towards very high frequencies for occupied seats which

*Table 6.3* Absorption coefficients of audience and chairs (reverberation chamber)

| *Type of seats* | *Frequency (Hz)* | | | | | | |
|---|---|---|---|---|---|---|---|
| | *125* | *250* | *500* | *1,000* | *2,000* | *4,000* | *6,000* |
| Audience seated on wooden chairs, two persons per m² | 0.24 | 0.40 | 0.78 | 0.98 | 0.96 | 0.87 | 0.80 |
| Audience seated on wooden chairs, one person per m² | 0.16 | 0.24 | 0.56 | 0.69 | 0.81 | 0.78 | 0.75 |
| Audience seated on moderately upholstered chairs, 0.85 m × 0.63 m | 0.72 | 0.82 | 0.91 | 0.93 | 0.94 | 0.87 | 0.77 |
| Audience seated on moderately upholstered chairs, 0.90 m × 0.55 m | 0.55 | 0.86 | 0.83 | 0.87 | 0.90 | 0.87 | 0.80 |
| Moderately upholstered chairs, unoccupied, 0.90 m × 0.55 m | 0.44 | 0.56 | 0.67 | 0.74 | 0.83 | 0.87 | 0.80 |

*Table 6.4* Absorption coefficients of unoccupied and occupied seating areas in concert halls (averages)

| Type of upholstery | | Frequency (Hz) 125 | 250 | 500 | 1,000 | 2,000 | 4,000 |
|---|---|---|---|---|---|---|---|
| Heavy | Unoccupied (seven halls) | 0.70 | 0.76 | 0.81 | 0.84 | 0.84 | 0.81 |
| | Occupied (seven halls) | 0.72 | 0.80 | 0.86 | 0.89 | 0.90 | 0.90 |
| Medium | Unoccupied (eight halls) | 0.54 | 0.62 | 0.68 | 0.70 | 0.68 | 0.66 |
| | Occupied (eight halls) | 0.62 | 0.72 | 0.80 | 0.83 | 0.84 | 0.85 |
| Light | Unoccupied (four halls) | 0.36 | 0.47 | 0.57 | 0.62 | 0.62 | 0.60 |
| | Occupied (six halls) | 0.51 | 0.64 | 0.75 | 0.80 | 0.82 | 0.83 |

Source: Adapted from Beranek and Hidaka (1998).

*Table 6.5* Residual absorption coefficients from concert halls (averages)

| Type of hall | Frequency (Hz) 125 | 250 | 500 | 1,000 | 2,000 | 4,000 |
|---|---|---|---|---|---|---|
| Group A: halls lined with wood, less than 3 cm thick, or with other thin materials (six halls) | 0.16 | 0.13 | 0.10 | 0.09 | 0.08 | 0.08 |
| Group B: halls lined with heavy materials, that is, with concrete, plaster more than 2.5 cm thick, etc. (three halls) | 0.12 | 0.10 | 0.08 | 0.08 | 0.08 | 0.08 |

Source: Adapted from Beranek and Hidaka (1998).

is missing in Table 6.4. Evidently, the influence of seat upholstery is particularly pronounced in the low-frequency range. Beranek and Hidaka (1998) had the opportunity of assembling reverberation data from several halls before and after the chairs were installed. From these values, they evaluated what they called the 'residual absorption coefficients', $\alpha_r$, that is, the total absorption of all walls, the ceiling, balcony faces, etc., except the floor, divided by their total area $S_r$. Since these data are interesting in their own right, we present their averages in Table 6.5. The residual absorption coefficients include the absorption of chandeliers, ventilation openings, and other typical installations, and they show remarkably small variances.

First group: 7.5 cm upholstery on front side of seat back, 10 cm on top of seat bottom, arm rest upholstered. Second group: 2.5 cm upholstery on front side of seat back, 2.5 cm on top of seat bottom, solid arm rests. Third group: 1.5 cm upholstery on front side of seat back, 2.5 cm on top of seat bottom, solid arm rests.

*Table 6.6* Absorption coefficients of audience and chairs (reverberation chamber)

| | *Frequency (Hz)* | | | | | |
|---|---|---|---|---|---|---|
| *Type of chairs* | *125* | *250* | *500* | *1,000* | *2,000* | *4,000* |
| *High-absorption chairs, with carpet,* | *0.37* | *0.55* | *0.78* | *0.83* | *0.88* | *0.88* |
| *row distance 1.2 m, unoccupied* | *(0.10)* | *(0.25)* | *(0.29)* | *(0.34)* | *(0.33)* | *(0.34)* |
| *High-absorption chairs, with carpet,* | *0.48* | *0.77* | *0.87* | *0.80* | *0.90* | *0.96* |
| *row distance 1.2 m, occupied* | *(0.14)* | *(0.19)* | *(0.26)* | *(0.37)* | *(0.34)* | *(0.33)* |
| *High-absorption chairs, with carpet,* | *0.79* | *0.69* | *0.97* | *1.06* | *0.99* | *1.00* |
| *row distance 0.76 m, unoccupied* | *(0.02)* | *(0.40)* | *(0.39)* | *(0.40)* | *(0.46)* | *(0.51)* |
| *High-absorption chairs, with carpet,* | *1.21* | *0.90* | *1.05* | *1.00* | *1.03* | *1.04* |
| *row distance 0.76 m, occupied* | *(0.0)* | *(0.31)* | *(0.37)* | *(0.45)* | *(0.43)* | *(0.46)* |
| *Low-absorption chairs, no carpet,* | *0.23* | *0.47* | *0.46* | *0.47* | *0.50* | *0.45* |
| *row distance 1.2 m, unoccupied* | *(0.0)* | *(0.0)* | *(0.03)* | *(0.06)* | *(0.06)* | *(0.15)* |
| *Low-absorption chairs, no carpet,* | *0.29* | *0.48* | *0.44* | *0.50* | *0.50* | *0.48* |
| *row distance 1.2 m, occupied* | *(0.04)* | *(0.08)* | *(0.11)* | *(0.11)* | *(0.13)* | *(0.17)* |
| *Low-absorption chairs, no carpet,* | *0.32* | *0.75* | *0.45* | *0.46* | *0.49* | *0.39* |
| *row distance 0.76 m, unoccupied* | *(0.04)* | *(0.0)* | *(0.18)* | *(0.21)* | *(0.24)* | *(0.35)* |
| *Low-absorption chairs, no carpet,* | *0.66* | *0.86* | *0.60* | *0.63* | *0.64* | *0.69* |
| *row distance 0.76 m, occupied* | *(0.0)* | *(0.03)* | *(0.16)* | *(0.19)* | *0.23* | *0.23* |

Source: Adapted from Choi et al. (2015).

Note: Upper entry, $\alpha_\infty$; lower entry (in brackets), $\beta$ in metre (see Equation 6.39).

More systematic measurements of the absorption of occupied or unoccupied blocks of chairs have been carried out by Choi et al. (2015). Concerning the dependence of the absorption on the size of blocks, these authors found the following relation:

$$\alpha = \beta \frac{P}{S} + \alpha_\infty \tag{6.39}$$

with $P$ denoting the perimeter of a block and $S$ its area. $\alpha_\infty$ is the absorption coefficient of an infinitely large block of chairs; it corresponds to what was denoted by $\alpha$ in Tables 6.3 and 6.4. The factor $\beta$, as well $\alpha_\infty$, depends on several parameters: on the type of chairs, on row spacing, on the presence of occupants, and of a carpet underneath the seats. Table 6.6

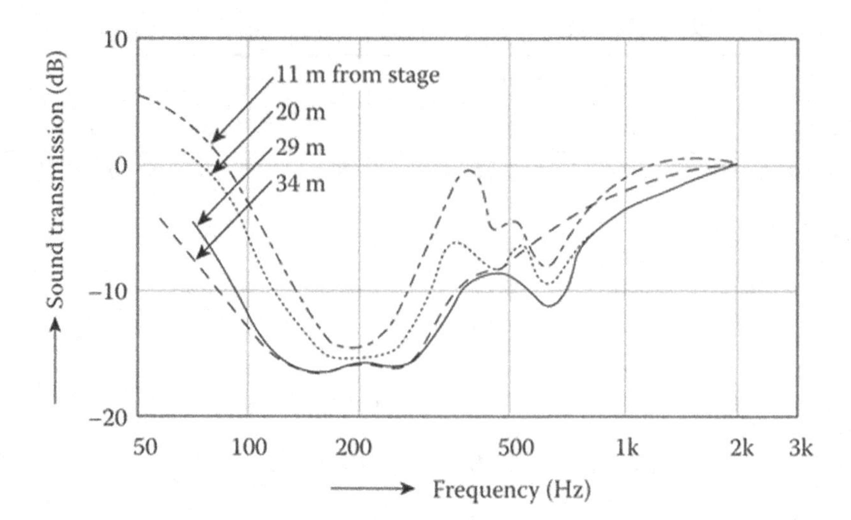


*Figure 6.16* Transmission characteristics of direct sound, measured at unoccupied seats of the main floor in Boston Symphony Hall. The numbers in the figure indicate the distance from the stage. (The level reduction due to spherical divergence has been subtracted from the data.)

Source: Schultz and Watters (1964).

(Choi et al. 2015) presents some of the results. The upper entry in each cell shows $\alpha_\infty$, while the lower one (in brackets) is the coefficient $\beta$ in Equation 6.39. All these data have been obtained with scale models of theatre chairs, of the listeners, and of a carpet, using a model reverberation chamber; the scaling factor was 1:10. Two types of model chairs were developed, namely, with low or with high absorption, and care has been taken to achieve a close agreement of their acoustical properties with those of typical full-scale chairs.

It has been known for a long time that the absorption of the impinging sound waves and, as a consequence, the reduction of the reverberation time are not the only acoustical effect brought about by audience. Another one is additional attenuation of sound waves travelling parallel or nearly parallel to the audience. It seems that two different causes are responsible for this effect. Attenuation in excess of the $1/r$ law of Equation 1.27 is actually observed whenever a wave propagates over an absorbent surface. If the sound velocity in the material behind (or rather below) the surface is smaller than that in air — which is the case for porous materials — a sound wave with grazing incidence will be partially refracted according to Equation 2.15 and enter the absorbent material. If this surface is not plane but shows a comb-like structure as with rows of occupied or upholstered seats, the regular arrangement gives rise to additional, frequency-dependent attenuation.

Even sound waves travelling over the empty seating rows of a large hall are subject to a characteristic attenuation, which is known as the 'seat dip effect' and has been observed by many researchers (Schultz and Watters 1964; Sessler and West 1964). The excess attenuation has typically a maximum in the range of 80–250 Hz, depending on the angle of sound incidence and other parameters. An example is shown in Figure 6.16. The minimum in these curves is usually attributed to a vertical $\lambda/4$ resonance of the space between seating rows. Some authors have found this 'seat dip' with occupied seats too; others have not. The missing energy is probably not always absorbed, but some of it may be redistributed in the hall. Figure 6.17 presents the level of reduction (shaded areas) caused by the audience, as measured

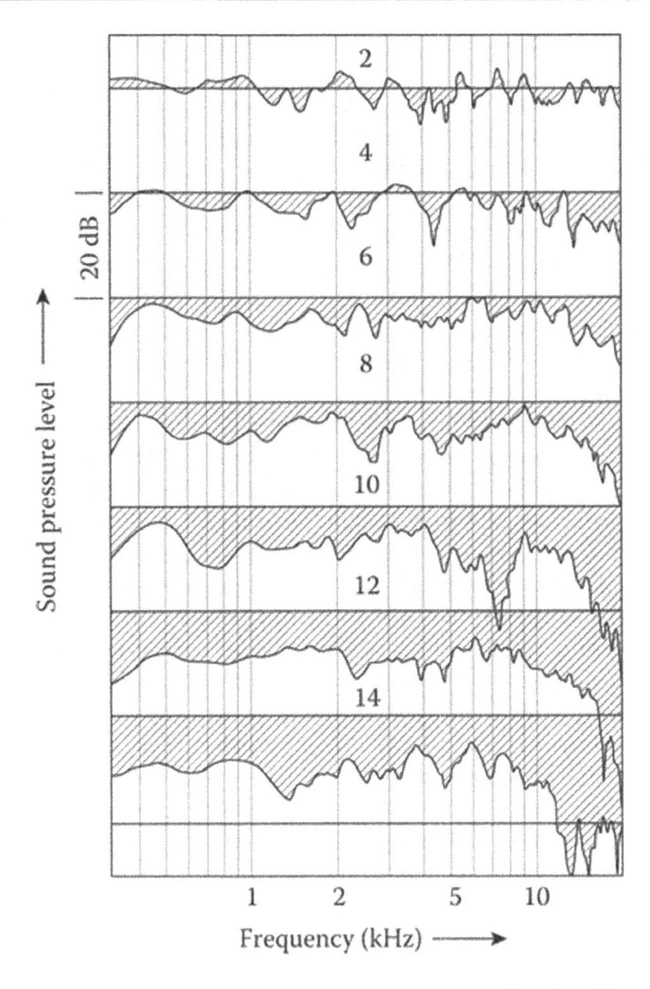


*Figure 6.17* Sound pressure level over audience, relative to the level at free propagation (horizontal lines). The figures indicate the number of the seating row. The source is 2.5 m before the first row and 1.4 m high.

Source: Mommertz (1993).

by Mommertz (1993) in a large hall with a horizontal floor, by employing maximum length sequence techniques (see Section 8.2). The numbers 2, 4, 6 . . . characterize the seating row in which the measuring microphone was located. An evaluation of these data shows that there is a linear level decrease from front to rear seats, indicating an excess attenuation of the audience of roughly 1 dB per metre in the range 500–2,000 Hz. It should be noted that not only the direct sound but also reflections from vertical side walls may be subject to this kind of attenuation.

The selective attenuation of sound propagating over the audience cannot be considered as an acoustical fault because it occurs in concert halls, which are renowned for their good acoustics, as well as in poorer ones. Since it is highly dependent on the arrangement of seats, there is no reliable method to predict this kind of attenuation. However, it can be avoided by sloping the audience area upwardly. This has the effect of exposing the listeners freely to the direct sounds without running the risk of grazing sound incidence (see Section 9.2).

## 6.8 ANECHOIC ROOMS

Certain types of acoustic measurements require an environment which is free of reflected sound waves. This is true for the free field calibration of microphones, or for the determination of directional patterns of sound sources, and so on. The same holds for psychoacoustic experiments. In all these cases, the accuracy and the reliability of results would be impaired by the interference of the direct sound with sound components reflected from the boundaries.

One way to avoid reflections — except those from the ground — would be to perform such measurements or experiments in the open air. This has the disadvantage, however, that the experimenter depends on favourable weather conditions, which implies the absence of not only rain but also wind. Furthermore, acoustic measurements in the open air can be badly affected by ambient noise.

A more convenient way is to use a so-called anechoic room or chamber, all the boundaries of which are treated in such a way that virtually no sound reflections are produced by them, at least in the frequency range of interest. How stringent the conditions are which have to be met by the acoustical treatment may be illustrated by a simple example. If all the boundaries of an enclosure have an absorption coefficient of 0.90, everybody would agree that the acoustics of this room is extremely 'dry', on account of its very low reverberation time. Nevertheless, the sound pressure level of a wave reflected from a wall would be only 10 dB lower than that of the incident wave! Therefore, the usual requirement for the walls of an anechoic room is that the absorption coefficient is at least 0.99 for all angles of incidence. This condition cannot be met with plane homogeneous layers of some absorbent materials; it can only be satisfied with a wall covering which achieves a stepwise or continuous transition from the air to a material with high internal energy losses.

In principle, this transition can be accomplished by a porous wall coating whose flow resistance increases in a well-defined way from the surface to the wall. It must be expected, however, that, at grazing sound incidence, the absorption of such a plane layer would be zero on account of total reflection: According to Equation 2.10 or 2.16, the reflection factor becomes $-1$ when the incidence angle $\theta$ approaches 90°. Therefore, it is more useful to achieve the desired transition by choosing a proper geometrical structure of the acoustical treatment than by varying the properties of the material. Accordingly, the absorbent material is applied in the form of pyramids or wedges which are fixed on the walls, the ceiling, and the floor of the test room. Hence, an incident sound wave will run into channels with absorbent walls whose cross sections steadily decrease in size, that is, into reversed horns. The apertures at the front of these channels are well matched to the characteristic impedance of the air, and thus, no significant reflection will occur.

This is only true, however, as long as the length of the channels, that is, the thickness of the lining, is at least about one-third of the acoustical wavelength. This condition can easily be fulfilled at high frequencies, but only with great expense, at frequencies of 100 Hz or below. For this reason, every anechoic room has a certain lower limiting frequency, usually defined as the frequency at which the absorption coefficient of its walls becomes less than 0.99.

As to the production of such a lining, it is easier to utilize wedges instead of pyramids. The wedges must be made of a material with suitable flow resistance and sufficient mechanical solidity, and the front edges of neighbouring wedges or packets of wedges must be arranged at right angles to each other (see Figures 6.18 and 6.19).

Since the floor, as well as the other walls, must be treated in the same way, a net of steel cables or plastic wires must be installed in order to give access to the space above the floor. The reflections from this net can be safely neglected at audio frequencies.

The lower limiting frequency of an anechoic room can be further reduced by combining the pyramids or wedges with cavity resonators which are located between the latter and

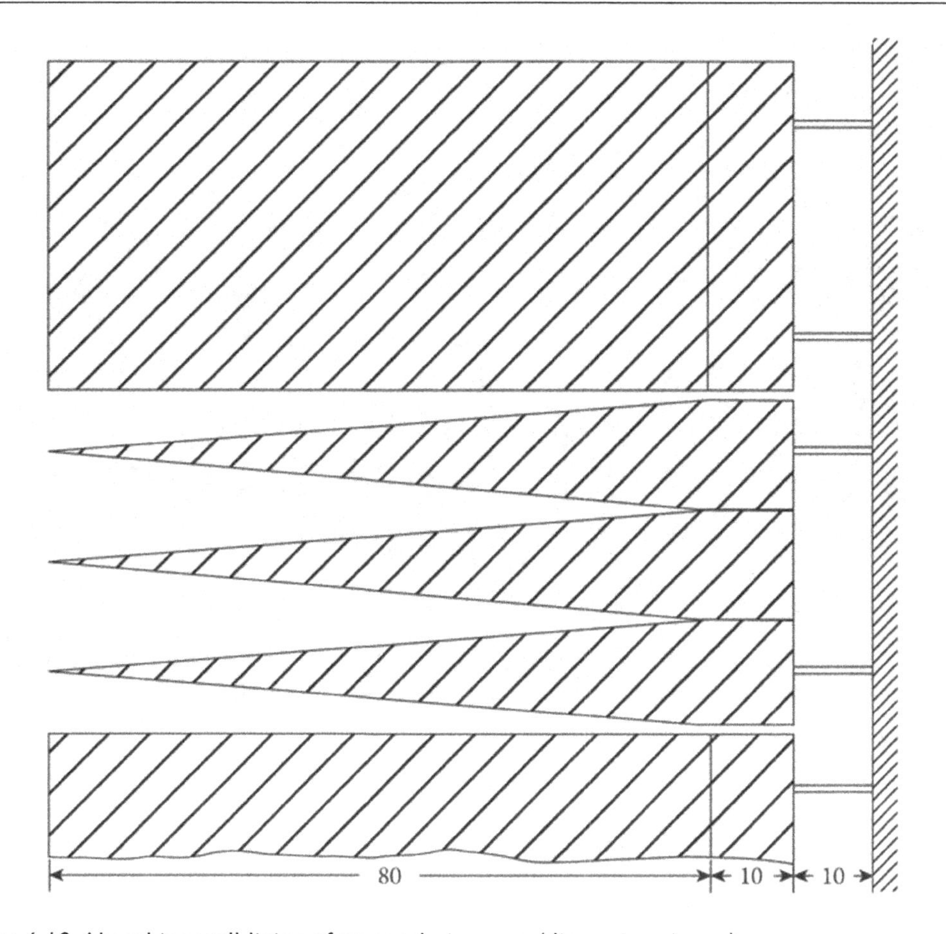


*Figure 6.18* Absorbing wall lining of an anechoic room (dimensions in cm).

the rigid wall. By choosing the apertures, that is, the lengths and widths and the depths of the resonators, carefully, the reflection can be suppressed at frequencies at which noticeable reflection would occur without the resonators.

In this way, for a particular anechoic room (Meyer et al. 1953), a lower limiting frequency of 80 Hz was achieved by a lining with a total depth of 1 m. The absorber material has a density of 150 kg $m^{-3}$ and a specific flow resistance $\Xi$ of about $10^5$ kg $m^{-3}s^{-1}$ (= $10^5$ Pa s $m^{-2}$). It is fabricated in wedges of 13 cm × 40 cm base area and 80 cm length, which terminate in rectangular blocks of the same base area and 10 cm length. Between these blocks, there are narrow gaps of 1 cm width which run into an air cushion of 10 cm depth between the absorber material and the concrete wall (see Figure 6.18). The latter acts as the volume of a large resonance absorber with many necks which are the gaps between the wedges. Three wedges with parallel edge are joined together in a packet; neighbouring packets are rotated by 90° with respect to each other. A view of the interior of this anechoic room is presented in Figure 6.19.

Anechoic rooms are usually tested by observing the way in which the sound amplitude decreases when the distance from a sound source is increased. This decrease should take place according to a $1/r$ law to simulate perfect outdoor conditions. In practice, with increasing distance, deviations from this simple law become more and more apparent in the form of random fluctuations. Since these fluctuations are caused by wall reflections, they can be used to evaluate the average absorption coefficient of the walls. Several methods have been

*Figure 6.19* Anechoic room fitted with 65 loudspeakers for synthesizing complex sound fields.

worked out to perform these measurements and to determine the wall absorption from their results (Diestel 1962; Delany and Bazley 1971).

## REFERENCES


Bass H.E., Sutherland L.C., Zuckerwar A.J., Blackstock D.T., Hester D.M. Atmospheric absorption of sound. Further developments. *J Acoust Soc Am* 1995; *97*: 680.

Beranek L.L. Audience and seat absorption in large halls. *J Acoust Soc Am* 1960; *32*: 661.

Beranek L.L. *Concert and Opera Halls: How They Sound*. Woodbury, NY: Acoustical Society of America, 1996.

Beranek L.L., Hidaka T. Sound absorption in concert halls by seats, occupied and unoccupied, and by the hall's interior surfaces. *J Acoust Soc Am* 1998; *104*: 3169.

Choi Y.J., Bradley J.S., Jeong D.U. Experimental investigation of chair type, row spacing occupants and carpet on theatre chair absorption. *J Acoust Soc Am* 2015; *137*: 105.

Delany M.E., Bazley E.N. The design and performance of free-field rooms. Proceedings of the Fourth International Congresses on Acoustics, Budapest, 1971, Paper 24A6.

Diestel H.G. Sound propagation in anechoic rooms (in German). *Acustica* 1962; *12*: 113.

Kath U., Kuhl W. Measurements on sound absorption of upholstered chairs with and without persons (in German). *Acustica* 1965; *15*: 127.

Maa D.Y. Wide-band sound absorber based on microperforated panels. *Chin J Acoust* 1985; *4*(3): 197 (original work in Chinese: Theory and design of microperforated panel sound-absorbing constructions. *Scientia Sinica* 1975: *55*).

Meyer E., Kurtze G., Severin H., Tamm K. A large anechoic room for sound waves and electromagnetic waves (in German). *Acustica* 1953; *3*: 409.

Mommertz E. Some measurements on the grazing sound propagation over audience and chairs (in German). *Acustica* 1993; *79*: 42.

Schultz T.J., Watters B.G. Propagation of sound across audience seating. *J Acoust Soc Am* 1964; *36*: 885.

Sessler G.M., West J.E. Sound transmission over theatre seats. *J Acoust Soc Am* 1964; *36*: 1725.

Chapter 7

# Subjective room acoustics

Previous chapters were devoted exclusively to the physical side of room acoustics, that is, the objectively measurable properties of sound fields in a room and the circumstances which are responsible for their origin. We could be satisfied with this aspect if the only problems at stake were those of noise abatement by reverberation reduction and, hence, by reduction of the energy density, or if we only had to deal with problems of measuring techniques.

In most cases, however, the 'final consumer' of room acoustics is the listener, for example, who wants to enjoy a concert or who attends a lecture or a theatre performance. This listener does not, by any means, require the reverberation time, at the various frequencies, to have certain values; neither does he insist that the sound energy at his seat should exhibit a certain directional distribution. Instead, he expects the room with its 'acoustics' to support the music being performed or to render speech easily intelligible (as far as this depends on acoustic properties).

The acoustical designer finds himself in a different situation. He must find ways to meet the expectations of the average or typical listener (whoever this is). To do this, he needs knowledge of the relationship between properties of the physical sound field on the one hand and the listener's subjective impression on the other (see Figure 7.1). For this purpose, several physical parameters have been isolated in the course of time, which are related more or less with certain aspects of the subjective listening impression. A second task is to find out in which way the physical sound field is determined by the architect's design, that is, by the size and shape of a hall and the material of its boundary. This was the subject of the initial chapters of this book. In this chapter, we deal with the subjective aspects, that is, with the left side of Figure 7.1. With these considerations, we clearly leave the region of purely physical fact and enter the realm of psychoacoustics.

The isolation of meaningful physical sound field parameters and the examination of their relevance have, in the past, been and still are the subject of numerous investigations — experimental investigations — since answers to these problems, which are not affected by the stigma of pure speculation, can only be obtained by systematic listening tests. Unfortunately, the results of this research do not form an unequivocal picture, in contrast to what we are accustomed to in the purely physical branch of acoustics. This is ascribed to the lack of a generally agreed vocabulary to describe subjective impressions and to the very involved physiological properties of our hearing organ, including the manner in which hearing sensations are processed by our brain. It can also be attributed to listeners' hearing habits and, last but not the least, to their personal aesthetic sensitivity — at least as far as musical productions are concerned. Another reason for our incomplete knowledge in this field is the vast number of sound field components, which may influence the subjective hearing impression. The experimental results which are available to date must therefore be considered in spite of the fact that many are unrelated or sometimes even inconsistent, and that every day new and surprising insights into psychoacoustic effects and their significance in room acoustics may be found.

DOI: 10.1201/9781003389873-7

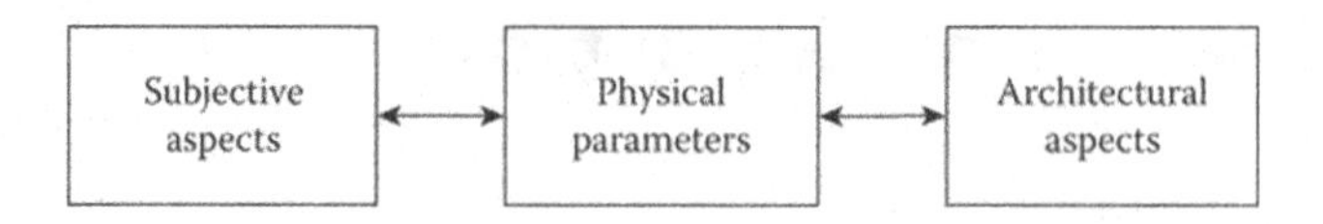


*Figure 7.1* Relationship between subjective, physical, and architectural aspects of room acoustics.

## 7.1 SOUND PRESSURE LEVEL, STRENGTH FACTOR

The first of parameters we introduce is the stationary sound pressure level (SPL) or energy density that a sound source of given output power produces in a room. For a long time, this was not regarded as an acoustical quality criterion because it was believed that this level just depends on the absorption area (or the reverberation time) of the room. More recently, however, the general attitude towards the overall level has changed, since high definition or clarity is of little use if the sound is too weak to be heard at a comfortable loudness. Moreover, the simple equation (5.5), which relates the energy density to the absorption area and the source power $P$, is valid for diffuse sound fields only, while the field within a real hall may show considerable deviations from this ideal condition.

If the stationary SPL in an enclosure is to reflect merely the properties of the enclosure and not of the source, it must be measured by using a non-directional sound source. Furthermore, the influence of the source is eliminated by subtracting a reference level $\mathrm{SPL_A}$. This is the level which the same sound source would produce in the free field (anechoic room) at a distance of 10 m. The result is the so-called 'strength factor':

$$G = \mathrm{SPL} - \mathrm{SPL_A} = 10\,\log_{10}(w / w_\mathrm{A}) \tag{7.1}$$

where $w$ and $w_A$ denote the corresponding energy densities. For the reference measurement, we have $w_\mathrm{A} = P / 4\pi c r_A^2$ according to Equation 5.43 with $r_\mathrm{A} = 10$ m. If the sound field in the considered hall were exactly diffuse, the energy density in the reverberant sound field of the room would be $w = 4P/cA$, with $A = 0.16\ V/T$ (see Section 5.1), and we could calculate the strength factor from the formula

$$G_\mathrm{exp} = 10\log_{10}\left(\frac{T}{V}\right) + 45\,\mathrm{dB} \tag{7.2}$$

($T$ in s and $V$ in m³). In contrast, Gade and Rindel (1985) found from measurements in 21 Danish concert halls that the strength factor in any hall shows a linear decrease from the front to the rear. This decrease corresponds to 1.2–3.3 dB per distance doubling, and the average of the strength factor $G$ falls short of the value of Equation 7.2 by 2–3 dB. Furthermore, the steady-state level does not depend in a simple way on the reverberation time or on the geometrical data of the hall. From these results, it may be concluded that the sound fields in real concert halls are not diffuse, and that the strength factor is, indeed, a useful figure of merit.

An alternative definition of the strength factor is based on the impulse response $g(t)$ of the room at a given point:

$$G = 10\log_{10}\left[\int_0^\infty [g(t)]^2\,\mathrm{d}t \Big/ \int_0^{t_\mathrm{d}} [g_\mathrm{A}(t)]^2\,\mathrm{d}t\right] \tag{7.3}$$

$g_\mathrm{A}(t)$ is the impulse response measured with the same sound source at 10 m distance in the free sound field. The upper integration limit $t_\mathrm{d}$ must be chosen in such a way that the second integral includes the direct sound but not any reflected components. Both definitions are equivalent.

## 7.2 SOME GENERAL REMARKS ON REFLECTIONS AND ECHOES

In the following discussion, we shall regard the sound transmission between two points of a room as formally represented by the impulse response of the transmission path. According to Equations 4.4 and 4.5, this impulse response is composed of the direct sound and numerous repetitions of the primary sound impulse which are caused by its reflections from the boundary of the room (see also Figure 4.8). Each of these reflections is specified by its level and its time delay, both with respect to the direct sound. Since our hearing is sensitive to the direction of sound incidence as well, this description must be completed by indicating the direction from which each reflection arrives at the receiving point. And finally, there may be differences in spectrum since, as already mentioned in Section 4.1, the various components of the impulse response are not exact replicas of the original sound signal, strictly speaking, because of frequency-dependent wall reflectivities and attenuation by audience.

What has been presented here is the physical description of a room impulse response and its components. From the subjective standpoint, however, there are great differences between reflections, depending on their strengths and their delays with respect of the direct sound. In particular, we have to distinguish slightly delayed reflections, so-called early reflections, from those with significantly longer delay. To illustrate this, the reader may be reminded of two common experiences: sometimes, a reflected sound can be perceived as an 'echo', that is, as a distinct repetition of the original sound signal. This is frequently observed outdoors, for instance, by hand-clapping in front of an extended building, provided there is no noise which would mask the echo. Fortunately, in closed rooms, such echoes are less familiar, since they are usually masked by the general reverberation of the room. Whether a reflection will be heard as an echo or not depends on several factors, most critically on the delay with respect to the original sound. In our hand-clapping experiments, there is no audible echo if the observer is too close to the reflecting wall, say, less than about 10 m, because then the delay time, which is the time the sound signal needs for its round trip from the hands to the building front and back to the observer's ear, is too short.

The second common experience concerns our ability to localize sound sources in closed rooms. Although in a room which is not too heavily damped, the sum of all reflected sound energies exceeds by far the directly received energy, our hearing can usually localize the direction of the sound source without any difficulty. Obviously, it is the sound signal to reach the listener first which subjectively determines the direction from which the sound comes. This effect is called the precedence effect or — according to L. Cremer — the 'law of the first wave front'. In Section 7.4, we shall discuss the conditions under which it is valid.

In the following sections, the subjective effects of sound fields with increasing complexity will be discussed. It is quite natural that the criteria of judgement become less and less detailed: in a sound field consisting of hundreds or thousands of reflections, we cannot investigate the effects of each reflection separately.

Many of the experimental results to be reported on have been obtained with the use of synthetic sound fields, as mentioned earlier: in an anechoic chamber, the reflections and the direct sound are 'simulated' by loudspeakers which have certain positions vis-à-vis the test subject; these positions correspond to the directional distribution of the reflections. Of course, all loudspeaker signals must be derived from the same signal source. The differences in strength are achieved by attenuators in the electrical lines feeding the loudspeakers, whereas their mutual delay differences are produced by electrical delay units. When necessary or desired,

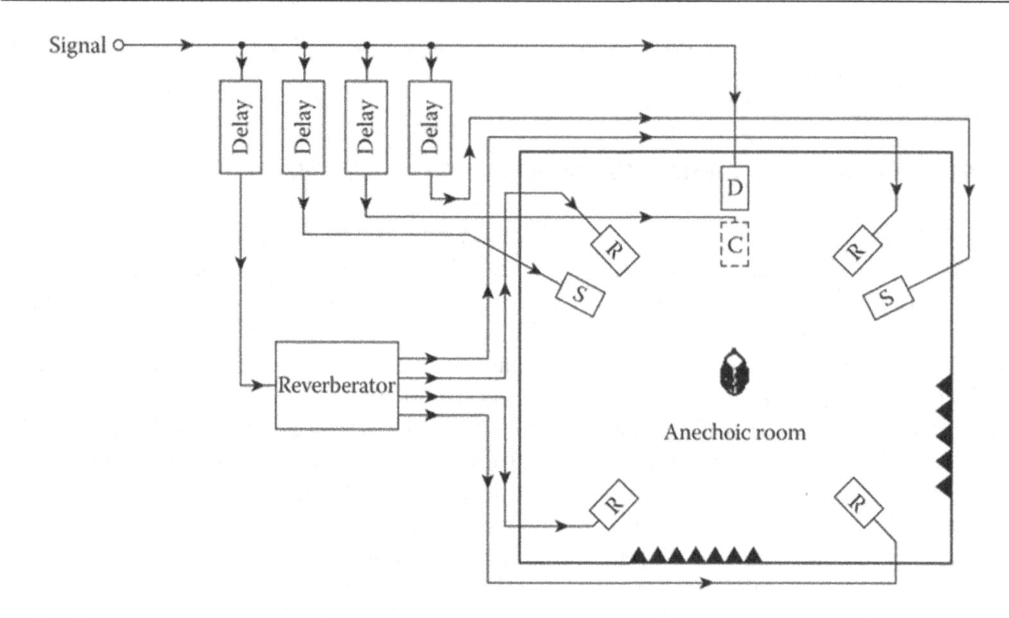


*Figure 7.2* Simulation of sound fields in an anechoic room. The loudspeakers are denoted by *D* = direct sound (*S* = side wall reflections, *C* = ceiling reflection [elevated], R = reverberation).

Source: Reichardt and Schmidt (1966, 1967).

reverberation with prescribed properties can be added to the signals. For this purpose, signals are passed through a so-called reverberator, which is also most conveniently realized with digital circuits nowadays. (Other methods of 'adding' reverberation to a given signal will be described in Section 11.5.) A typical setup for psychoacoustic experiments related to room acoustics is depicted in Figure 7.2; it allows simulation of the direct sound, two side-wall reflections, and one ceiling reflection. The reverberated signal is reproduced by four additional loudspeakers. A more flexible loudspeaker arrangement employed for investigations of this kind is shown in Figure 6.19.

## 7.3 THE PERCEPTION OF REFLECTIONS AND ECHOES

We begin with sound fields consisting of the direct sound and one or a few repetitions of it, that is, reflections. There are two questions to be raised, namely:

1. Under what condition is a reflection perceivable at all, without regard to the way in which its presence is manifested, and under what condition is it masked by the direct sound?
2. Under what condition does the presence of a reflection rate act as a disturbance of the listening impression, for instance, an echo or a change of timbre ('colouration')?

In this section, we deal with the first question; the second one is discussed in the next section. Most of the reported results are due to a research group of the University of Göttingen (Burgtorf 1961; Burgtorf and Oehlschlägel 1964; Seraphim 1961) and are published in *Acustica*, Vol. 11–14.

We postulate that there is a critical level separating the levels at which a reflection is audible from those at which it is completely masked. This 'absolute threshold of perception', or simply 'audibility threshold', is a function not only of the time delay with respect to the direct sound but also of the direction of its incidence and of the kind of test signal. Through all the further discussions, we assume the listener is facing the source of the direct sound.

To find this threshold, a subject is presented with two alternate sound field configurations that differ in the presence or absence of a specified reflection. The test subject is asked to indicate solely whether he or she notices a difference or not. (One has to make sure, of course, that the test subjects do not know beforehand to which configuration they are listening at a given moment.) The answers of the subjects are evaluated statistically; the level at which 50% of the answers are positive is regarded as the threshold of absolute perceptibility.

The first situation we consider is an impulse response consisting of just two components: the direct sound and one reflection. For a speech level of 70 dB, and for frontal incidence of the direct sound, as well as of the reflected component, the audibility threshold turns out to be (Burgtorf 1961)

$$\Delta L = -0.575 t_0 - 6 \text{ dB} \tag{7.4}$$

where $\Delta L$ is the pressure level of the reflected sound signal relative to the sound pressure of the direct sound and $t_0$ is its time delay in milliseconds. Take, for example, a reflection delayed by 60 ms with respect to the direct sound. According to Equation 7.1, it is audible even when its level is 40 dB below that of the direct signal.

The picture is different when the reflection arrives from a lateral direction. Figure 7.3 plots for three different signals (continuous speech, a short syllable, and a white noise pulse with a duration of 50 ms), the angle dependence of the audibility thresholds when the reflection is delayed by 50 ms. It is remarkable to what extent the thresholds depend on the type of sound signal. In any case, however, the masking effect of the direct sound is most effective for both

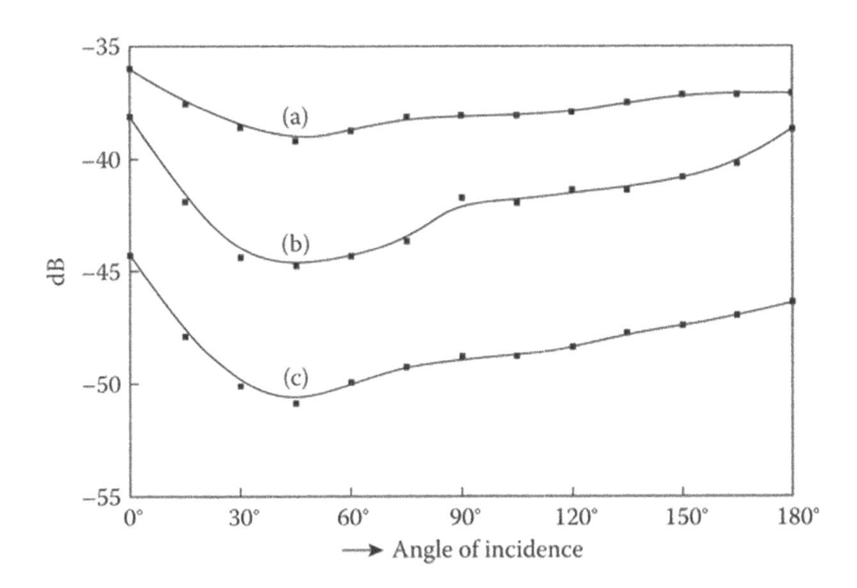


*Figure 7.3* Threshold of perception of a reflection with 50 ms delay, obtained with (a) continuous speech, (b) a short syllable, and (c) noise pulses of 50 ms duration. Abscissa is the horizontal angle at which the reflection arrives. The sound pressure level of the direct signal is 75 dB.

Source: Burgtorf and Oehlschlägel (1964).

components arriving from the same direction, or in other words, our hearing is more sensitive to reflections arriving from lateral directions than to those arriving from the front or the rear. It should be added that reflections arriving from above are also masked more effectively by the direct sound than are lateral reflections.

If the sound signal is a music sample, our hearing is generally much less sensitive to reflections. This is the general result of investigations carried out by Schubert (1969), who measured the audibility threshold with various music motifs. One of his typical results is presented in Figure 7.4, which plots the average threshold taken over six different music samples. With increasing delay time, it falls much less rapidly than according to Equation 7.4; its maximum slope is about –0.13 dB/ms. As with speech, the threshold is noticeably lower for reflections arriving from lateral directions than with frontal incidence. Furthermore, added reverberation renders the detection of a reflection more difficult. Obviously, reverberated sound components cause additional masking, at least with continuous sound signals.

If there is more than one reflection, the number of parameters to be varied increases rapidly. Fortunately, each additional reflection does not create a completely new situation for our hearing. This is demonstrated in Figure 7.5, which shows the audibility threshold for a variable reflection which is added to a masking sound field consisting of the direct sound plus one, two, three, or four reflections at fixed delay times and levels. The fixed reflections are indicated as vertical lines, and their length is a measure of the strength of the reflections. In this study, it was assumed that all the reflections arrive from the same direction as the direct sound. If all the reflections and the direct sound arrive from different directions, the thresholds are different from those of Figure 7.5 in that they immediately begin to fall and then jump back to the initial value whenever one of the fixed reflections is arriving.

Apart from the absolute thresholds of perception, the subjective difference limen for reflections is also of interest. According to Reichardt and Schmidt (1966, 1967), variations of the

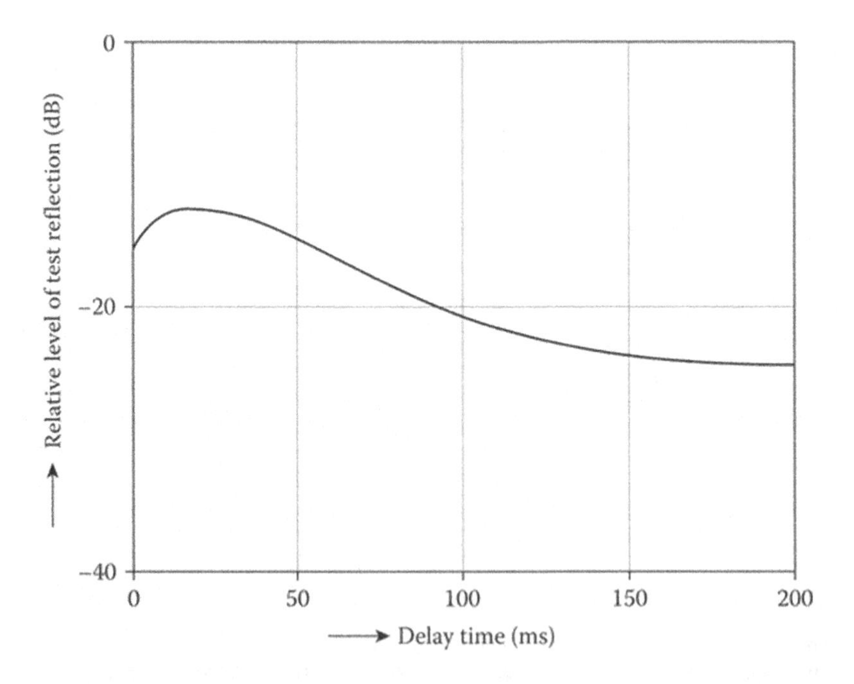


*Figure 7.4* Threshold of perception of a reflection as a function of its delay (frontal incidence of reflection). The threshold is the average taken over six different music samples.

Source: Schubert (1969).

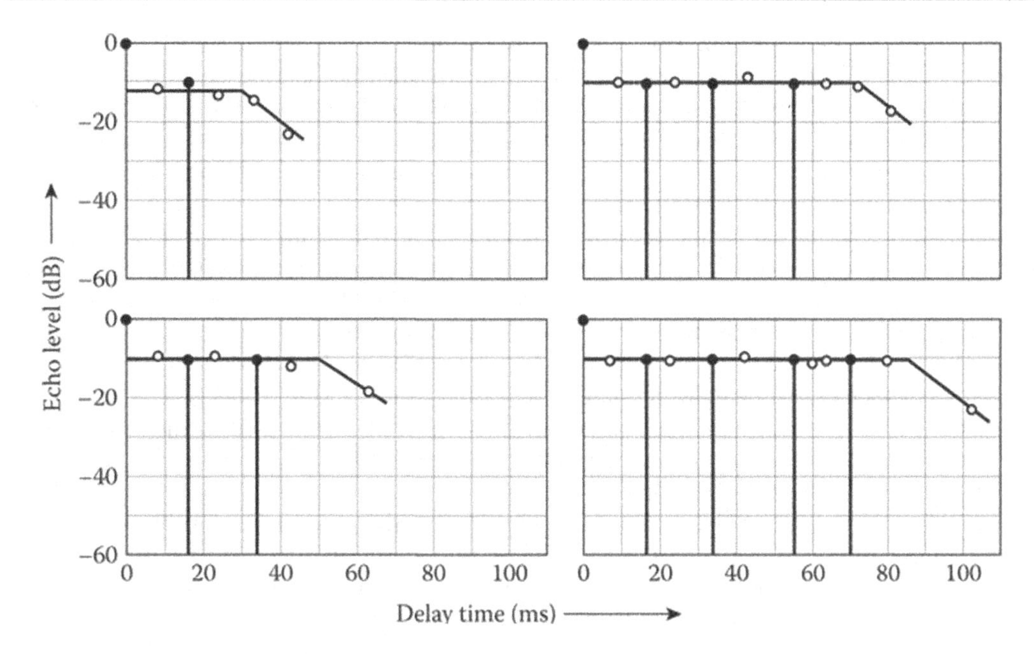


*Figure 7.5* Threshold of perception of a delayed reflection being added to a sound field consisting of direct sound plus one, two, three, or four reflections at fixed delay times and relative levels, as indicated by vertical lines. The test signal is speech. All components arriving from the front.

Source: Seraphim (1961).

reflection level as small as about ±1.5 dB can be detected by our hearing if music is used as a test signal. In contrast, the auditory detection of differences in delay times is afflicted with great uncertainty.

## 7.4 ECHOES AND COLOURATION

A reflection which is perceived at all does not necessarily reach the consciousness of a listener. At low levels, it manifests itself only by an increase in loudness of the total sound signal, by a change in timbre, or by an increase in the apparent size of the sound source. But at higher levels, a reflection can be heard as a separate event, that is, as a repetition of the original sound signal. This effect is commonly known as an 'echo', as already mentioned in Section 7.2. But what outdoors usually appears as a funny experience may be rather unpleasant in a concert hall or in a lecture room, in that it distracts the listeners' attention. In severe cases, an echo may significantly reduce our enjoyment of music or impair the intelligibility of speech, since subsequent speech sounds or syllables are mixed up and the text is confused.

In the following text, the term 'echo' will be used for any sound reflection which is subjectively noticeable as a distinct repetition of the original sound signal, and we are discussing the conditions under which a reflection will become an echo. Thus, we are taking up again the second question raised at the outset of the foregoing section.

From his outdoor experience, the reader may know that the echo produced by sound reflection from a house front disappears when he approaches the reflecting wall and when his distance from it becomes less than about 10 m, although the wall still reflects the sound. Obviously, it is the reduction of the delay time between the primary sound and its repetition which makes the echo vanish. This shows that our hearing has only a restricted ability to resolve subsequent acoustical events, a fact which is sometimes attributed to some kind of 'inertia' of hearing. Like the absolute threshold of perception, however, the echo disturbance depends not only on the delay of the repetition but also on its relative strength, its direction, on the type of sound signal, on the presence of additional components in the impulse response, and on other circumstances.

Systematic experiments to find the critical echo level of reflections are performed in much the same way as those investigating the threshold of absolute perception, but with a different instruction given to the test subjects. It is clear that there is more ambiguity in fixing the critical echo levels than in establishing the absolute perception threshold, since an event which is considered as disturbing by one person may be found tolerable by others.

Results of such experiments were published as early as in 1950 by Haas (1951), who used continuous speech as a primary sound signal. This signal was presented to the test subjects with two loudspeakers: the input signal of one of them could be attenuated (or amplified) and delayed with respect to the other.

Figure 7.6 shows one of Haas's typical results. It plots the percentage of subjects who felt disturbed by an echo as a function of the time delay between the undelayed signal (primary sound) and the delayed one (reflection). The numbers next to the curves indicate the level of the artificial reflection in decibels relative to that of the primary sound. The rate of speech was 5.3 syllables per second; the listening room had a reverberation time of 0.8 s. At a delay time of 80 ms, for instance, only about 20% of the observers felt irritated by the presence of a reflection with a relative level of −3 dB, but the percentage was more than 80% when the level was +10 dB. Further investigations concerned the dependence of critical delay times on the speech rate and the reverberation of the listening room. A critical delay of about 70 ms can be regarded typical for most situations.

Figure 7.6 shows a striking feature: if the relative level of the reflection is raised from 0 to +10 dB, there is only a small increase in the percentage of observers feeling disturbed by

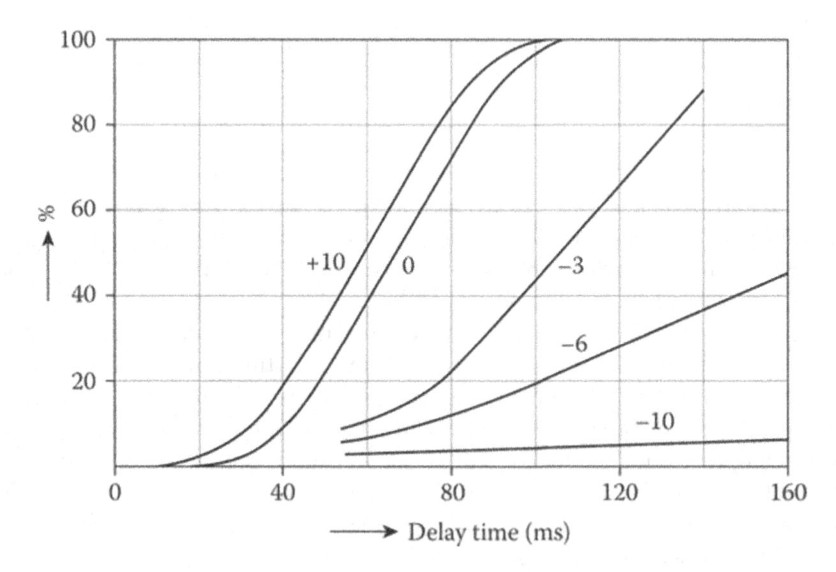


*Figure 7.6* Percentage of listeners disturbed by a delayed speech signal in a room with a reverberation time of 0.8 s. Speaking rate is 5.3 syllables per second. The relative echo levels (in dB) are indicated by numbers next to the curves.

Source: Haas (1951).

the reflected sound signal. Hence, no disturbance is expected to occur for a reflection with time delay of, say, 20 ms even if its energy is 10 times the energy of the direct sound. This finding is frequently referred to as the 'Haas effect'. In a slightly modified set of experiments, Meyer and Schodder (1952) applied a different criterion for the annoyance by a reflection: to restrict the range of possible judgements, the test subjects were not asked to indicate the level at which they felt disturbed by an echo; instead, they were to indicate the level at which they heard both the delayed signal and the undelayed one equally loud. Since in these tests the undelayed signal reached the test subject from the front, whereas the delayed one came from a lateral angle of 90°, the test subjects could also be asked to indicate the reflection level that made the total sound signal seemingly arrive from halfway between both directions. Both criteria of judgement led to the same results. One of them is shown in Figure 7.7, where the critical level difference between primary sound and reflection is plotted as a function of the delay time. It confirms the results obtained by Haas and renders them somewhat more precise. For our hearing, the primary sound determines the perception of direction even when the reflection — provided it has a suitable delay time — is stronger by up to 10 dB. If the reflection is split up into several small reflections of equal strengths and with successive mutual delay times of 2.5 ms, leaving constant the total reflection energy, the curve shown in Figure 7.7 is shifted upwards by another 2.5 dB. This result shows that many small reflections separated by short time intervals of the order of milliseconds cause about the same disturbance as one single reflection, provided the total reflected energy and the (centre) delay time are the same for both configurations. These findings have important applications in the design of public address systems.

Muncey et al. (1953) have performed similar experiments for speech as well as for various kinds of music. Their investigations clearly showed that our hearing is less sensitive to echoes in music than in speech. The annoyance of echoes in very slow music, for example, organ music, is particularly low. In Figure 7.8, the critical echo level (50% level) for fast string and organ music is plotted as a function of time delay.

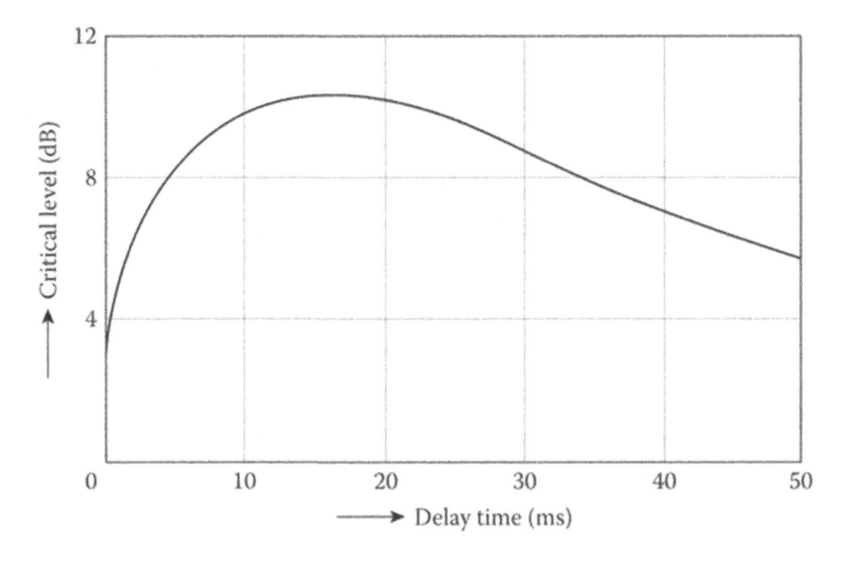


*Figure 7.7* Critical reflection level (relative to the direct sound level) as a function of delay time. At this level, both components, the direct and the delayed, are heard equally loud.

Source: Meyer and Schodder (1952).

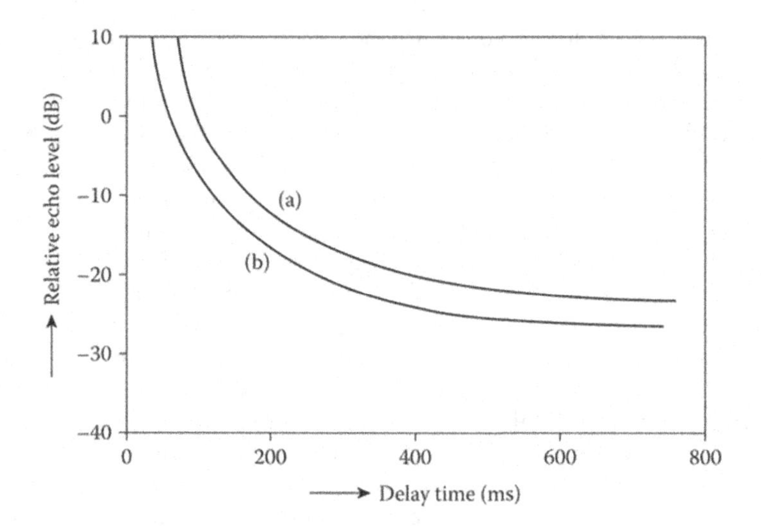


*Figure 7.8* Critical reflection level (relative to the direct sound level) as a function of delay time for (a) organ music and (b) string music. Note the different time scale in this figure.

Source: Muncey et al. (1953).

From these results, we can draw the following practical conclusions: In general, the law of the first wave front can be considered valid; exceptions, that is, erroneous localizations and disturbing echoes, will occur only in special situations, for example, when most of the room boundaries, except for a few remote portions of wall, are lined with an absorbent material or when certain portions of wall are concavely curved and, hence, produce exceptionally strong reflections by focusing the sound. Often, this situation is encountered in halls equipped with an electroacoustic system for sound amplification (see Chapter 10).

So far, our discussion has been restricted to the somewhat-artificial case that the sound field consists of the primary or direct sound, followed by just one single repetition. However, the impulse responses of virtually all real rooms have a more complicated structure, and it is clear that the presence of numerous reflections must influence the way we perceive one of them in particular. Hence, from a practical point of view, it would be desirable to have a criterion to indicate whether a certain peak in a measured impulse response or 'reflectogram' hints at an audible echo and should be removed by suitable constructive measures. Such a criterion was proposed by Dietsch and Kraak (1986). Its description is provided in Section 8.3.

We come back now to the effects of a single strong reflection. If its delay is small, say, smaller than about 30 ms, its superposition onto the direct sound can cause a characteristic change of timbre, called 'colouration', which may be quite disturbing no matter whether we are listening to music or speech. This is easy to understand: the impulse response of a transmission system which produces this kind of signal reads

$$g_1(t) = \delta(t) + q\delta(t - t_0) \tag{7.5}$$

with $q < 1$. The squared absolute value of its Fourier transform after Equation 1.66 is

$$|G_1(f)|^2 = |1 + q\,\exp(2\pi i f t_0)|^2 = 1 + q^2 + 2q\,\cos(2\pi f t_0) \tag{7.6}$$

This is the squared absolute transfer function of a comb filter, that is, a frequency filter with a regular succession of maxima and minima; the spacing of maxima is $1/t_0$ (see Figure 7.9).

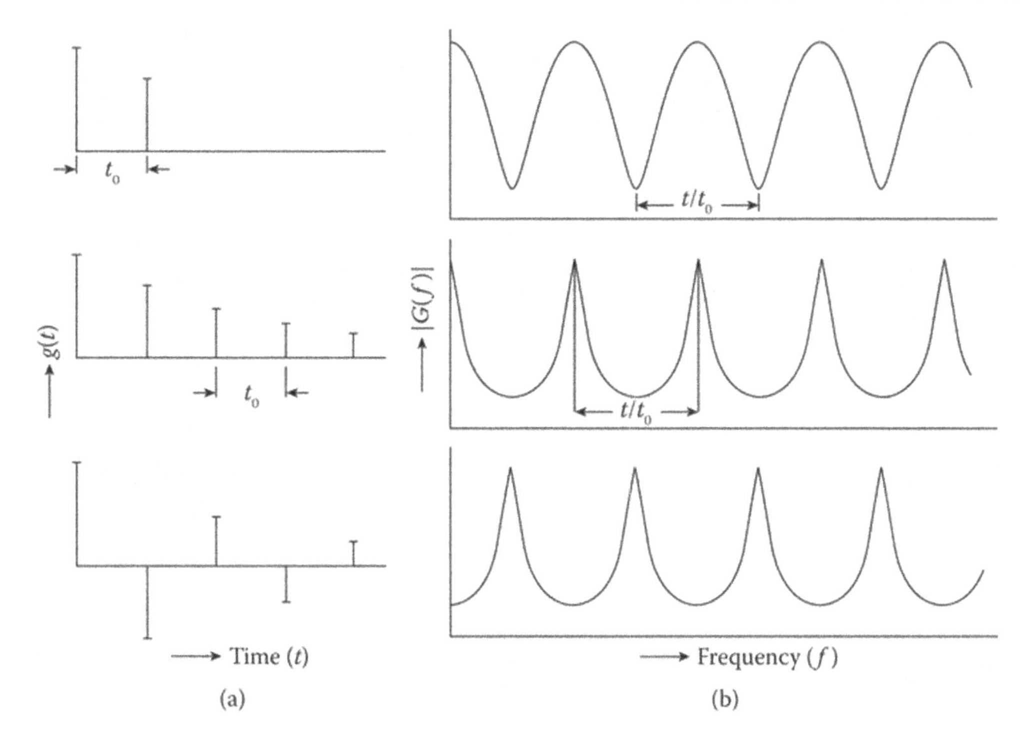


*Figure 7.9* Impulse responses (a) and absolute values of transfer function (b) of various comb filters ($q$ = 0.7).

The ratio of maximum to minimum values is $(1 + q)^2/(1 - q)^2$. A similar effect has an infinite and regular succession of reflections; its impulse response is

$$g_2(t) = \sum_{n=0}^{\infty} q^n \delta(t - nt_0) \tag{7.7}$$

and its squared absolute spectrum is

$$|G_2(f)|^2 = \left[1 + q^2 - 2q\cos(2\pi f t_0)\right]^{-1} \tag{7.8}$$

It has the same spacing of maxima as that in Equation 7.6; the same holds for the relative height of the peaks. However, the peaks are much sharper here than with a single reflection, except for $q << 1$.

Whether such a comb filter will produce audible colourations or not depends again on the delay time $t_0$ and on the relative heights of the maxima (Atal et al. 1962). Beginning from very low values of the delay time or distance $t_0$, the absolute threshold for audible colourations grows with increasing delay time $t_0$, that is, with decreasing distance $1/t_0$ between subsequent maxima on the frequency axis (see Figure 7.10). When $t_0$ exceeds a certain value, say, 25 ms or so, the regularity of impulse responses does not subjectively appear as colouration, that is, as a change of the timbre of sound, but rather, the sounds have a rattling or buzzing character, that is, the listener becomes aware of the regular repetitions of the signal as a phenomenon occurring in the time domain (echo or flutter echo). This is because our ear is not just a sort of frequency analyzer but also sensitive to the temporal structure of the sound signals.

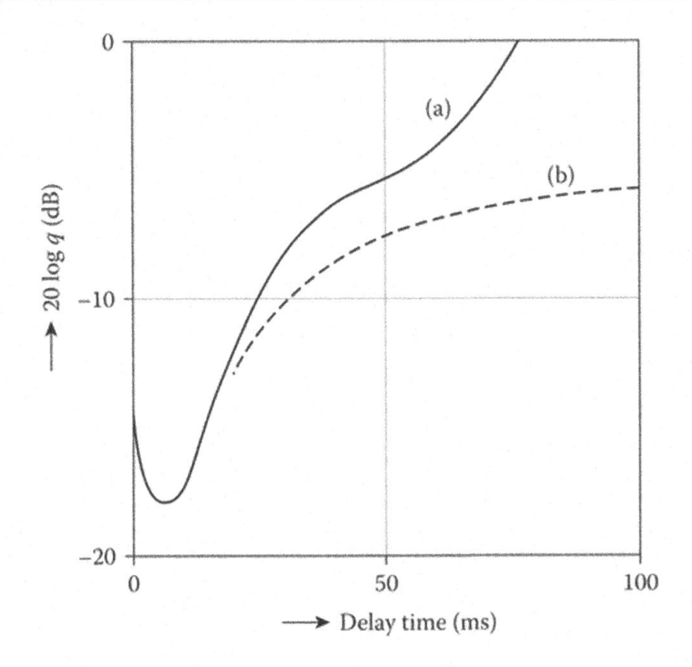


*Figure 7.10* Critical values of the factor $q$ resulting in just audible colouration of white noise passed through comb filter based on (a) Equation 7.6 and (b) Equation 7.8.

Source: Atal et al. (1962).

In Section 8.3, we shall discuss an objective criterion for the perceptibility of sound colouration or of a flutter echo which is based on thresholds of the kind shown in Figure 7.10.

## 7.5 PARAMETERS CHARACTERIZING THE 'EARLY ENERGY'

From the somewhat-artificial impulse responses we have considered in the past section, we now turn to the impulse responses of real rooms which consist of countless components exposing a broad spectrum of strengths and delays. As may be seen from Figure 4.8, they show the general tendency of decreasing strength with growing delay. This is easy to understand, because 'later' reflections have undergone more wall reflections, each of which causes some energy loss. Furthermore, the average rate of reflections grows in proportion to the square of the delay time. Regarding the subjective effects of the reflections, one has to realize that all of them are strongly interwoven because of complicated masking effects. Therefore, instead of studying the contributions of single reflections to the acoustical overall impression of a room, one has tried to condense the huge amount of information into certain parameters which summarize the situation in a more compact but nevertheless subjectively meaningful way. The result is a number of parameters which are generally believed to be related to particular acoustical aspects of a room.

In the preceding sections, it was shown that a reflection is not perceived subjectively as a separate event as long as its delay and its relative strength do not exceed certain limits. Its only effect is to make the sound source appear somewhat more extended and to increase the apparent loudness of the direct sound. Thus, these 'early reflections' give support to the sound source and improve the intelligibility of speech presented in that room. Reflections which reach the listener with longer delays are noticed as echoes in unfavourable cases; in favourable cases, they contribute to the reverberation of the room. In principle, any reverberation has the tendency to smooth the time structure of a signal and to mix up the spectral characteristics of successive sounds. This may be highly desirable for the presentation of music; at the same time, it impairs the intelligibility of speech.

Most of the following criteria characterize in some way the relative energy contained in the early part of the impulse response.

### 7.5.1 Definition

The first attempt to assess a proper value of the amount of 'early energy' is due to Meyer and Thiele (1956), who introduced a parameter named *definition* (originally *Deutlichkeit*). It is directly derived from the impulse response *g(t)*:

$$D = \left[ \int_0^{50\,\mathrm{ms}} [g(t)]^2 \, dt \Big/ \int_0^{\infty} [g(t)]^2 \, dt \right] \cdot 100\% \tag{7.9}$$

Both integrals must include the direct sound, the arrival of which at the observer determines the time $t = 0$. Obviously, $D$ will be 100% if the impulse response does not contain any components with delays in excess of 50 ms.

To validate $D$ as a descriptor of speech intelligibility, one has to correlate its values with the result of articulation tests, carried out in the same room with the same arrangement of the sender and the receiver. Such tests can be performed in the following way: A sequence of meaningless syllables ('logatoms') is read aloud in the environment under test. To obtain representative results, it is advisable to use phonetically balanced material (from so-called 'PB lists') for this purpose, that is, sets of syllables with properly distributed initial consonants, vowels, and final consonants. Listeners placed at various positions are asked to write down what they have heard. The percentage of syllables which have been correctly understood is considered to be a relatively reliable measure of speech intelligibility, called 'syllable intelligibility'.

The relationship between definition and the syllable intelligibility is shown in Figure 7.11. The plotted values of $D$ have been obtained by averaging over the frequency range 340–3,500 Hz. Obviously, the objective measure definition $D$ is highly correlated with the intelligibility and hence is a useful descriptor of the latter.

### 7.5.2 Clarity index

A quantity which is formally similar to definition but intended to characterize the transparency of music in a concert hall is the clarity index $C$ (originally *Klarheitsmaß*), as introduced by Reichardt et al. (1974). It is defined by

$$C = 10 \log_{10} \left[ \int_0^{80ms} [g(t)]^2 \, dt \Big/ \int_{80ms}^{\infty} [g(t)]^2 \, dt \right] \mathrm{dB} \tag{7.10}$$

The higher limit of delay time (80 ms compared with 50 ms in Equation 7.9) makes allowance for the fact that, with music, a reflection from a room wall is less perceptible than it is with speech signals. By subjective tests with synthetic sound fields, these authors have determined the values of $C$ preferred for the presentation of various styles of orchestral

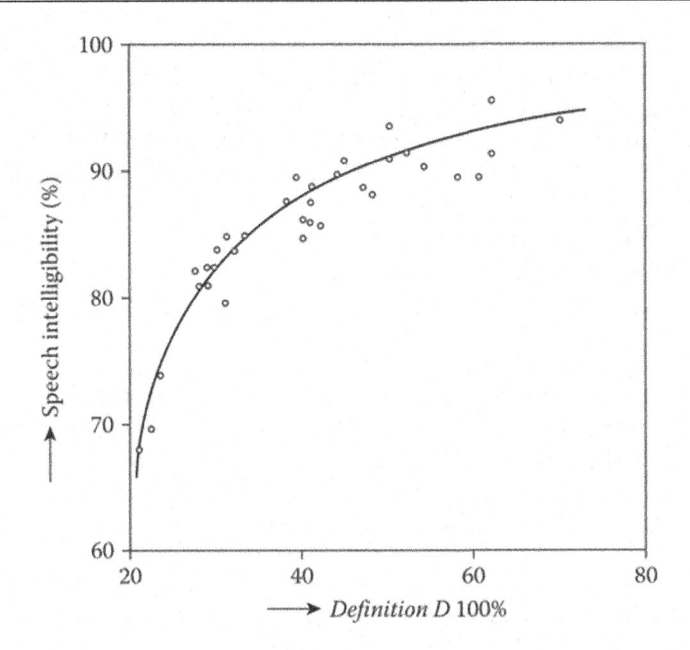


*Figure 7.11* Relationship between syllable intelligibility and *definition*.

music. They found that C = 0 dB indicates that the subjective clarity is quite good even for fast musical passages, whereas a value of C = −3 dB seems to be still tolerable. Nowadays, C (often referred to as C80 or $C_{80}$) is widely accepted as a useful criterion for the clarity and transparency of musical sounds in concert halls. According to an investigation of concert halls in Europe and the USA carried out by Gade (1994), its typical range is from about −5 to +3 dB. Bradley (1986) found that $C_{80}$ is, too, a reliable descriptor of speech intelligibility.

The assumption of a sharp delay limit separating useful from non-useful reflections is certainly a crude approximation to the way in which repetitions of sound signals are processed by our hearing. From a practical point of view, it has the unfavourable effect that, in critical cases, a small change in the arrival time of a strong reflection may result in a significant change in *D* or *C*. Therefore, several authors have proposed a gradual transition from 'early' to 'late' reflections by calculating the useful energy with a continuous weighting function $a(t)$:

$$E_u = \int_0^\infty a(t)\cdot\left[g(t)\right]^2 \mathrm{d}t \tag{7.11}$$

For a linear transition, $a(t)$ is given by

$$a(t) = \begin{cases} 1 & \text{for } 0 \le t < t_1 \\ \dfrac{t_2 - t}{t_2 - t_1} & \text{for } t_1 \le t \le t_2 \\ 0 & \text{for } t > t_2 \end{cases} \tag{7.12}$$

A reasonable choice of $t_1$ and $t_2$ would be about 35 and 100 ms, for example.

### 7.5.3 Centre time

No delay limit whatsoever is involved in the 'centre time', which was proposed by Kürer (1969) and is defined as the first moment of the squared impulse response:

$$t_c = \frac{\int_0^\infty [g(t)]^2 t \, dt}{\int_0^\infty [g(t)]^2 \, dt} \tag{7.13}$$

Obviously, a reflection contributes more to $t_s$ the longer it is delayed with respect to the direct sound. High transparency or speech intelligibility is indicated by low values of the centre time $t_c$, and vice versa. The high (negative) correlation between measured values of $t_c$ and intelligibility scores is demonstrated in Figure 7.12.

For speech, the centre time should not exceed about 80 ms, while for music, values of about 150 ms are still tolerable.

### 7.5.4 Speech transmission index

Quite a different approach to quantifying the speech intelligibility from objective sound field data is based on the modulation transfer function (MTF) already introduced in Section 5.6. It describes the smoothing effect of reverberation on the envelope of speech signals as well as that on their spectral components. For strictly exponential sound decay with a reverberation time $T = 6.91\,\delta$, the complex MTF reads (see Equation 5.40)

$$m = \frac{1}{1 + i\Omega / 2\delta} \tag{7.14}$$

where $\Omega$ is the angular modulation frequency. The absolute value of $m(\Omega)$,

$$|m(\Omega)| = \left[1 + \left(\frac{\Omega T}{13.8}\right)^2\right]^{-1/2} \tag{7.15}$$

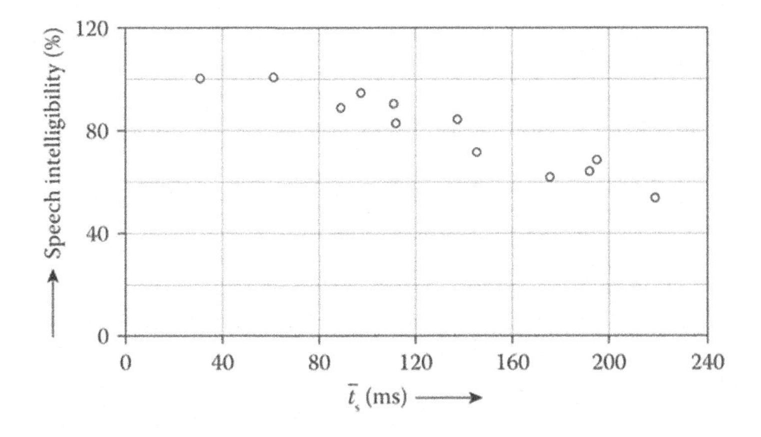


*Figure 7.12* Relationship between speech intelligibility and centre time $t_c$.
Source: Kürer (1969).

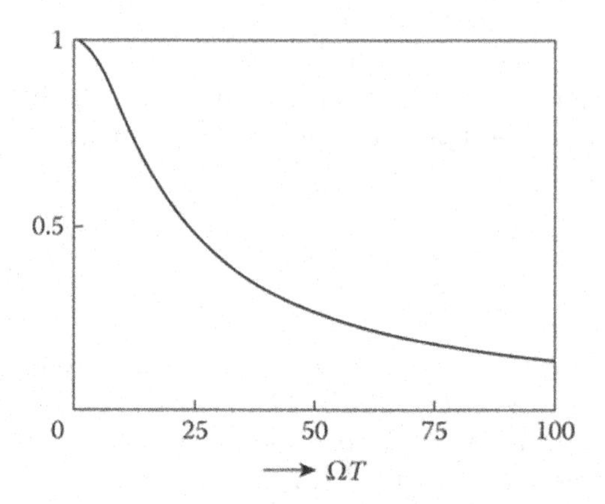


*Figure 7.13* Modulation transfer function (absolute value) of a room with exponential sound decay ($T$ = reverberation time; $\Omega$ = angular modulation frequency).

is plotted in Figure 7.13. It shows that very slow variations in a signal's envelope are not levelled out to any noticeable extent, but very rapid fluctuations are almost completely eliminated by the reverberant tail. However, real sound decay usually does not follow a simple exponential law; hence, the MTFs differ more or less from that shown in Figure 7.13. Furthermore, the MTF depends on the frequency or, more generally, on the spectral composition of the sound signal.

Houtgast and Steeneken (1973) and Steeneken and Houtgast (1980) have developed a practical procedure to convert numerous MTF data measured in 7 octave bands and at 14 modulation frequencies into one single figure of merit, which they called the 'speech transmission index' (STI). For collecting these data, random noise is used as a test signal which is filtered with octave filters, the mid-frequencies of which are ranging from 125 to 8,000 Hz; the filtered noise is modulated with 14 different modulation frequencies. From the modulation indices measured in this way, the final quantity is obtained by forming a weighted average. Houtgast and Steeneken have shown in numerous experiments that STI is very closely related to the results of articulation tests carried out with various types of speech signals (see Figure 7.14). Obviously, to guarantee sufficient speech intelligibility, STI must be at least 0.5.

A simpler and less time-consuming version of this criterion is the 'rapid speech transmission index', RASTI (Houtgast and Steeneken 1984). It is obtained by applying only four modulation frequencies in the octave band centred at 500 Hz and five modulation frequencies in the 2,000 Hz octave band. The 13 modulation frequencies range from 0.7 to 11.2 Hz. Each of the nine values of the modulation index $m$ is converted into an 'apparent signal-to-noise ratio':

$$(S/N)_{app} = 10\log_{10}\left(\frac{m}{1-m}\right) \tag{7.16}$$

These figures are averaged after eliminating those exceeding the range of ±15. The final parameter is obtained by forming the average $\left(\overline{S/N}\right)_{\text{app}}$ and normalizing it in such a way that the result is a figure between 0 and 1:

$$\text{RASTI} = \frac{1}{30}\left[\left(\overline{S/N}\right)_{\text{app}} + 15\right] \tag{7.17}$$

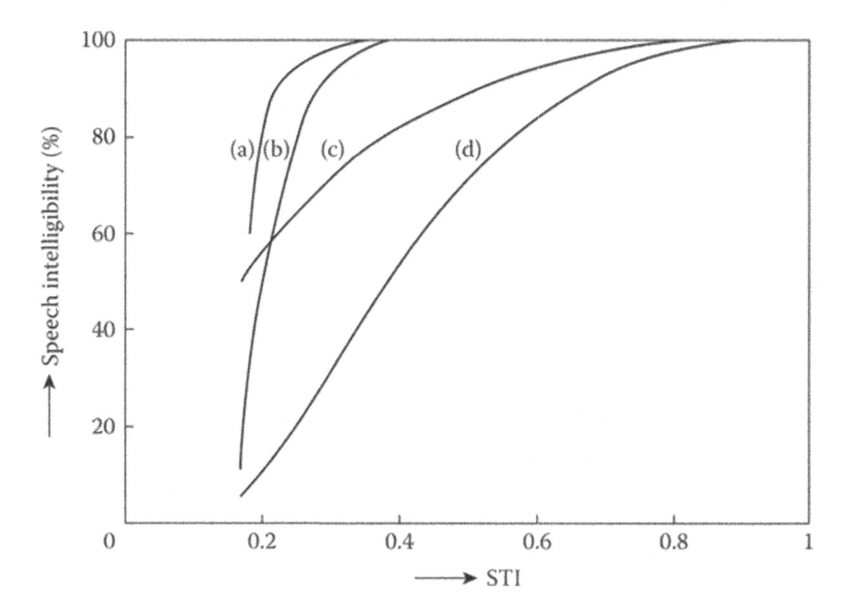


*Figure 7.14* Relationship between the speech transmission index (STI) and speech intelligibility, the latter obtained with (a) numbers and spell alphabet, (b) short sentences, (c) diagnostic rhyme test, and (d) logatoms.

Source: Houtgast and Steeneken (1973, 1984).

*Table 7.1* Relationship between speech transmission quality and the RASTI values

| *Quality score* | *RASTI* |
|---|---|
| Bad | <0.32 |
| Poor | 0.32–0.45 |
| Fair | 0.45–0.60 |
| Good | 0.60–0.75 |
| Excellent | >0.75 |

For practical RASTI measurements, both octave bands are emitted simultaneously, each with a complex power envelope containing five modulation frequencies. The automated analysis of the received sound signal is performed in parallel. With these provisions, it is possible to keep the duration of one measurement as low as about 12 s. Table 7.1 shows the relationship between five classes of speech quality and corresponding intervals of RASTI values.

We conclude this subsection with two remarks which more or less apply to all the preceding criteria. First, it is evident that they are highly correlated among each other. If, for example, a particular impulse response is associated with a short centre time $t_s$, its evaluation according to Equation 7.9 will yield a high value of *definition D*, and vice versa. Therefore, there is no need to measure more than one or two of them.

Secondly, if the sound decay in the room under consideration would strictly obey an exponential law, all the parameters defined earlier could be directly expressed by the reverberation time, as has already been done in Equation 7.15. Hence, they would not yield any information beyond the reverberation time. In real situations, however, the exponential law is a useful but nevertheless crude approximation to a much more complicated decay process. Especially in its early portions, the impulse response of a room is anything but a smooth time function (see, for example, the upper part of Figure 8.7). Furthermore, the pattern of reflections usually varies from one observation point to another; accordingly, these parameters, too, may vary over a wide range within one hall and are quite sensitive to geometrical and acoustical details of a room. Therefore, they are well suited to describe differences of listening conditions at different seats in a hall, whereas the reverberation time does not significantly depend on the place where it has been measured.

### 7.5.5 Support

For musicians performing in a concert hall, the demands on the acoustical conditions are different from those of the listeners in the audience. In the first place, musicians want to get some acoustic response of the hall, which is a matter of reverberance, as will be discussed in the next section. Furthermore, they need acoustical contact with their co-players and acoustical support by their environment, that is, by the stage, its ceiling, and the walls. Both factors are indispensable for good ensemble playing, which, in turn, is needed for correct synchronism and helps achieve correct intonation and balance.

According to Gade (1989), an objective parameter for this condition is what he calls the 'early support', defined by

$$ST_e = 10\log_{10}\left[\int_{20\ ms}^{100\ ms}[g(t)]^2\,dt \Big/ \int_0^{\Delta t}[g(t)]^2\,dt\right]\text{dB} \tag{7.18}$$

As earlier, $g(t)$ is the impulse response of the room measured at several places on the stage with a distance of 1 m between the sound source and the receiver. The time interval $\Delta t$ must be chosen so that the second integral comprises the direct sound but no reflections whatsoever. Apart from this, Gade defined also a quantity named 'late support', the definition of which is similar to Equation 7.18, with the difference that the limits of the first integral are 100 and 1,000 ms. Both support parameters are measured in the four octaves: 250, 500, 1,000, and 2,000 Hz. From this definition, it follows immediately that the boundaries surrounding the stage and also the ceiling should be made of reflecting material, which means the surface of these boundaries must not be porous and their weight should be sufficiently high so that even low-frequency sound components will be reflected (see Equation 6.3). This is particularly important for temporarily used orchestra shells. Typical values of $ST_e$ in existing concert halls are in the range of −15 to −10 dB.

## 7.6 REVERBERATION AND REVERBERANCE

If we disregard all details of the impulse response of a room, we finally arrive at the general decay the sound energy undergoes after an impulsive excitation or after a sound source has abruptly been stopped. As discussed in Chapters 4 and 5, the duration of this decay is characterized by the reverberation time or decay time, at least if the energy decay obeys an exponential law in its gross appearance. A more general notion which is rather to characterize the subjective aspect of reverberation is 'subjective reverberation time' or 'reverberance'. However, there is no sharp distinction between both these concepts. Clearly, the reverberance of an environment depends not only on its reverberation time but also on other features. This can be illustrated by Figure 7.15: it shows two schematic decay processes.

Both of them indicate low reverberance, although for different reasons. In Figure 7.15a, the reason is the short decay time. In case of Figure 7.15b, the cause of low reverberance is the low level at which the decay onsets, or in other words, the decay process is masked by the strong direct sound.

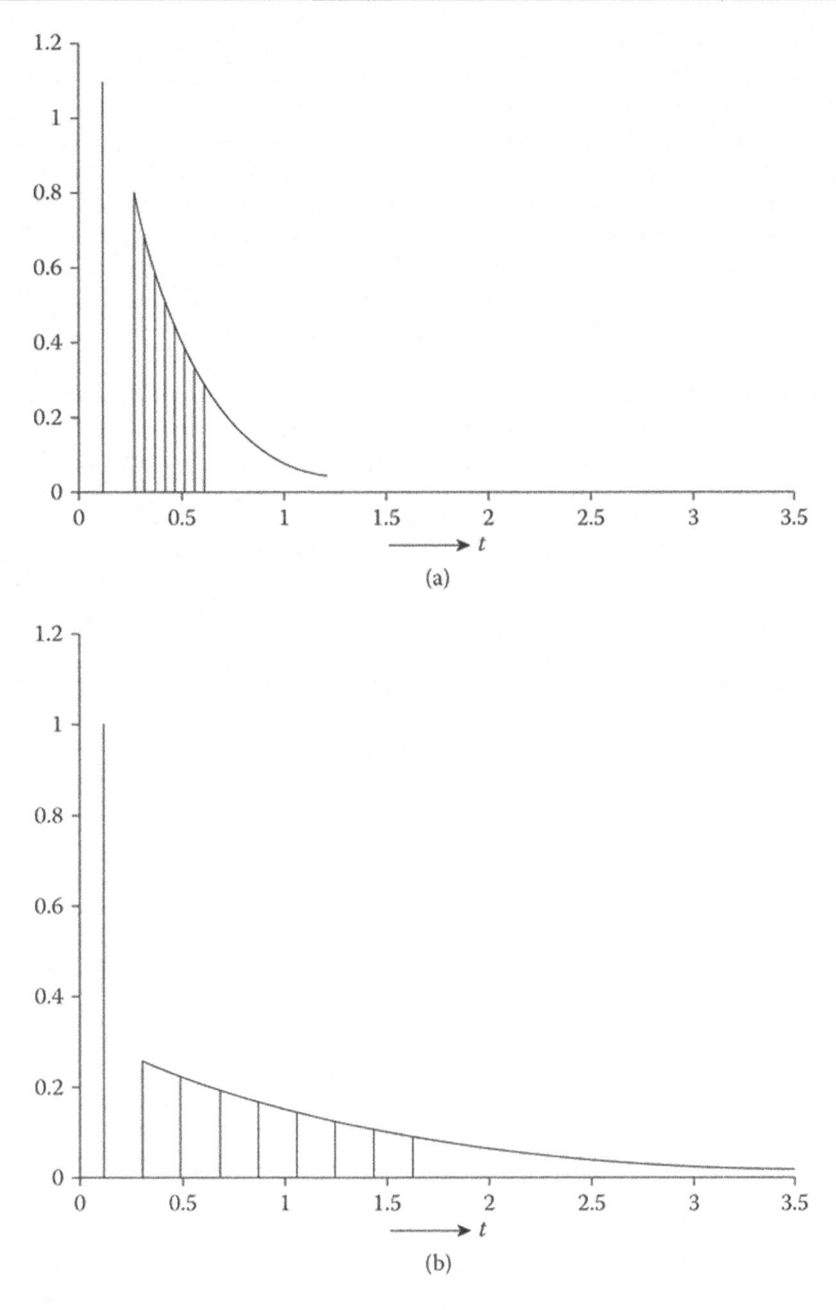


*Figure 7.15* Two reflection diagrams (schematic) indicating low reverberance: (a) short decay time and (b) low level of reflections.

Historically, the outstanding role of reverberation was first recognized by Sabine (1923), whose famous investigations, carried out during the last years of the 19th century, mark the origin of modern room acoustics. Sabine defined the reverberation time and developed several methods to measure it, and he was the first to formulate the laws of reverberation. Furthermore, he investigated the sound-absorbing power of numerous materials.

Nowadays, the reverberation time (or decay time) is still considered as the most important objective quantity in room acoustics, although it has been evident for some time that it characterizes only one particular aspect of the sound field and needs to be supplemented by additional data if a full description of the listening conditions is to be obtained. This predominance of reverberation time as an acoustical quality criterion has at least three reasons. First, it can be measured and predicted with reasonable accuracy and moderate expenditure. Secondly, the reverberation time of a room does not depend significantly on the observer's position in a room, a fact which is also underlined by the simple structure of the formulae by which it can be calculated from room data (see Section 5.1). Hence, it is well suited to characterize the overall acoustic properties of a hall, neglecting details which may vary from one place to another. And finally, abundant data on the reverberation time of existing halls are available nowadays, including their frequency dependence. They can serve as a yardstick to tell us which values are generally accepted and which are not.

Before discussing the important question of which reverberation times are desirable or optimal for the various types of rooms and halls, a remark on the just audible differences in reverberation time, that is, on the difference limen of reverberation time, may be in order. By presenting exponentially decaying noise impulses with variable decay times, bandwidths, and centre frequencies to numerous test subjects, Seraphim (1958) found that the relative difference limen for decay time is about 4%, at least in the most important range of decay times. Although these results were obtained under somewhat artificial conditions, they show at least that there is no point in giving reverberation times with a greater accuracy than about 0.05 or 0.1 s.

In principle, preferred ranges of the reverberation time can be determined by systematic listening tests, that is, by presenting speech or music samples to a test audience. To lead to meaningful results, such tests should be performed, strictly speaking, in environments (real enclosures or synthetic sound fields) which allow variations of the reverberation time under otherwise-unchanged conditions. Just comparing sound recordings from different halls may lead to interesting results which, however, are not very reproducible.

A more empirical approach consists of collecting the reverberation times of halls, which are generally considered as acoustically satisfactory or even excellent for the purpose they have to serve (lectures, drama theatre, operatic performances, orchestra or chamber music, etc.). It should be noticed, however, that subjective opinions on acoustical qualities, and hence the conclusions drawn from them, are afflicted with several factors of uncertainty, such as the question of who is eligible for constructive criticism in this field. Certainly, musicians have the best opportunity of comparison, since they often perform in different concert halls. However, musicians have a very special standpoint (meant literally as well as metaphorically), which does not necessarily agree with that of a typical listener. On the other hand, acousticians and sound recording engineers usually have a professional attitude towards acoustical matters and may frequently concentrate their attention on special properties which are insignificant to the 'average' concert listeners. The latter, however, for example, the concert subscribers, often lack the opportunity to compare different concert halls, or else, they are not very critically minded in acoustical matters, or their opinion is influenced from the point of view of local patriotism. Furthermore, there are — and, again, this applies particularly to musical events and their appropriate surroundings — individual differences in taste which cannot be discussed in scientific terms. And finally, it is quite possible that there are certain trends of 'fashion' towards longer or shorter reverberation times. All these uncertainties make it understandable that it is impossible to specify one single optimum value of reverberation time for each room type or type of presentation; instead, only ranges of favourable values can be set up.

We begin with rooms mainly used for speech, such as lecture rooms, congress halls, parliaments, theatres for dramatic performances, and so on. As mentioned earlier, no reverberation at all is required for such rooms in principle, since any noticeable sound decay has the tendency to blur the syllables and, thus, to reduce speech intelligibility. On the other hand, a highly absorbing treatment of all walls and of the ceiling of a room would not only remove virtually all the reverberation but, at the same time, would also prevent the formation of useful reflections which increase the loudness of the perceived sounds and which are responsible for the relative ease with which communication is possible in enclosures as compared to outdoor communication. Furthermore, the lack of any audible reverberation in a closed space creates an unnatural and uncomfortable feeling, as can be observed when entering an anechoic room, for instance. Obviously, one subconsciously expects to encounter some reverberation which bears a certain relation to the size of the room. For this reason, the reverberation time in rooms of this kind should not fall short of, say, 0.5 s, except for very small rooms, such as living rooms or small broadcasting or television studios. For larger rooms, such as drama theatres, values of about 1.2 s are still tolerable.

As is well known, low-frequency signal components contribute very little to speech intelligibility. Therefore, it is advisable to provide for sufficient low-frequency absorption by applying suitably designed absorbers — typically resonance absorbers (see Section 6.3) — to the boundary of such rooms in order to reduce the reverberation time and also the stationary sound level at low frequencies.

Now, we shall turn to the reverberation times, which are considered favourable for concert halls. In order to discover these values, we depend completely on subjective opinions concerning existing halls, at least as long as there are no results available of systematic investigations with synthetic or simulated sound fields. As has been pointed out before, there is always some divergence in the opinions about a certain concert hall; furthermore, they are not always constant in time. Old concert halls particularly are often commented on enthusiastically, probably more than is justified by their real acoustical merits. (This is true especially for those halls which were destroyed by war or other catastrophes.) In spite of all these reservations, it is a matter of fact that certain concert halls enjoy a high reputation for acoustical reasons. This means, among other things, that at least their reverberation time does not give cause for complaint. On the whole, it seems that the optimum values for occupied concert halls are in the range from about 1.6 to 2.1 s at mid-frequencies. Table 7.2 lists the reverberation times of several old and new concert halls, both for low frequencies (125 Hz) and for the medium-frequency range (500–1,000 Hz).

At first glance, it may seem curious that which is good for speech — namely, a relatively short reverberation time — should be bad for music. This discrepancy can be resolved by

*Table 7.2* Reverberation time of some concert halls, fully occupied

| *Name and location* | *Volume (m³)* | *Seating capacity* | *Year of completion (reconstruction)* | $T_{125}$ | $T_{500-1,000}$ |
|---|---|---|---|---|---|
| Großer Musikvereinssaal, Vienna | 14,600 | 2,000 | 1870 | 2.3 | 2.05 |
| St Andrew's Hall, Glasgow | 16,100 | 2,130 | 1877 | 2.1 | 2.2 |
| Chiang Kai Shek Memorial Hall, Taipei | 16,700 | 2,077 | 1987 | 1.95 | 2.0 |
| Symphony Hall, Boston | 18,800 | 2,630 | 1900 | 2.2 | 1.8 |
| Concertgebouw, Amsterdam | 19,000 | 2,200 | 1887 | 2.3 | 2.2 |
| Neues Gewandhaus, Leipzig | 21,000 | 1,900 | 1884 (1981) | 2.0 | 2.0 |
| Neue Philharmonie, Berlin | 24,500 | 2,230 | 1963 | 2.4 | 1.95 |
| Concert Hall De Doelen, Rotterdam | 27,000 | 2,220 | 1979 | 2.3 | 2.2 |

bearing in mind that, when listening to speech, we are interested in perceiving each element of the sound signal, since this increases the ease with which we can understand what the speaker is saying. When listening to music, it would be rather disturbing to hear every detail, including the bowing noise of the string instruments or the airflow noise of flutes. Furthermore, it is impossible to achieve perfect synchronism among the various players of an orchestra, let alone differences in intonation. It is these inevitable imperfections which are hidden or masked by reverberation, at least up to a certain degree. What is even more important, reverberation of sufficient duration improves the blending of musical sounds and increases their loudness and richness as well as the continuity of musical line. The importance of all these effects for musical enjoyment becomes obvious if one listens to music in an environment virtually free of reverberation, for example, to a military band or a light orchestra playing outdoors: the sounds are brittle and harsh, and it is obvious that it is of no advantage to be able to hear every detail. Furthermore, the loudness of music heard outdoors is reduced rapidly as the distance increases from the sound source.

But perhaps the most important reason that relatively long reverberation times are adequate for music is simply the fact that listeners are accustomed to hearing music in environments which happen to have reverberation times of the order of magnitude mentioned. This applies equally well to composers who unconsciously take into account the blending of sounds which is produced in concert halls.

As regards the frequency dependence of the reverberation time, it is generally considered tolerable, if not as favourable, to have an increase in the reverberation time towards lower frequencies, beginning at about 500 Hz. (This can also be seen from Table 7.2.) From the physical point of view, such an increase is quite natural, since the sound absorption by the audience is generally lower at low frequencies than it is at medium and high frequencies (see, for example, Tables 6.3 and 6.4). Concerning the subjective sensation, it is often believed that a slightly increased reverberation time at low frequencies is responsible for what is called the 'warmth' of musical sounds. On the other hand, there are quite a number of concert halls without this increase or even with a slightly lower reverberation time and which are nevertheless considered to be excellent acoustically.

The optimum range of reverberation time as indicated earlier refers to the performance of orchestral and choral music. Smaller ensembles perform often in halls having seating capacities of 400–700 and reverberation times ranging from 1.5 to 1.7 s (Hidaka and Nishihara 2004). These shorter decay times are not only dictated by the smaller size of a chamber music hall but are also more adequate to the character of the kind of music performed in it.

The dependence of optimum reverberation time on the style of music has been investigated by Kuhl (1954) in a remarkable round robin experiment. In this experiment, three different pieces of music were recorded in many concert and broadcasting studios with widely varying reverberation times. These recordings were later replayed to a great number of listeners — musicians as well as acousticians, music lovers and music historians, recording and other engineers, all in all, individuals who were regarded as competent in that field in one way or another. They were asked to indicate whether the reverberation times in the different recordings, whose origin they did not know, appeared too short or too long. The pieces of music played back were the first movement of Mozart's Jupiter Symphony, the fourth movement of Brahms's 4th Symphony, and the Danse Sacrale from Stravinsky's Le Sacre du Printemps. The final result was that a reverberation time of 1.5 s was considered to be most appropriate for the Mozart Symphony as well as for the Stravinsky piece, whereas 2.1 s was felt to be most suitable for the Brahms Symphony. In the first two pieces, there was almost complete agreement in the listeners' opinions; in the Brahms Symphony, however, there was considerable divergence of opinion.

*Table 7.3* Reverberation time of some opera theatres, fully occupied

| *Name and location* | *Volume ($m^3$)* | *Seating capacity* | *Year of completion (reconstruction)* | $T_{125}$ | $T_{500-1,000}$ |
|---|---|---|---|---|---|
| La Scala, Milano | 10,000 | 2,290 | 1778 (1946) | 1.5 | 1.2 |
| Covent Garden, London | 10,100 | 2,180 (60) | 1858 | 1.2 | 1.1 |
| Festspielhaus, Bayreuth | 11,700 | 1,800 | 1876 | 1.7 | 1.5 |
| National Theater, Taipei | 11,200 | 1,522 | 1987 | 1.6 | 1.4 |
| Staatsoper, Wien | 11,600 | 1,658 (560) | 1869 (1955) | 1.5 | 1.3 |
| Staatsoper, Dresden | 12,500 | 1,290 | 1878 (1985) | 2.3 | 1.7 |
| Neues Festspielhaus, Salzburg | 14,000 | 2,158 | 1960 | 1.7 | 1.5 |
| Metropolitan Opera House, New York | 30,500 | 3,800 | 1966 | 2.25[a] | 1.8[a] |

[a] With 80% occupancy.

These results should certainly not be overemphasized, since the conditions under which they have been obtained were far from ideal in that they were based on monophonic recordings, replayed in rooms with some reverberation. But they show clearly that no hall can offer optimum conditions for all types of music, and they may explain the large range of 'optimum' reverberation times for concert halls.

In opera houses, the listener should be able to enjoy the full sound of music as well as to understand the text, at least partially. Therefore, one would expect that these somewhat-contradictory requirements can be reconciled by a compromise as far as the reverberation time is concerned, and that, consequently, the optimum of the latter would be somewhere about 1.5 s. As a matter of fact, however, the reverberation times of well-renowned opera theatres scatter over a wide range (see Table 7.3). Traditional theatres have reverberation times close to 1 s only, whereas more modern ones show a definite trend towards longer values. One is tempted to explain these differences by a changed attitude of the listeners, who, nowadays, seem to give more preference to a full and smooth sound of music than to the intelligibility of the text, whereas earlier opera goers presumably just wanted to be entertained by the plot. This trend is supported by the tendency to perform operas in the original language and to display the translated text on a projection board. There is still another possible reason for the tendency towards longer reverberation times: old theatres were designed in such a way as to seat as many spectators as possible, whereas the architects of more modern ones (including the famous Festspielhaus in Bayreuth, which was specially designed to stage Richard Wagner's operas) tried to follow more or less elaborate acoustical concepts.

The rehearsal room of an orchestra cannot be expected to offer the same acoustical conditions as the hall where the orchestra performs. It has typically a volume of 1,500–2,500 m$^3$, and its reverberation time should not exceed about 1.2 s in order to ensure transparency of the produced sounds and to keep the sound level within tolerable limits. By providing for variable sound absorption in the form of curtains, absorbing or reflecting screens, and so on, musicians should be given the opportunity to do some experimentation in order to find the optimum conditions themselves.

The question of optimum reverberation times is even more difficult to answer if we turn to churches and other places of worship which cannot be considered merely under the heading of acoustics. It depends on the character of the service whether more emphasis is given to organ music and liturgical chants or to the sermon. In the first case, longer reverberation times are to be preferred, but in the latter, the reverberation time should certainly not exceed 2 s. Churches with still shorter reverberation times are often not well accepted by the congregation for reasons which have nothing to do with acoustics. This shows that the churchgoers' acoustical expectations are not only influenced by rational arguments, such as that of speech intelligibility, but also by tradition.

As mentioned earlier, the reverberation time is a meaningful measure for the duration of the decay process if the latter is exponential, that is, if the decay level decreases linearly with time. If, on the contrary, a logarithmic decay curve is bent and, consequently, each section of it has its own decay rate, the question arises as to which of these sections is most significant for the 'reverberance' of a room.

To answer this question, Atal et al. (1965) passed speech and music samples through an artificial reverberator consisting of a combination of computer-simulated comb filters (see Section 10.5) with adjustable, non-exponential decay characteristics. The signals modified in this way were presented with earphones to test subjects, who were asked to compare them with exponentially reverberated signals in order to find the subjectively relevant decay rate, that is, the reverberance of the non-exponential decays. It turned out that the reverberance of a sound field is highly correlated with the 'initial reverberation time', which corresponds to the slope of a decay curve during the first 160 ms.

These findings can be explained by the fact that the smoothing effect of reverberation on the irregular level fluctuations of continuous speech or music is mainly achieved by the initial portion of the decay process, while its later portions add up to some general 'background', which is not felt subjectively as carrying much information on the signal. Only final or other isolated chords present the listener with the opportunity of hearing the complete decay process, but these chords occur too rarely for them to influence to any great degree the overall impression which a listener gains of the hall's reverberance.

Nowadays, it has become common to characterize the rate of initial sound decay by the 'early decay time' (EDT). This is the time in which the first 10 dB fall of a decay process occurs, multiplied by a factor 6. It is mostly shorter than the Sabine reverberation time. Listening tests based on binaural impulse responses recorded in different concert halls confirmed that the perceived reverberance is closely related to EDT.

Lee et al. (2011) investigated whether subjective reverberance corresponds better with loudness decay than with level decay. Time-dependent loudness includes both spectral masking and the temporal effects of vibrations in the cochlea. Corresponding models from psychoacoustics are available from the models developed by E. Zwicker and H. Fastl and by B. Moore and B. Glasberg which have been implemented in the two parts of the standard (ISO 532-1 2017; ISO 532-2 2017). This work has shown that loudness decay rates are, indeed, correlated to a higher degree with subjective reverberation. Furthermore, loudness decay rates also serve to analyze reverberation not only on the basis of impulse responses or decay curves but also on the basis of room response to music or speech signals. Accordingly, the subjective reverberance can be divided into two components, namely, the sound decay in the signal transients and its effect on signal modulation, and the decay at the end of the final chord. The overall reverberation time does not show substantial variations with room shape. This is so because the decay process as a whole is made up of numerous reflections with different delays, strengths, and wall portions where they originated. On the contrary, the 'early decay time' is strongly influenced by early reflections and, therefore, depends noticeably on a listener's position; furthermore, it is sensitive to details of the room's geometry. In this respect, it resembles, to some extent, the parameters discussed in the preceding sections.

## 7.7 SPACIOUSNESS OF SOUND FIELDS

The preceding discussions of this chapter predominantly referred to the temporal structure of the impulse response of a room and to the auditive sensations associated with them. A subjective effect not mentioned so far, which is nevertheless very important, at least for concert halls, is the acoustical 'sensation of space' which a listener usually experiences in a room. It is caused by the fact that the sound in a closed room reaches the listener from quite different directions and that our hearing, although not able to distinguish these directions, processes them into an overall impression, namely, the sensation or feeling of space.

It is evident that this sensation is not brought about just by reverberation. If music recorded in a reverberant room is replayed through a single loudspeaker in a non-reverberant environment, it never suggests acoustically the illusion of being in a room of some size, no matter if the reverberation time is long or short. Likewise, if the music is replayed through several loudspeakers which are placed at equal distance but in different directions seen from the listener, and which are fed by identical signals, the listener will not feel more enveloped by the sound. Instead, all the sound seems to arrive from a single imaginary sound source, a so-called 'phantom source', which can easily be located and which seems, at best, somewhat more extended than a single loudspeaker. The same effect occurs if the sound field is produced by a great number of loudspeakers arranged in an anechoic room over a hemisphere (see Figure 6.19) which are fed with the same electrical signal. A listener at the centre of the hemisphere does not perceive a 'spacious' or 'subjectively diffuse' sound field, but instead, he perceives a phantom sound source immediately overhead. Even the usual two-channel stereophony employing two similar loudspeaker signals, which differ in a certain way from each other, cannot provide a full acoustical impression of space, since the apparent directions of sound incidence remain restricted to the region between both loudspeakers.

The 'sensation of space' has attracted the interest of acoustic researchers for many years, but only since the late 1960s has real progress been made in finding the cause of this subjective property of sound fields. The different authors used expressions like 'spatial responsiveness', 'spatial impression', 'ambience', 'apparent source width', 'subjective diffusion', 'Räumlichkeit', 'spaciousness', 'listener envelopment', and others to circumscribe this sensation. Assuming that all these verbal descriptions signify about the same sensation, we prefer the term 'spaciousness' or 'spatial impression' in the following.

For a long time, it was common belief among acousticians that spaciousness was a direct function of the uniformity of the directional distribution (see Section 4.3) in a sound field: the more uniform this distribution, the higher the degree of spaciousness. This belief was supported by the fact that the ceiling and walls of many famous concert halls are highly structured by cofferings, niches, pillars, statuettes, and other projections which supposedly reflect the sound in a diffuse manner rather than specularly.

It was the introduction of synthetic sound fields as a research tool which led to the insight that the uniformity of the stationary directional distribution is not the primary cause of spaciousness. According to Damaske (1967), spatial impression can be created with quite a few synthetic reflections, provided they reach the listener's head from lateral directions, and that the signals they carry are mutually incoherent (as is usually the case in a large hall).

That only the lateral reflections (i.e. not reflections from the front, from overhead, or from the rear) contribute to spaciousness has been observed by several earlier authors, and many subsequent publications have confirmed this finding. A very extensive investigation into spaciousness has been carried out by Barron (1971). He found that only a reflection with a delay time in the range from 5 to 80 ms contributes to spaciousness and that its contribution is proportional to its energy and to cos $\theta$, where $\theta$ is the angle between the direction of sound incidence and an imagined axis through the ears of a listener who is facing the sound source. Furthermore, its effect is independent of other reflections and of the presence or absence of reverberation. Based on these results, the 'lateral energy fraction', LEF (Barron and Marshall 1981) was proposed as an objective measure for the spatial impression:

$$LEF = \int_{5ms}^{80ms} \left[g(t)\cos\theta\right]^2 dt \Big/ \int_{5ms}^{80ms} \left[g(t)\right]^2 dt \tag{7.19}$$

(In the original definition of LEF, the numerator contains the factor cos θ in the first power only. This modification is a concession to the ease of practical measurements; its influence on the result is insignificant.) The LEF is usually averaged over four octave bands with mid-frequencies 125, 250, 500, and 1,000 Hz. This accounts for the fact that the low- and mid-frequency components contribute most to the sensation of spaciousness. Theoretically, the LEF may vary from 0 (only frontal sound incidence, $\theta = 0$) to 1 (sound incidence exactly from the side $\theta = \pm 90°$). In most halls, the LEF lies in the range of 0.1 to 0.3.

Another method of characterizing the laterality of reflected sounds makes use of the fact that sound impinging on a listener's head from its vertical symmetry plane will produce equal sound pressures at both his ears, whereas a sound wave from outside the symmetry plane will produce different ear signals. Generally, the similarity or dissimilarity of two signals is measured by their cross-correlation function or by the correlation coefficient as defined in Equation 1.59. In the following, we apply that expression to the impulse responses $g_r$ and $g_l$ measured at the right and the left ear, respectively. Since these signals are transient, time averaging would be meaningless and is therefore replaced by integration over the time. Then, the correlation coefficient reads

$$\Psi_{rl}(\tau) = \int_{t_1}^{t_2} g_r(t) g_l(t+\tau) dt \Big/ \left\{ \int_{t_1}^{t_2} \left[g_r(t)\right]^2 dt \int_{t_1}^{t_2} \left[g_l(t)\right]^2 dt \right\}^{1/2} \tag{7.20}$$

Usually, the limits $t_1$ and $t_2$ are set to 0 and 100 ms in order to restrict the integration over the arrival times of the 'early reflections'; hence, $\Psi_{rl}$ is a 'short time' correlation factor. The maximum of its absolute value within the range $|\tau| < 1$ ms is called the 'interaural cross-correlation', IACC (Damaske and Ando 1972):

$$\mathrm{IACC} = \max\left\{\left|\Psi_{rl}\right|\right\} \text{ within } -1 \text{ ms} < \tau < 1 \text{ ms} \tag{7.21}$$

Of course, this quantity is negatively correlated with the spatial impression, and high values of the IACC mark a low degree of spaciousness, and vice versa.

It should be mentioned that several versions of the IACC are in use which differ in the integration limits in Equation 7.20, or in the way the impulse responses are frequency-filtered before processing them according to Equation 7.20. Beranek (1996) prefers the use of an IACC version which comprises three octave bands with mid-frequencies, 0.5, 1, and 2 kHz. This limitation accounts for the fact that at low frequencies, the difference between $g_r$ and $g_l$ is negligibly small. Beranek reports that the values of this quantity are as low as 0.3 in excellent concert halls.

Laterality of the early reflections is certainly the most decisive factor on which the impression of spaciousness depends, but it is not the only one. Several researchers have observed that the spatial impression increases monotonically with the listening level. Furthermore, there seems to be some agreement nowadays that spatial impression is not a 'one-dimensional' sensation but consists of at least two different components which are more or less independent. The most prominent of these are the 'apparent source width' (ASW) and 'listener envelopment' (LEV) (Bradley and Soulodre 1995). If this view is adopted, it is reasonable to attribute Damaske's and Barron's results to the ASW and, hence, to consider the objective quantities LEF and IACC as predictors of this particular aspect of spaciousness.

This leaves us with the question of which objective parameters of the sound field are responsible for the remaining component, namely, the sense of 'listener envelopment'. This question was the subject of careful listening experiments performed by Bradley and Soulodre (1995) using a synthetic sound field similar to that shown in Figure 7.2. With this system, the reverberation time, the early-to-late energy ratio expressed by the clarity index $C_{80} = C$ (see Equation 7.10), the $A$-weighted SPL, and the angular distribution of the reverberated signals could be independently varied, while the IACC and the LEF were kept constant through all combinations.

It turned out that the angular distribution had the largest effect on spaciousness. The wider it was, the higher is the LEV. Another strong influence on the perceived 'listener envelopment' was that of the overall sound level, while the relative amount of early energy ($C_{80}$) and the reverberation time was found to be less significant. A correlation analysis of the collected data revealed that the best objective predictor for LEV is the strength (see Equation 7.3) of the late lateral reflections

$$LG_{80}^{\infty} = 10\log_{10}\left\{\int_{80ms}^{\infty}\left[g(t)\cos\theta\right]^2 dt \Big/ \int_0^{t_d}\left[g_A(t)\right]^2 dt\right\} \text{dB} \tag{7.22}$$

averaged over four octave bands with mid-frequencies 125, 250, 500, and 1,000 Hz, thus accounting for the fact that high-frequency components do not contribute much to the sense of LEV. $g_A(t)$ is the impulse response received in the free field in a 10 m distance; $t_d$ is chosen such that the integral contains the direct sound. The range from missing to maximum envelopment corresponds to a range in $LG_{80}^{\infty}$ from –20 to 2 dB.

We return to the measures of ASW, namely, the LEF and the IACC. Surprisingly, both quantities are not very highly correlated to each other, which may be a consequence of the different frequency ranges in which they are usually determined. Another puzzling finding is that these parameters show considerable fluctuations with the position of the receiver (microphones or dummy head), as has been revealed by careful experiments carried out by de Vries et al. (2001). These fluctuations are caused by interfering components of the wave field and are not accompanied with corresponding variations in the subjectively perceived ASW. Even within one seat width, the fluctuations of the measured parameter are quite strong, although it is inconceivable that a listener in a concert hall will perceive any change in spatial impression when he moves his head by 10 or 20 cm. This shows us that still more research on significant spaciousness parameters is needed.

## 7.8 ASSESSMENT OF CONCERT HALL ACOUSTICS

Although nowadays quite a number of parameters are at the acoustician's disposal to quantify the listening conditions in a concert hall or particular aspects of them, the situation is unsatisfactory in that important questions are still open: Do these parameters yield a complete description of the acoustics of a hall? To what extent are they correlated with each other? Which relative weight is to be given to each of them? Is it possible, for instance, to compensate insufficient reverberation by a large amount of early lateral energy? And finally, is it possible to condense the values of these parameters into one single figure of merit?

Conventional attempts to correlate the acoustical quality of a concert hall with an objective measure or a set of such measures have not been very satisfactory because they have concentrated on one particular aspect only or were based on plausible but unproven assumptions. Furthermore, they relied on the listeners' memory, since the immediate comparison of different halls or positions in a hall was not possible at that time.

For assessing the acoustical quality of concert halls, the following steps should be carried out:

1. The hall under test is 'excited' either by real artists (an orchestra, a choir, or a chamber music ensemble) or by loudspeakers positioned on the stage. The second method is to be preferred since it is more flexible and allows testing the hall with defined and reproducible music samples. A still more refined method has been developed by Pätynen et al. (2008). To excite all halls to be tested with exactly the same music samples, these authors employed a 'loudspeaker orchestra' consisting of 34 calibrated loudspeakers placed at defined positions on the stage. These are fed with samples of orchestra music, the parts of which had been recorded separately one player at a time in an anechoic chamber. The sounds produced by the instruments are picked up with 22 microphones positioned evenly around the player. Since these recordings must be combined to the complete music sample afterwards, particular care must be taken to achieve correct synchronism and intonation. Quite a different way to prepare the test samples was chosen by Soulodre and Bradley (1995), who convolved anechoic music motifs with the measured impulse responses of the halls to be examined.
2. The opinions of qualified listeners about the acoustics of a hall or various perceptual aspects of it can be elicited *in situ*, for instance, by interviewing them after the performance or by handing formal questionnaires over to them. Alternatively (and preferably), the sounds at typical positions in the hall are picked up with microphones and digitally stored for later analysis.
3. In the latter case, the recorded signals are presented to listeners in an anechoic room. It is important that the test persons are carefully selected and have a certain familiarity with the sort of music which is usually performed in concert halls. The listeners are asked to quantify in some way the differences between successive presentations. This can be done by assigning scores to the replayed sound signals using a number of bipolar rating scales with verbally labelled extremes such as 'dull–brilliant', 'cold–warm', and so on. Another possibility is having the subjects indicate which of two presentations they prefer, and thus to set up a preference scale on the basis of these judgements.
4. Finally, the results are analyzed using modern psychometric methods. The aim of this procedure is to isolate from this material the relevant perceptual aspects and their relative significance.

It goes without saying that all electro-acoustic components, such as microphones, recorders, and loudspeakers, which are used for the aforementioned operations must be of excellent quality. Equally important is that the system is capable of transmitting and reproducing all subjective effects brought about by the complex structure of the original sound field. One way to achieve this is to use a binaural system which transplants the signals which a listener's ears would receive in the original sound field to the ears of a remote listener as correctly as possible. Very often, a 'dummy head' is used for recording the original sounds. This is an artificial head with microphones built into the ear channels. The purpose of this device, which usually includes the shoulders and part of the trunk, is to diffract the arriving sound waves in a way which is representative for the majority of human listeners. Nowadays, several types of dummy heads are available which meet this requirement. An example is shown in Figure 7.16. The signals recorded in this way are stored using a digital storage medium.

*Figure 7.16* Dummy head.

Source: Courtesy of Institute for Hearing Technology and Acoustics, RWTH Aachen University, Germany.

In principle, binaural signals could be replayed and presented to the test listeners by means of high-quality headphones. Unfortunately, this kind of reproduction is often plagued by 'in-head localization', that is, for some reason, the listener has the impression that the sound source is located within the rear part of his head. On the other hand, replacing the earphones by two loudspeakers, as in the usual stereophony, creates another problem, namely, that the right-hand loudspeaker will inevitably send a cross-talk signal to the left ear, and vice versa. This can be avoided by a filter which 'foresees' this effect and adds proper cancellation signals to the input signals. This technique, nowadays referred to as 'crosstalk cancellation' (CTC), was invented and first demonstrated by Schroeder and Atal (1963). Figure 7.17 depicts the principle of CTC. A more detailed description of this reproduction technique may be found in the work by Masiero (2012). It should be mentioned that the cancellation effect only occurs if the listener keeps his head at a fixed position and orientation. This limitation can be avoided by 'dynamical cross-talk cancellation' (Lentz and Behler 2004) — a method which employs four regularly arranged loudspeakers which can be combined into six different CTC configurations. At a time, only one of these configurations is in operation; their control is effected by a head-tracking device.

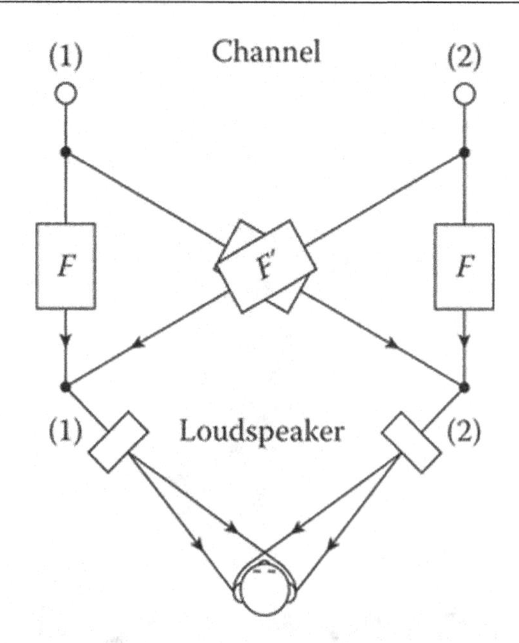


*Figure 7.17* Principle of cross-talk cancellation in loudspeaker reproduction of binaural signals. *F* and *F'* are digital filters.

In a final step, the judgements elicited from the test listeners are subjected to a mathematical procedure called factor analysis. The objective of this process is to extract the most prominent aspects of judgement from the experimental material, the so-called factors. In the following, an attempt is made to give at least an idea of this technique. A more detailed explanation of these techniques and their application to room-acoustical problems may be found in the book by Cremer and Müller (1982).

At first, the data are arranged in a matrix C consisting of *n* columns, which refer to the objects (hall, seats), and *m* rows containing the attributes (e.g. 'brilliance of sounds', or 'intimacy'):

$$\mathbf{Z} = \begin{pmatrix} z_{11} & z_{12} & z_{13} \ldots\ldots & z_{1n} \\ z_{21} & z_{22} & z_{23} \ldots\ldots & z_{2n} \\ \vdots & \vdots & \vdots \ldots\ldots & \vdots \\ z_{m1} & z_{m2} & z_{m3} \ldots\ldots & z_{mn} \end{pmatrix} \tag{7.23}$$

From this matrix, the 'correlation matrix' C is derived:

$$\mathrm{C} = \frac{1}{n-1}\mathbf{Z}\mathbf{Z}' \tag{7.24}$$

where **Z'** denotes the transposed matrix, that is, the matrix obtained from **Z** by interchanging lines and columns, and *n* is the number of 'objects' (e.g. of halls or different locations within a hall). The matrix C is quadratic, with $m \times m$ elements, where $m$ denotes the number of attributes. Next, the matrix C is orthogonalized, that is, one must find a coordinate transformation **W** — the principal axis transformation — by which the matrix C can be converted into an equivalent matrix **D**, the only non-zero elements of which appear on its main diagonal. According to the rules of matrix algebra, this transformation reads:

$$\mathbf{WCW}^{-1} = \mathbf{D} = \begin{pmatrix} \lambda_1 & 0 & 0 \ldots\ldots & 0 \\ 0 & \lambda_2 & 0 \ldots\ldots & 0 \\ \vdots & \vdots & \vdots \ldots\ldots & \vdots \\ 0 & 0 & 0 \ldots\ldots & \lambda_m \end{pmatrix} \quad (7.25)$$

The matrix $\mathbf{W}^{-1}$ is the inverse of **W**; the numbers $\lambda_1, \ldots, \lambda_m$ are the so-called eigenvalues of matrix **C**. By this operation, the space of the original attributes (or preferences) is transformed into a perceptual space, the Cartesian coordinate axes of which are the 'factors' $F_1$, $F_2, \ldots, F_m$, that is, independent or 'orthogonal' aspects of perception. Their relative significance is determined from the eigenvalues $\lambda_1, \ldots, \lambda_m$. Usually, the number of significant factors is 3 or 4. The meaning of these factors or scales is principally unknown beforehand, but sometimes they can be circumscribed vaguely by verbal labels such as 'resonance' or 'proximity'.

As a simple example, let us consider a perceptual space which consists of just two significant factors. This space is a plane with rectangular coordinates $F_1$ and $F_2$, as shown in Figure 7.18. If the preceding factor analysis was based on preference tests, the individual preference scale of each listener can be represented as a vector in this plane. The figure shows just one of them. The projections of the 'object points' A, B, C, etc. on this vector indicate this listener's personal preference rating of these halls. According to the angle $\alpha$ which it includes with the axis $F_1$, this particular listener gives in his preference judgement a weight of $\cos\alpha$ to the factor $F_1$ and of $\sin\alpha$ to the factor $F_2$. Similarly, if bipolar rating scales have been used, these scales can be presented as vectors in such a diagram.

Investigations of concert halls using bipolar rating scales have been carried out by several authors. Wilkens and Plenge (1974) and Wilkens (1977) collected 30 test samples by accompanying the Berlin Philharmonic Orchestra on a concert tour where it presented the same programme (Mozart, Bartok, Brahms) in six different halls. For the reproduction of the

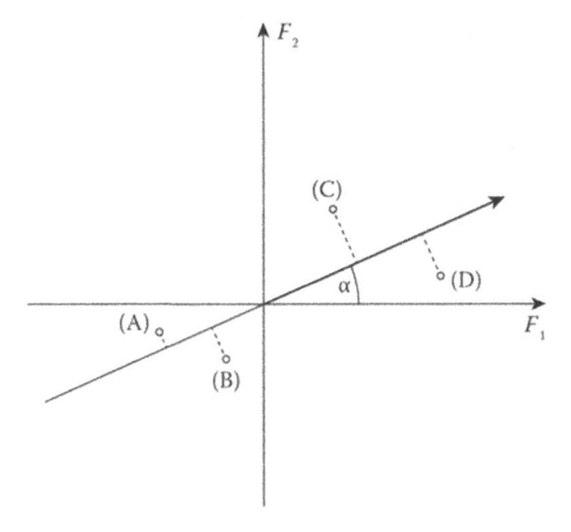


*Figure 7.18* Two-dimensional perceptual space with factors $F_1$ and $F_2$. The vector represents the preference scale of one particular listener. The points (A), (B), (C), and so on indicate different concert halls or seats.

samples recorded with a dummy head, Wilkens and Plenge used headphones and found three factors to be relevant, which they labelled as follows:

$F_1$: strength and extension of the source (47%)
$F_2$: clearness or distinctness (28%)
$F_3$: timbre (14%)

The numbers in parentheses denote the relative significance of the three factors. They add up to 89% only, which means that there are still more, although less significant, factors. It is interesting to note that there seem to be two groups of listeners: one group which prefers loud sounds (high values of $F_1$), and another one giving more preference to clearness and distinctness of sounds (high values of $F_2$). (A similar division of the subjects into two groups with different preferences has also been found in an investigation conducted by Barron [1988]). The members of the first group preferred 'reverberance', while the other group gave preference to 'intimacy'.)

In the course of Wilkens's research project, physical sound field parameters have been measured at the same positions in which the sound recordings have been made. A subsequent correlation analysis revealed that $F_1$ is highly correlated with the strength factor $G$ from Equation 7.1, whereas $F_2$ shows high (negative) correlation with Kürer's 'centre time' $t_s$, defined in Equation 7.13. Finally, the factor $F_3$ seems to be related to the frequency dependence of the 'early decay time'.

At about the same time, another group of researchers, namely, Schroeder et al. (1974), studied the acoustics of 22 European concert halls, using a different method to collect the test samples. They replayed a 'dry', that is, reverberation-free, motif of Mozart's Jupiter Symphony with two loudspeakers located onstage of the tested concert hall and re-recorded them with a dummy head placed at various listener positions. Afterwards, the samples prepared in this way were presented to the test subjects in an anechoic room at constant level employing a CTC system described earlier in this section. The subjects were asked to judge preference between pairs of presentations. They were allowed to switch back and forth between both recordings. Application of factor analysis indicated four factors, $F_1$–$F_4$, with relative significance of 45%, 16%, 12%, and 7%, respectively. In Figure 7.19, which is analogous to Figure 7.18, part of results by Siebrasse (1973) are plotted in the plane of factors $F_1$ and $F_2$. The vectors representing the individual preference scales have different lengths since they have non-vanishing components also in $F_3$ and $F_4$ directions. The fact that they all are directed towards the right side suggests the conclusion that $F_1$ is a 'consensus factor', whereas the components in the $F_2$ direction reflect differences in the listener's personal taste. This holds even more with regard to the factors $F_3$ and $F_4$.

In order to find the objective sound field parameter with the highest correlation to the consensus factor $F_1$, Gottlob (1973) analyzed the impulse responses observed at the same places where the music samples had been recorded. The reverberation time proved to be important only when it deviated substantially from an optimum range centred on 2 s. If halls with inadequate reverberation times are excluded, $F_1$ shows high correlation with several parameters, of which the IACC (see Section 7.7) seems to be of particular interest, since it is virtually independent of the reverberation time. Another highly correlated quantity is the width of the concert hall. Both the IACC and the width of the hall are negatively correlated with $F_1$, that is, narrow concert halls are generally preferred. This again proves the importance of early lateral reflections, which are particularly strong in narrow halls leading to a low value of the IACC.

There may be several possible reasons for the differences between the results of both studies as, for instance, the different ways of collecting the test samples and of presenting them to

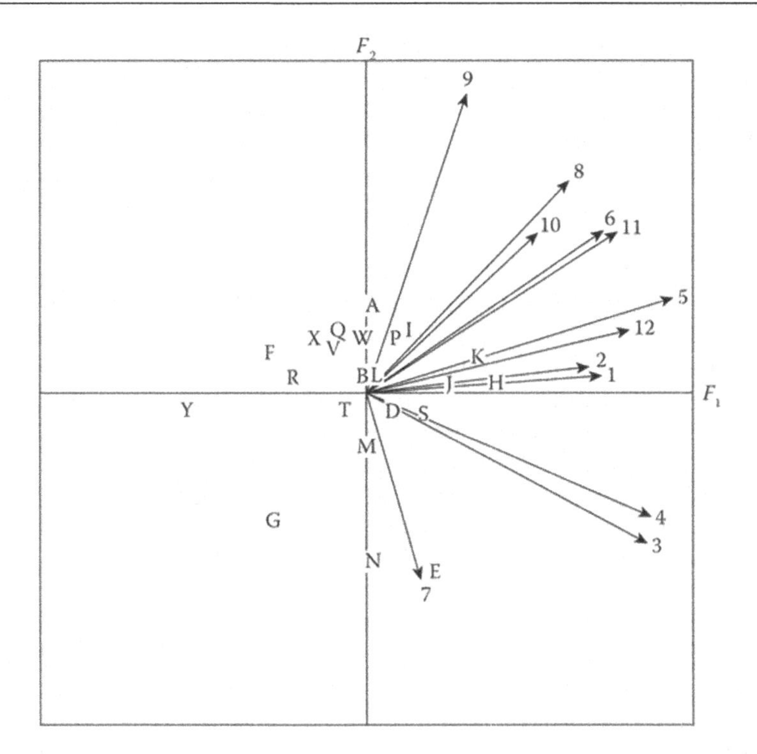


*Figure 7.19* Individual preference scales of 12 test subjects (arrows) and 22 concert halls (points A, B, . . . ,Y) represented in the $F_1$–$F_2$ plane of a four-dimensional perceptual space.

Source: Siebrasse (1973).

the subjects, furthermore, the different concert halls tested and the use of different assessment schemes. And it is not surprising, after all, that in the second study, the strength or loudness of the sound signal does not show up as a relevant attribute, since care had been taken to present all test samples at equal level to the listeners.

More recent investigations of this kind have been carried out by Lokki et al. (2011, 2012), summarized in Kuusinen et al. (2017). It differed from the previous studies in four major points:

1. To excite all halls to be tested in exactly the same way, these authors employed Pätynen's 'loudspeaker orchestra' as described earlier, arranged on the stage of the hall.
2. Instead of directly judging the sound field in the hall, the impulse responses of each transmission path connecting a loudspeaker with a selected listener position was measured, using a six-channel intensity probe (see Section 8.1). In this way, it was possible to estimate the direction of sound incidence at each instant.
3. From the output signals of this probe, 24 spatial impulse responses were derived. Each of them was convolved with the anechoic instrument recordings prepared as described earlier and fed to one of 24 loudspeakers distributed in a listening room which formed constituting the final 3D sound reproducing system. By this somewhat-involved procedure, which is described in a condensed form in a publication by Lokki (2014), the test listeners — the 'assessors' — were enabled to switch between different halls or different seats and thus to compare directly their auditory impressions.

4. Each assessor was encouraged to develop his individual set of perceptual attributes for the sensory evaluation of the tested halls. A subsequent clustering process led to certain consensus groups of attributes, such as the 'reverberance' or 'loudness' group, or the 'clarity' group. By further analysis of the data, the so-called 'sensory profiles' could be established, that is, rankings of the examined halls according to each of those attribute groups. Additionally, assessors were asked to arrange the concert halls according to the results of preference rating.

Similar to Lokki's approach to find an overall perceptual scheme, Weinzierl et al. (2018) presented the 'Room Acoustical Quality Inventory', a vocabulary describing the music and speech perception in performance venues. A focus group with room acoustical experts collected a list of perceptual attributes of room acoustical impression. Afterwards, 190 test subjects rated their acoustical impression of 35 binaurally simulated rooms from two listening positions stimuli from an orchestra, a solo trumpet, and speech. The resulting set of perceptual dimensions and attributes serves as basis for quantitative quality analysis of the overall room-acoustic perception. The evaluation of the assessors' responses to the processed musical excerpts led to many interesting insights which cannot be reported on here in detail. One remarkable result is the fact that the test listeners can be divided into two groups on account of their preferences. The first one preferred more clear and intimate sounds with high definition, whereas the members of the other one favoured halls producing loud and more reverberant, enveloping sounds with strong bass. These results agree to some extent with those obtained by the Wilkens and the Schroeder group. Furthermore, they underline the importance of early lateral sound reflections for good concert hall acoustics (Kuusinen et al. 2014). Concerning global preference ratings, the highest correlation was found with the 'perceived proximity'.

The discussions of this section would be incomplete without mentioning the work of Ando (1983, 2007), who developed an acoustic rating system for concert halls or for seats within a concert hall. His system is based on numerous listening experiments carried out with synthetic sound fields similar to that described in the beginning of this chapter (see Figure 7.2) — in contrast to the aforementioned researchers, who collected their material in real concert halls. Ando and his collaborators isolated four 'orthogonal', that is, independent sound field parameters which they believed to contain all the necessary information on the sound field. Two of them were temporal parameters, namely, the reverberation time and the 'initial time delay gap' $\Delta t_1$. (The latter is the delay of the first reflection in a room with respect to the direct sound and was first introduced by Beranek [1962].) It is remarkable that both of them depend on the effective duration $\tau_e$ of the autocorrelation function of the presented sound signal, that is, on the character of the performed music (see Section 1.7). The other parameters are the sound pressure level $L$ and the IACC. For each of these parameters, Ando determined the 'most preferred value' by paired comparison tests. His results are listed in Table 7.4.

*Table 7.4* Most preferred values of the sound pressure level, the initial time delay gap, the reverberation time, and the interaural autocorrelation coefficient

| *Parameter* | *Symbol* | *Most preferred value* |
|---|---|---|
| Initial time delay gap | $(\Delta t_1)_p$ | $(1-\log_{10} A_r)\cdot\tau_e$ |
| Reverberation time | $T_p$ | $23\cdot(\Delta_1)_p$ |
| Sound pressure level | $L_p$ | 79 dB (A) |
| IACC | $(IACC)_p$ | 0 |

In this table, $A_r$ is the total pressure amplitude of all reflections, given by

$$A_r^2 = \int_{\varepsilon}^{\infty} [g(t)]^2 \, dt \Big/ \int_{0}^{\varepsilon} [g(t)]^2 \, dt \tag{7.26}$$

The character $\varepsilon$ signifies a time interval which is just small enough to exclude the direct sound from the first integration. The most preferred IACC is zero, in which case both ear signals are completely uncorrelated.

From numerous preference tests, Ando derived a (negative) figure of merit which he called the 'total subjective preference':

$$S_a = -\sum_{i=1}^{4} \alpha_i |X_i|^{3/2} \tag{7.27}$$

The quantities $X_1$, $X_2$, and $X_3$ are to characterize the deviations of the actual parameters from their 'most preferred values' $p_p$, $(\Delta t_1)_p$, and $T_p$. They are given by the following expressions along with the weighting coefficients $\alpha_i$:

$$X_1 = L - L_p, \qquad \alpha_1 = \begin{cases} 0.07 \text{ for } X_1 \geq 0 \\ 0.04 \text{ for } X_1 < 0 \end{cases} \tag{7.28}$$

$$X_2 = \log_{10}\left(\frac{\Delta t_1}{(\Delta t_1)_p}\right), \qquad \alpha_2 = \begin{cases} 1.42 \text{ for } X_2 \geq 0 \\ 1.11 \text{ for } X_2 < 0 \end{cases} \tag{7.29}$$

$$X_3 = \log_{10}\left(\frac{T}{T_p}\right), \qquad \alpha_3 = \begin{cases} 0.45 + 0.74A \text{ for } X_3 \geq 0 \\ 2.36 - 0.42A \text{ for } X_3 < 0 \end{cases} \tag{7.30}$$

$$X_4 - \text{IACC}, \quad \alpha_4 = 1.45 \tag{7.31}$$

Ando's 'total subjective preference' $S_a$ vanishes if all the parameters assume their most preferred values; negative values of $S_a$ indicate certain acoustical deficiencies.

Beranek (1996) has modified this rating scheme by adding two further components — 'warmth' (equal to low-frequency divided by mid-frequency reverberation time) and an index characterizing the surface diffusivity of a concert hall, determined by visual inspection. He applied it to his collection of data on concert halls and opera theatres. On the other hand, Beranek attributed each of these concert halls to one of six categories of acoustical quality, ranging from 'fair' (category C) to 'superior' (category A+), by evaluating numerous interviews with musicians and music critics, and found satisfactory agreement between these subjective categories and the rating numbers derived from the objective data.

In a way, Ando's approach seems to answer the questions raised at the beginning of this section. However, his results show only partial agreement with those of previous investigations from which the initial time delay gap did not emerge as a significant acoustical criterion. Even more important, Ando's investigations did not show evidence that there are two groups of listeners with clearly different preferences. For this reason, it may be very difficult, if not impossible, to arrange concert halls in an order according to their acoustical quality. We should not be too sorry about this consequence. In fact, if all designers of concert halls decided to follow the guidelines of this system, they would arrive at auditoria with similar acoustics. One may doubt whether this is a desirable goal of room acoustical efforts, since differences in listening environments are just as enjoyable as different architectural solutions to a building or different interpretations of a musical work.

In recent research in the field of psychoacoustics, auditory models of binaural processing have proven to be a powerful tool for assessing localization, diffuseness, and spaciousness. Based on the Jeffress model of cross-correlation (Jeffress 1948), various auditory modelling approaches have been developed by Lindemann (1986), Dau et al. (1996), and others (see overview in Blauert and Braasch [2020]). These models also open up new possibilities for room acoustic research. van Dorp Schuitman (2011) presented a method for objective measurement of the acoustic properties of a room. It is based on the binaural non-linear model of human hearing introduced by Breebaart et al. (2001) and can be used to predict human responses in listening experiments on reverberation, clarity, apparent source width (ASW), and listener envelopment (LEV). The results of his psychoacoustic model, 'AMARA', correlate better with the perceptual results from the listening tests than the conventional parameters from ISO 3382. Furthermore, the psychoacoustic parameters vary less. Referring to de Vries et al. (2001) and his array of measurements in the Concertgebouw Amsterdam for clarity according to ISO 3382, fluctuations in the range of a concert hall seat by more than the just noticeable difference (JND) can be observed. AMARA's psychoacoustically predicted clarity fluctuates less than its JND, which is more like the listening experience in a concert hall venue where slight head movements do not result in large clarity fluctuations. We have to admit that the investigations which have been reported in the preceding sections do not combine to form a perfectly well-rounded picture of subjective concert hall acoustics; their results are not free of inconsistencies and even contradictions and, hence, are to be considered as preliminary only. Nevertheless, it is evident that important insights have been gained into what is relevant for good concert hall acoustics. Beranek et al. (2011) is an example of how experiences from consulting on concert hall projects are reported and reflected in the further development of evaluations based on physical phenomena and their effects on subjective perception in music performance. These insights, along with the auralization techniques to be described in Section 9.7, will certainly help acoustical designers to create concert halls with satisfactory or even excellent listening conditions and to avoid serious mistakes.

## REFERENCES

Ando Y. Calculation of subjective preference at each seat in a concert hall. *J Acoust Soc Am* 1983; *74*: 873.

Ando Y. Concert hall acoustics based on subjective preference theory. In: Rossing T.D. (ed.), *Springer Handbook of Acoustics*. New York: Springer, 2007.

Atal B.S., Schroeder M.R., Kuttruff H. Perception of coloration in filtered Gaussian noise short-time spectral analysis by the ear. Proceedings of the Fourth International Congress on Acoustics, Copenhagen, 1962, Paper H31.

Atal B.S., Schroeder M.R., Sessler G.M. Subjective reverberation time and its relation to sound decay. Proceedings of the Fifth International Congress on Acoustics, Liege, 1965, Paper G32.

Barron M. The subjective effects of first reflections in concert halls. *J Sound Vibr* 1971; *15*: 475.

Barron M. Subjective study of British symphony concert halls. *Acustica* 1988; *66*: 1.

Barron M., Marshall A.H. Spatial impression due to early lateral reflections in concert halls. *J Sound Vibr* 1981; 77(2): 211.

Beranek L.L. *Music, Acoustics and Architecture*. New York: Wiley, 1962.

Beranek L.L. *Concert and Opera Halls: How They Sound*. Woodbury, NY: Acoustical Society of America, 1996.

Beranek, L.L., Gade A.C., Bassuet, A., Kirkegaard, L., Marshall, H., Toyota, Y. Concert hall design—present practices. *Build Acoust* 2011; *18*: 159.

Blauert J., Braasch J. *The Technology of Binaural Understanding*. Cham: Springer, 2020.

Bradley J.S. Predictors of speech intelligibility in rooms. *J Acoust Soc Am* 1986; *80*: 837.
Bradley J.S., Soulodre G.A. Objective measures of listener envelopment. *J Acoust Soc Am* 1995; *98*: 2590.
Breebaart J., van de Par S., Kohlrausch A. Binaural processing model based on contralateral inhibition. I. Model setup. *J Acoust Soc Am* 2001; *110*: 1074.
Burgtorf W. The perceptibility of delayed sound signals (in German). *Acustica* 1961; *11*: 97.
Burgtorf W., Oehlschlägel H.K. Angle-dependent perceptibility of delayed sound signals (in German). *Acustica* 1964; *14*: 254.
Damaske P. Subjective investigations of sound fields (in German). *Acustica* 1967/68; *19*: 199.
Damaske P., Ando Y. Interaural cross-correlation for multichannel loudspeaker reproduction. *Acustica* 1972; *27*: 232.
Cremer L., Müller H.A. *Principles and Applications of Room Acoustics*, Vol. 1, Chapter III.2. London: Applied Science, 1982.
Dau T., Püschel D., Kohlrausch A. A quantitative model of the effective signal processing in the auditory system. I. Model structure. *J Acoust Soc Am* 1996; *99*: 3615.
de Vries D., Hulsebos E.M., Baal J. Spatial fluctuations in measures for spaciousness. *J Acoust Soc Am* 2001; *110*: 947.
Dietsch L., Kraak W. An objective criterion for the detection of disturbing echos in the presentation of music and speech (in German). *Acustica* 1986; *60*: 205.
Gade A.C. Investigations of musicians' room acoustic conditions in concert halls. *Acustica* 1989; *69*: 193 and 249.
Gade A.C. Acoustic properties of concert halls in the US and in Europe. Proceedings of the Sabine Centennial Symposium, Acoustical Society of America, Cambridge, MA, NY, 1994, p. 191.
Gade A.C., Rindel J.H. The dependence of the sound level on the distance in concert halls (in German). Fortschr. d. Akustik — DAGA '85, Bad Honnef, DPG-GmbH, 1985, p. 435.
Gottlob D. *Comparison of objective parameters with results of subjective investigations on concert halls (in German)*. Dissertation, University of Göttingen, 1973.
Haas H. On the influence of a single reflection on the perception of speech (in German). *Acustica* 1951; *1*: 49.
Hidaka T., Nishihara N. Objective evaluation of chamber-music halls in Europe and Japan. *J Acoust Soc Am* 2004; *116*: 357.
Houtgast T., Steeneken H.J.M. The modulation transfer function in room acoustics as a predictor of speech intelligibility. *Acustica* 1973; *28*: 66.
Houtgast T., Steeneken H.J.M. A multi-language evaluation of the RASTI-method for estimating speech intelligibility. *Acustica* 1984; *54*: 186.
ISO 532-1:2017 Acoustics — Methods for calculating loudness — Part 1: Zwicker method.
ISO 532-2:2017 Acoustics — Methods for calculating loudness — Part 2: Moore-Glasberg method.
Jeffress L.A. A place theory of sound localization. *J Comp Physiol Psychol* 1948; *41*: 35.
Kuhl W. Experiments on the optimum reverberation time of large music studios (in German). *Acustica* 1954; *4*: 618.
Kürer R. Isolation of single number criteria from impulse measurements in room acoustics (in German). *Acustica* 1969; *21*: 370.
Kuusinen A., Lokki T. Wheel of concert hall acoustics. *Acta Acust United Acust* 2017; *103*(2): 185.
Kuusinen A., Pätynen J., Tervo S., Lokki T. Relationships between preference ratings, sensory profiles, and acoustical measurements in concert halls. *J Acoust Soc Am* 2014; *135*: 239.
Lee D., Cabrera D., Martens W.L. Equal reverberance contours for synthetic room impulse responses listened to directly: Evaluation of reverberance in terms of loudness decay parameters. *Build Acoust* 2011; *18*(1–2): 189.
Lentz T., Behler G. Dynamic cross-talk cancellation for binaural synthesis in virtual reality environments. Proceedings of the 117 Convention Audio Engineering Society, San Francisco, CA, 2004.
Lindemann W. Extension of a binaural cross-correlation model by means of contralateral inhibition. II. The law of the first wave front. *J Acoust Soc Am* 1986; *80*: 1623.
Lokki T. Tasting music like wine: Sensory evaluation of concert halls. *Physics Today*, January 2014.
Lokki T., Pätynen J., Kuusinen A., Vertanen H., Tervo S. Concert hall acoustics assessment with individually elicited attributes. *J Acoust Soc Am* 2011; *130*: 835.

Lokki T., Patynen J., Kuusinen A., Vertanen H., Tervo S. Disentanglement preference ratings of concert hall acoustics using subjective sensory profiles. *J Acoust Soc Am* 2012; *132*: 3148.
Masiero B. *Individualized binaural technology – measurement, equalization and perceptual evaluation.* PhD thesis, RWTH Aachen University, 2012.
Meyer E., Schodder G.R. On the influence of sound reflections on the localisation and loudness of speech (in German). *Nachr Akad Wissensch Göttingen Math-Phys Kl* 1952; *6*: 31.
Meyer E., Thiele R. Room acoustical investigations in numerous concert halls and broadcasting studios using more recent measurement techniques (in German). *Acustica* 1956; *6*: 425.
Muncey R.W., Nickson A.F.B., Dubout P. The acceptability of speech and music with a single artificial echo. *Acustica* 1953; *3*: 168.
Pätynen J., Tervo S., Lokki T. A loudspeaker orchestra for concert hall studies. The Seventh International Conference on Auditorium Acoustics (Institute of Acoustics Oslo 2008), p. 45.
Reichardt W., Abdel Alim O., Schmidt W. Dependence of the limit between useful and useless transparency on the kind of music, of the reverberation time and the onset time of sound decay (in German). *Appl Acoust* 1974; *7*: 243.
Reichardt W., Schmidt W. The audible differences of the spatial impression with music (in German). *Acustica* 1966; *17*: 75.
Reichardt W., Schmidt W. The detectibility of changes in sound field parameters for music (in German). *Acustica* 1967; *18*: 274.
Sabine W.C. *The American architect 1900. See also Collected Papers on Acoustics.* Cambridge: Harvard University Press, 1923.
Schroeder M.R., Atal B.S. Computer simulation of sound transmission in rooms. *IEEE Int Conv Rec* 1963; *7*: 150.
Schroeder M.R., Gottlob D., Siebrasse K.F. Comparative study of European concert halls: Correlation of subjective preference with geometric and acoustic parameters. *J Acoust Soc Am* 1974; *56*: 1195.
Schubert P. The perceptibility of reflections with music (in German). *ZeitschrHochfrequenztechn Elektroakust* 1969; *78*: 230.
Seraphim H.P.S. Investigations on the difference limen of exponential decay of noise impulses (in German). *Acustica* 1958; *8*: 280.
Seraphim H.P.S. The perceptibility of several reflections of speech sound (in German). *Acustica* 1961: *11*: 80.
Siebrasse K.F. *Comparative subjective investigations on the acoustics of concert halls (in German).* Dissertation, University of Göttingen, 1973.
Soulodre G., Bradley J. Subjective evaluation of new room acoustic measures. *J Acoust Soc Am* 1995; *98*: 1995.
Steeneken H.J.M., Houtgast T. A physical method for measuring speech-transmission quality. *J Acoust Soc Am* 1980; *67*: 318.
van Dorp Schuitman J. *Auditory modelling for assessing room acoustics.* Dissertation, Technical University Delft, The Netherlands, 2011.
Wilkens H. A multidimensional description of subjective assessment of concert hall acoustics. *Acustica* 1977; *38*: 10.
Wilkens H., Plenge G. The correlation between subjective and objective data of concert halls. In: Mackenzie R. (ed.), *Auditorium Acoustics.* London: Applied Science, 1974.
Weinzierl S., Lepa S., Ackermann D. A measuring instrument for the auditory perception of rooms: The room acoustical quality inventory (RAQI). *J Acoust Soc Am* 2018; *144*(3): 1245.

Chapter 8

# Measuring techniques in room acoustics

The starting point of modern room acoustics is marked by attempts to define physical sound field parameters suited to quantify the subjective acoustic impression of a listener sitting in a hall and attending a presentation. As shown in *Figure 7.1*, these parameters or a set of them can be thought of as a link between the world of subjective perceptions on the one hand and the realm of architecture responsible for the geometrical and other room data on the other. As we have seen in Chapter 7, several useful parameters of this kind have evolved in the course of time. It is the object of this chapter to describe how such quantities can be measured and which kind of equipment is used for this purpose.

Room acoustic measurements are necessary for research and design purposes and are also used as a diagnostic tool for existing rooms. Suppose there are frequent and persistent complaints about the acoustics of a particular hall. If the problem cannot be solved on the basis of a simple inspection, acoustical measurements are needed to reveal and to characterize the deficiencies and to find their cause. This holds not only if the complaining people are concert- or operagoers but also, at least to the same extent, for performers, in particular, for musicians. Although it is sometimes difficult to translate the comments of musicians into the language of acousticians, their opinions must be taken seriously; a concert hall is not very likely to get a good reputation if the performing artists are continuously dissatisfied. Another typical situation is that the performance of an electroacoustic reinforcement system is not satisfactory. Then, the examination of the sound field can indicate whether the location or the directivities of the loudspeakers should be improved, or whether it is rather a matter of readjusting the signal equalizer (if there is any). And finally, it may be wise to take at least some measurements in a new or refurbished hall prior to its opening or reopening when minor corrections are still possible.

Some measuring procedures are not directly related to subjective impressions but concern the acoustic properties of materials, especially the absorption of wall and ceiling materials, of seats, and so on. Knowledge of such data is essential for any planning in room acoustics. Many of them can be found in published collections, but often, the acoustical consultant is faced with new products or with specially designed wall linings for which no absorption data are available. This holds even more with regard to the scattering efficiency of acoustically 'rough' surfaces (see Section 8.8).

## 8.1 GENERAL REMARKS ON INSTRUMENTATION

Viewed from our present state of the art, the equipment which Sabine (1922) had at his disposal for his famous investigation of reverberation appears quite modest: he excited the room under test with a few organ pipes and used his ear as a measuring instrument together

DOI: 10.1201/9781003389873-8

with a simple stopwatch. With the development and introduction of the electrical amplifier in the 1920s, almost all measuring techniques became electrical. In acoustics, the ear was replaced with a microphone, and the fall in sound level was observed with an electromechanical level recorder. Nowadays, the room under test is usually excited with an electric loudspeaker which is fed from an electronic signal generator. More recently, the introduction of the digital computer has triggered off a second revolution in measuring techniques: all kinds of signal processing (filtering, storing, and processing the measured signals as well as the presentation of results) can now be done with digital equipment, which is generally more powerful, precise, and flexible and less expensive than the more traditional equipment.

For field measurement of the reverberation time, the requirements concerning the uniformity of the sound field are not very stringent, since the various sound components will anyway be mixed during the decay process. Therefore, it is often sufficient to excite the room with a pistol shot which produces sufficient sound power even if there is some background noise. The same holds for wooden hand clappers or bursting balloons which yield strong excitation of a room, especially at low frequencies. The frequency dependence of the decay process can be examined by using band-pass filters in the receiver side with variable passband. Frequently applied are octave ($f_2/f_1 = 2$) and third-octave filters ($f_2/f_1 = 5/4$), where $f_2$ and $f_1$ are the upper and the lower frequency limits of a passband.

Matters are different when it comes to measuring the correct impulse response of a room. Then, the excitation signals must be exactly reproducible. This is achieved by use of an electrical loudspeaker, driven by a power amplifier. Any directionality of the sound source must be avoided; otherwise, the result of the measurement would depend not only on the position of the source but also on its orientation. The frequency characteristics of the loudspeaker should be as flat as possible, and also, the phase distortions must be small. The same holds, of course, for any non-linear distortions. One way to achieve omnidirectional sound radiation, at least at lower frequencies, is to employ a combination of 12 or 20 equal and equally fed loudspeaker systems mounted on the faces of a regular polyhedron (dodecahedron or icosahedron). A dodecahedron loudspeaker is shown in Figure 8.1. It should be noted that even this kind of source shows some directionality at elevated frequencies. For this reason, two polyhedral loudspeaker systems of different size are sometimes employed when a wide frequency range is to be covered.

Another useful device is the driver of a powerful horn loudspeaker with the horn replaced by a metal tube. Its open end radiates the sound uniformly, provided its diameter is small compared with the wavelength. The resonances of the tube can be suppressed by inserting some damping material into it. Nearly omnidirectional radiation of powerful acoustical wide-band impulses can also be achieved by specially designed electrical spark gaps.

To pick up the sound in the enclosure, pressure-sensitive high-quality microphones with omnidirectional characteristics are most commonly used. For certain measurements, however, a gradient receiver, that is, figure-of-eight microphones, or a dummy head must be employed. Receivers with still higher directivity are needed to determine the directional distribution of sound energy at a certain position. Examples of directional microphones are arrays of equal microphones, or a pressure microphone placed in the focus of a concave mirror, or the 'line microphone', which is the combination of a usual pressure microphone with a slotted tube (Kuttruff 2007). Although the experimenter should have a certain idea of the sensitivity of his microphones, absolute calibration is usually not required, since virtually all sound measurements in room acoustics are relative.

To measure the sound intensity in a wave field, both the sound pressure and the particle velocity must be determined, according to Equation 1.34. For the latter, miniaturized hot-wire anemometers can be used. Such a device consists essentially of two thin parallel platinum wires, which are heated to 200–400°C. When it is exposed to a local airflow directed from one wire to the other, the thermal balance in the vicinity of both wires is disturbed,

*Figure 8.1* Dodecahedron loudspeaker for room acoustical measurements.

Source: Courtesy of the Institute of Hearing Technology and Acoustics, RWTH Aachen University, Germany.

causing a difference between the resistance of both wires which is proportional to the velocity of the flow (= particle velocity). The combination of a velocity sensor with a pressure microphone is known as a p-u probe (since many authors use the symbol $u$ for the particle velocity). Although originally developed for measuring the intensity in a sound field, it can be used as well as a device for the measurement of wall impedances (see Section 8.6). A more indirect method of measuring the particle velocity employs a so-called p-p probe consisting of two pressure microphones with equal sensitivity and phase characteristics in some fixed distance $d$. If $d$ is very small compared to the acoustical wavelength, the sound pressure at both microphones differs by

$$\Delta p \approx \left(\frac{\partial p}{\partial x}\right) \cdot d = -i\omega\rho_0 v_x \tag{8.1}$$

(The latter expression has been obtained by applying Equation 1.2 and replacing the time derivative with the factor $i\omega$.) In this formula, the $x$-axis is parallel to the line connecting both microphones. The distance $d$ is usually controlled by solid spacers, the dimensions of which can be adapted to the desired frequency range. To determine the complete intensity vector, each of its components must be measured, either by changing the orientation of the velocity probe or by using a combination of three separate velocity probes (Fahy 1995).

After proper pre-amplification, the output signal of the microphone may be stored until further evaluation in the laboratory. This can be achieved with a magnetic tape recorder, for instance. The most convenient method, however, is to apply the microphone signal immediately to a portable digital computer, where it is stored in the computer's hard disk or directly processed to yield the parameters which one is looking for. This requires, of course, that the computer is fitted out with an analogue-to-digital converter which covers a sufficiently wide range of amplitudes.

## 8.2 MEASUREMENT OF THE IMPULSE RESPONSE

According to system theory, all properties of a linear transmission system are contained in its impulse response or, alternatively, in its transfer function, which is the Fourier transform of the impulse response. Since a room can be considered as an acoustical transmission system, the impulse response yields a complete description of the changes a sound signal undergoes when it travels from one point in a room to another, and almost all the parameters we discussed in Chapter 7 can be derived from it, at least in principle. Parameters related to spatial or directional effects can be based upon the 'binaural impulse response' picked up at both ears of a listener or with a dummy head. From these remarks, it is clear that the determination of impulse responses is one of the most fundamental tasks in experimental room acoustics. It requires high-quality standards for all measuring components, which must be free of linear or non-linear distortions, including phase shifts. By its very definition, the impulse response of a system is the signal obtained at the system's output after its excitation by a vanishingly short impulse (yet with non-vanishing energy), that is, by a Dirac or delta impulse (see Section 1.4). Since we are interested only in frequencies below, say, 10 kHz, this signal can be approximated by a short rectangular impulse, the duration of which is smaller than about 20 μs. However, it should be kept in mind that there exist no loudspeakers which are completely free of linear distortions. This means the loudspeaker transforms a Dirac impulse $\delta(t)$ applied to its input terminal into a different sound signal $g_{LS}(t)$. The response $g'(t)$ of the room to this latter signal is the room impulse response $g(t)$ convolved with $g_{LS}(t)$ according to Equation 1.63 or 1.64:

$$g'(t) = \int_{-\infty}^{\infty} g(\tau) g_{LS}(t-\tau) d\tau = g(t) * g_{LS}(t) \tag{8.2}$$

The linear distortion caused by the loudspeaker can be eliminated by 'deconvolution', that is, by undoing the convolution of Equation 8.2, which is most easily performed in the frequency domain: Let $G(f)$ and $G_{LS}(f)$ denote the Fourier transforms of $g(t)$ and $g_{LS}(t)$, respectively. Then, the Fourier transform of Equation 8.2 reads

$$G'(f) = G(f) \cdot G_{LS}(f) \tag{8.3}$$

Dividing $G'(f)$ by $G_{LS}$ yields $G(f)$, the transfer function of the room, from which the corrected room impulse response $g(t)$ can be recovered by inverse Fourier transformation (see Equation 1.41). It may be practical to avoid the influence of the loudspeaker by pre-emphasis, that is, by passing the test signal through a filter with the transfer function $[G_{LS}(f)]^{-1}$ prior to feeding it to the loudspeaker.

In practice, however, results obtained in this way are more or less impaired by the omnipresent background noise, for instance, by traffic noise intruding into the room from the outside, or by noise from technical equipment, and so on. One way to overcome this difficulty is repeating the measurement several or many times and adding the results. If $N$ is the number of measurements, the total energy of the collected impulse responses grows in proportion to $N^2$, while the energy of the resulting noise grows only proportionally to $N$, provided the noise is random. Hence, the signal-to-noise ratio, expressed in decibels, is increased by $10 \cdot \log_{10} N$ decibels.

A less time-consuming method employs test signals which are stretched in time and, hence, can carry more energy than a short impulse. Suppose the system to be tested, that is, the room, is excited by an arbitrary signal $s(t)$. Its response is given by

$$s'(t) = \int_{-\infty}^{\infty} s(\tau)g(t-\tau)\mathrm{d}\tau = \int_{-\infty}^{\infty} g(\tau)s(t-\tau)\mathrm{d}\tau \tag{8.4}$$

or expressed in the frequency domain

$$S'(f) = S(f) \cdot G(f) \tag{8.5}$$

The impulse response $g(t)$ can be recovered from the output signal $s'(t)$ by applying the same recipe as described earlier, namely, by 'deconvolution'. Accordingly, its Fourier transform $G(f)$ is obtained as $S'/S$, provided $S$ is non-zero within the whole frequency range of interest. Figure 8.2 shows the principle of this method, which is known as dual-channel analysis. If $s(t)$ is a random signal, for instance, white noise, averaging over several or many single measurements is required to arrive at a reliable result. The execution of several Fourier transformations offers no difficulties using current computer technology. If the signal $s(t)$ is known beforehand, its spectrum $S(f)$ must be calculated only once for all and can be used in all subsequent measurements.

Particularly well-suited for this technique are excitation signals with 'flat', that is, frequency-independent, power spectra. The frequency spectrum of such a signal has the form (see Equation 1.48)

$$S(f) = \exp\left[i\psi(f)\right].$$

with $\psi(f)$ denoting the phase spectrum. Accordingly, $S(f)^{-1} = \exp[-i\psi(f)] = S^*(f)$, where the asterisk indicates the conjugate complex quantity. Hence, we obtain from Equation 8.5

$$G(f) = S'(f) \cdot S^*(f) \tag{8.6}$$

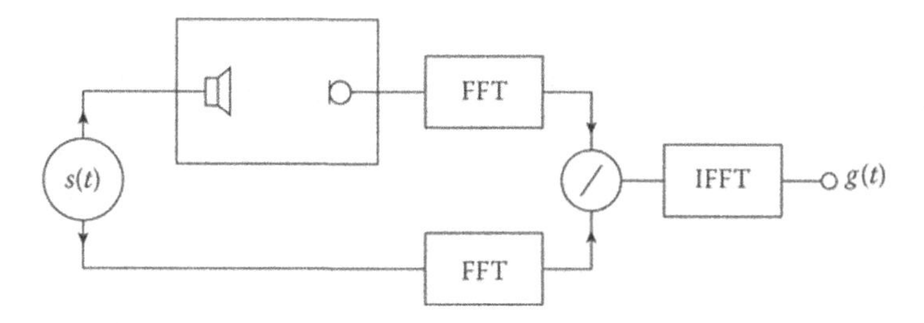


*Figure 8.2* Principle of dual-channel analysis (*FFT* = fast Fourier transform; *IFFT* = inverse FFT).

which corresponds to

$$g(t) = s'(t) * s(-t) \tag{8.7}$$

in the time domain, since $s(-t)$ is the inverse Fourier transform of $S^*(f)$. This means that the measured signal must be passed through a filter, the impulse response of which is the time-inversed excitation signal. This technique is also known as 'matched filtering'.

One signal of this kind is the Dirac impulse, as already discussed. Other test signals with the required property are trains of equally spaced Dirac impulses, the signs of which alternate according to a particular pattern. The best-known kind of such binary signals are 'maximum length sequences' (MLS). Their application to room acoustics was first described by Alrutz and Schroeder (1983).

Maximum length sequences are periodic pseudorandom signals, consisting of the elements +1 and −1. The number of elements per period is

$$L = 2^m - 1 \tag{8.8}$$

where $m$, the order of the sequence, is a positive integer. An MLS can be generated by a digital $m$-step shift register with the outputs of certain stages fed back to the input (MacWilliams and Sloane 1976). An example with $L = 7$ ($m = 3$) is

$$\{s_k\} = \cdots +1, -1, +1, -1, -1, -1, +1 \cdots \tag{8.9}$$

Of particular interest are the correlation properties of an MLS. Generally, the circular autocorrelation function of a discrete signal $s_k$ is defined, in analogy to Equation 1.56, by

$$(\phi_{ss})_n = \sum_{k=0}^{L-1} s_k s_{k+n} \tag{8.10}$$

where the subscript $k + n$ has to be taken modulo $L$, the periodicity of the sequence. (As in Section 2.7, the modulo operation yields the remainder after dividing a given number by an integer.) Applied to the sequence (8.9), this yields

$$(\phi_{ss})_n = \begin{Bmatrix} L & \text{for } n=0 \\ -1 & \text{for } n \neq 0 \end{Bmatrix} \tag{8.11}$$

Figure 8.3 shows the MLS for $m = 3$ ($L = 7$) along with its (circular) autocorrelation function. From the latter, we can conclude that the power spectrum of an MLS is nearly flat. (The power spectrum of $(\Phi_{ss})_n + 1$ is exactly constant). In contrast, the phase angle is randomly distributed in the interval from 0 to $2\pi$.

What about the choice of $\Delta t$, the time interval between two impulses? In accordance with the sampling theorem, it should be shorter than half the period of the highest frequency $f_m$ encountered in g(t), that is, $\Delta t < 1/2f_m$. In order to avoid time-aliasing, that is, overlap of different portions of the room impulse response, the period of the sequence must exceed the duration of the impulse response to be measured. (In practice, the impulse response ends when it drops below the level of the background noise.) Suppose the sampling frequency $1/\Delta t$ has the usual value of 44.1 kHz and we choose $n = 18$, then the period of the sequence $L$ has the duration of about 6 s. Hence, for most purposes in room acoustics, m = 17 or 18 would be sufficient.

The main advantage of maximum length sequences is the possibility of recovering the impulse response from the measured sequence without leaving the time domain. This is

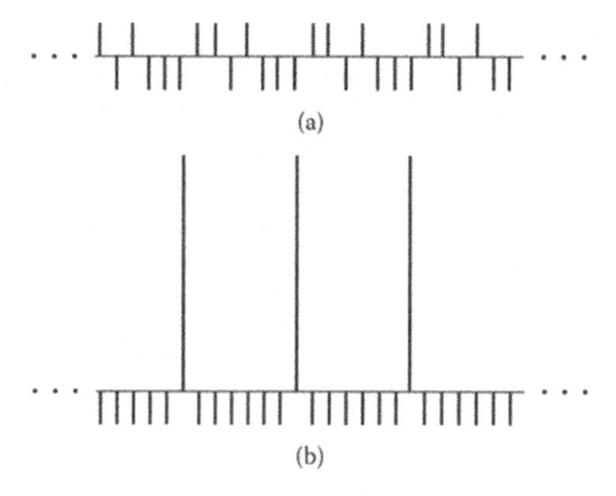


*Figure 8.3* Maximum length sequence with $n$ = 3: (a) the sequence, (b) its autocorrelation function.

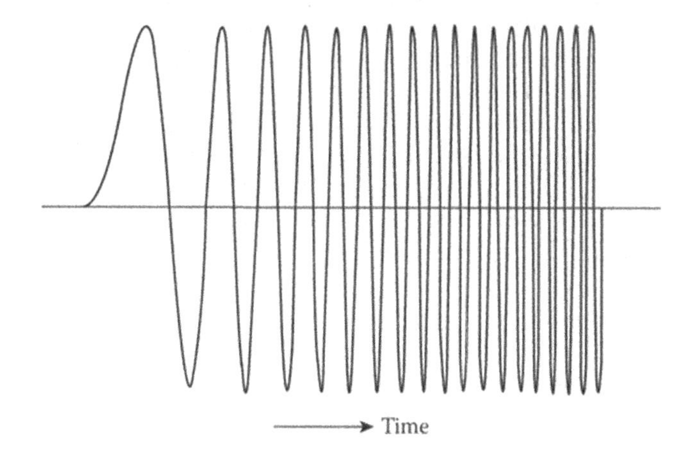


*Figure 8.4* Sine sweep signal.

because these sequences are closely related to Hadamard matrices (Borish and Angell 1983), which have particular symmetry and recursive properties and thus permit carrying out the required calculations in a very time-efficient way, since the only operations needed are additions and subtractions.

Another class of currently more widely used excitation signals are sine sweep signals, that is, sinusoidal signals with continuously varying frequency (Berkhout et al. 1980). The simplest example of such signals is the linear sine sweep,

$$s(t) = A\sin\left(bt^2\right) \tag{8.12}$$

(see Figure 8.4) with the instantaneous frequency,

$$f(t) = \frac{1}{2\pi}\frac{\mathrm{d}\left(bt^2\right)}{\mathrm{d}t} = \frac{1}{\pi}bt \tag{8.13}$$

It is easy to show that the frequency spectrum of this signal $s(t)$ has constant magnitude. Hence, the impulse response of the room can be retrieved immediately by 'matched filtering' according to Equation 8.7, that is, by applying the output signal of the measuring microphone to a filter, the impulse response of which is the time-reversed excitation signal, $s(-t)$. In this case, it is particularly easy to see how matched filtering works: the spectral components received last are processed first, and vice versa. However, due to modern computer technology, it is more practical to perform this operation in the frequency domain (see Equation 8.6).

The duration of the sine sweep must be limited; hence, start and end frequencies, $f_1$ and $f_2$, must be defined. Then, the constant $b$ can be determined from these frequencies and the duration $t_s$ of the sweep:

$$b = \pi \frac{f_2 - f_1}{t_s} \tag{8.14}$$

Truncating the sine sweep at finite frequencies introduces some ripple at the extremities of the amplitude spectrum which can be minimalized by providing for a soft switch-on and switch-off at the frequencies $f_1$ and $f_2$. The constant $b$ must be chosen small enough to make the duration $t_s$ of the sweep sufficiently long. In any case, $t_s$ must be longer than the duration of the impulse response to be measured.

It is noteworthy that a very fast sweep corresponds in the limit $b \to \infty$ to a Dirac function; hence, the response to the sweep is the impulse response of the enclosure. If, on the contrary, $b$ becomes vanishingly small, the response is, by its very definition, the frequency transfer function.

Apart from the linear sine sweep after Equation 8.12, the logarithmic (or rather exponential) sweep is also in use:

$$s(t) = A \sin\left[\exp(bt)\right] \tag{8.15}$$

With the instantaneous frequency:

$$f(t) = b \cdot \exp(bt) \tag{8.16}$$

Its power spectrum is not constant but drops proportionally to $1/f$ with increasing frequency corresponding to 3 dB per octave. One of the assets of this signal is that it gives more emphasis to the low-frequency range and, therefore, has a closer correspondence with the human perception of pitch. To get the exact impulse response, the $1/f$ dependence of the spectrum must be eliminated, for instance, by pre-emphasis.

Comparisons of the various methods of measuring the impulse response of a room have been published (Müller and Massarani 2001; Stan et al. 2002). Sine sweeps as excitation signals are more advantageous because of a higher signal-to-noise ratio and the possibility of eliminating any distortion products. The longer processing times are no longer of concern because of today's powerful digital processors. However, when it comes to measurements in occupied rooms or in other noisy environments, maximum length sequences may still be superior to sweep techniques, since their disturbance in a populated space is less than with tonal excitation signals.

## 8.3 EXAMINATION OF THE IMPULSE RESPONSE

As mentioned before, the impulse response of a room or, more precisely, of a particular transmission path within a room is the most characteristic objective feature of its 'acoustics'. Some of the information it contains can be found by direct inspection of a 'reflectogram', by which term we mean the graphical representation of the impulse response. For a more quantitative evaluation, the parameters discussed in Chapter 7 may immediately be extracted from the impulse response. (The measurement of reverberation time will be postponed to the next section.)

From the visual inspection of a reflectogram, the experienced acoustician may learn quite a bit about the acoustical merits and faults of the place for which it has been measured. One important question is, for instance, to what extent the direct sound will be supported by shortly delayed reflections, and how these are distributed in time. Furthermore, strong and isolated peaks with long delays which hint at the danger of echoes are often detected in this way.

The examination of a reflectogram can be facilitated — especially that of a band-limited reflectogram — by removing insignificant details beforehand. In principle, this can be effected by rectifying and smoothing the impulse response. This process, however, introduces some arbitrariness into the obtained reflectogram with regard to the applied time constant: if it is too short, the smoothing effect may be insufficient; if it is too long, important details of the reflectogram will be suppressed. One way to avoid this uncertainty is to apply a mathematically well-defined procedure to the impulse response, namely, to form its 'envelope'. Let $s(t)$ denote any signal, then its envelope is defined as

$$e(t) = \sqrt{\left([s(t)]^2 + [\breve{s}(t)]^2\right)} \tag{8.17}$$

Here, $\breve{s}(t)$ denotes the Hilbert transform of $s(t)$:

$$\breve{s}(t) = \frac{1}{\pi}\int_{-\infty}^{\infty} \frac{s(t-\tau)}{\tau} d\tau \tag{8.18}$$

Probably, the most convenient way to calculate the Hilbert transform is by exploiting its spectral properties. Let $S(f)$ denote the Fourier transform of $s(t)$, then the Fourier transform of $\breve{s}(t)$ is

$$\breve{S}(f) = -iS(f)\,\text{sign}(f) = \begin{Bmatrix} -iS(f) & \text{for } f > 0 \\ iS(f) & \text{for } f < 0 \end{Bmatrix} \tag{8.19}$$

Hence, a function $s(t)$ may be Hilbert-transformed by computing its spectral function $S$, modifying the latter according to Equation 8.19 and transforming the result back into the time domain. There are also efficient methods to compute the Hilbert transform directly in the time domain.

Figure 8.5 shows the original impulse response (upper trace) along with its squared envelope (lower trace). It is obvious that the latter shows the significant features much more clearly.

A reflectogram may be further modified by smoothing its envelope in order to simulate the integrating properties of our hearing. For this purpose, the envelope $e(t)$ (or the rectified impulse response $|g(t)|$) is convolved with $\exp(-t/\tau)$. This corresponds to applying an electrical signal $|g(t)|$ or $e(t)$ to a simple RC network with the time constant $\tau$ = RC. A reasonable choice of the time constant is 25 ms.

Figure 8.6 shows three examples of measured 'reflectograms', obtained at several positions in a lecture room, which was excited by short tone bursts with a centre frequency of 3,000

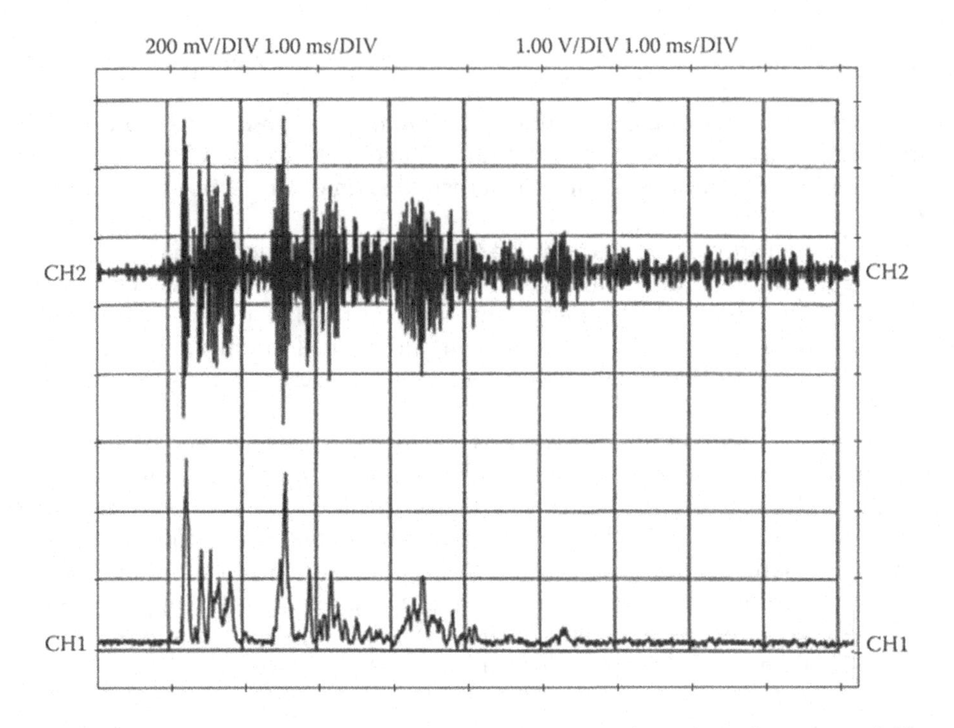


*Figure 8.5* Room impulse response (upper trace) and its squared envelope (lower trace). The total range of abscissa is 400 ms.

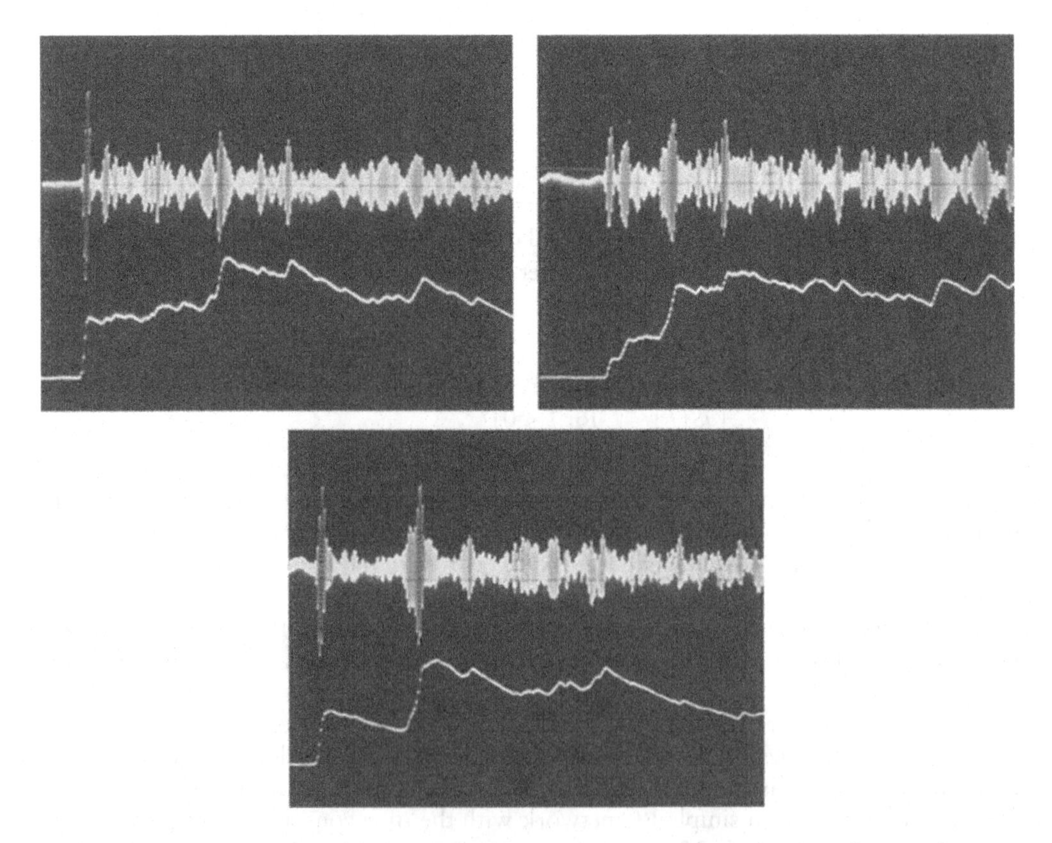

*Figure 8.6* Reflectograms measured at different positions in a lecture room (centre frequency 3,000 Hz; impulse duration about 1 ms). The total length of all recordings is 190 ms. Upper traces: original response of the room to the excitation signal. Lower traces: same, after rectifying and smoothing with time constant $\tau$ = 25 ms.

Hz and a duration of about 1 ms. The lower trace of each recording shows the result of the smoothing described earlier. The total length of each trace corresponds to a time interval of 190 ms. The upmost reflectogram was taken at a position close to the sound source; consequently, the direct sound is relatively strong. The most outstanding feature in the lowest recording is the strong reflection delayed by about 40 ms with regard to the direct sound. It is not heard as an echo, as it still lies within the integration time of our ear (see Section 7.3).

Just looking at a reflectogram or its modification may be very suggestive; nevertheless, it does not permit a safe decision as to whether a particular peak indicates the risk of an audible echo or not. This can only be achieved by applying an objective echo criterion, for instance, that proposed by Dietsch and Kraak (1986). It is based on the function

$$t_s(\tau) = \int_0^\tau |g(t)|^n t\mathrm{d}t \Big/ \int_0^\tau |g(t)|^n \mathrm{d}t \tag{8.20}$$

with $g(t)$ denoting as before the impulse response of the room. $n$ is a real number to be determined by experiment. $t_s(\tau)$ is a monotonous function of $\tau$ approaching a limiting value $t_s(\infty)$ as $\tau \rightarrow \infty$. The latter is the first moment of $|g(t)|^n$, and the function $t_s(\tau)$ indicates its temporal build-up (Figure 8.7). The quantity used for rating the strength of an echo — the *'echo criterion'* (EC) — is the difference quotient of $t_s(\tau)$:

$$\mathrm{EC} = \text{maximum of} \frac{\Delta t_s(\tau)}{\Delta \tau} \tag{8.21}$$

where $\Delta\tau$ can be adapted to the character of the sound signal. The dependence of the EC on the directional distribution of the various reflections is accounted for by recording the impulse responses with both microphones of a dummy head and adding their energies. By numerous subjective tests, carried out both with synthetic sound fields and in real halls, the mentioned authors determined not only suitable values for the exponent $n$ and for $\Delta\tau$ but also the critical values $\mathrm{EC}_{crit}$, which must not be exceeded to ensure that not more than 50% of the listeners will hear an echo (Table 8.1). It should be noted that echo disturbances are mainly due to somewhat-elevated spectral components. For practical purposes, however, it seems sufficient to employ test signals with a bandwidth of 1 or 2 octaves.

Another source of undesirable subjective effects is regular trains of reflections. They cannot be detected just by visual inspection of a reflectogram because they may be hidden by many other, non-periodic reflections. Usually, periodic components in room impulse responses are caused by repeated reflections of the sound between parallel walls, or generally in rooms with a very regular shape, for instance, in rooms with a circular or regularly polygonal ground plan. At relatively short repetition times, they are perceived as colouration, at least under certain conditions. Even a one single dominating reflection may cause audible colouration, especially of music, since the corresponding transmission function has a regular structure or substructure (see Section 7.4).

The standard technique for testing the randomness of a signal is autocorrelation analysis. In this procedure, all the irregularly distributed components are swept together into a single

*Table 8.1* Echo criterion of Dietsch and Kraak (1986): characteristic data

| *Type of signal* | *n* | *$\Delta\tau$ (ms)* | *$EC_{crit}$* | *Bandwidth of test signal (Hz)* |
|---|---|---|---|---|
| Speech | 2/3 | 9 | 1.0 | 700–1,400 |
| Music | 1 | 14 | 1.8 | 700–2,800 |

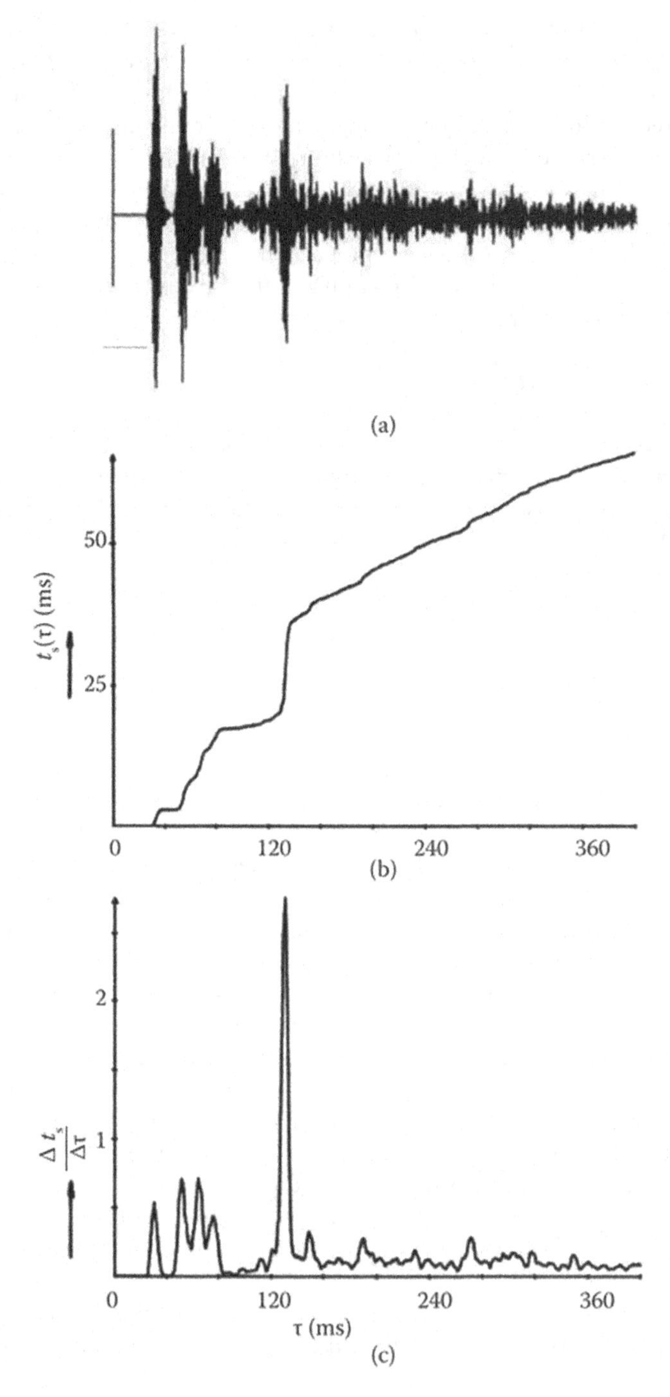


*Figure 8.7* First 400 ms (a) of a room impulse response, (b) of the associated build-up function $t_s(\tau)$, and (c) of the difference quotient $\Delta t_s(\tau)/\Delta\tau$ (with $\Delta\tau$ = 5 ms). EC is 2.75 in this example.

central peak, whereas the remaining reflections will form side components or even satellite maxima in the autocorrelogram. Since the impulse response *g(t)* is a non-stationary signal, we use the autocorrelation function according to Equation 1.56:

$$\phi_{gg}(\tau)=\int_{-\infty}^{\infty}g(t)\,g(t+\tau)\,dt=\int_{-\infty}^{\infty}g(-t)\,g(\tau-t)\,dt \tag{8.22}$$

The second expression is just the convolution of the impulse response with its time-reversed replica. It tells us that the autocorrelation function can be obtained in real time by exciting the room with its time-reversed impulse response, which has been measured beforehand. The autocorrelation functions shown in Figure 8.8 have been obtained with this simple 'playback'

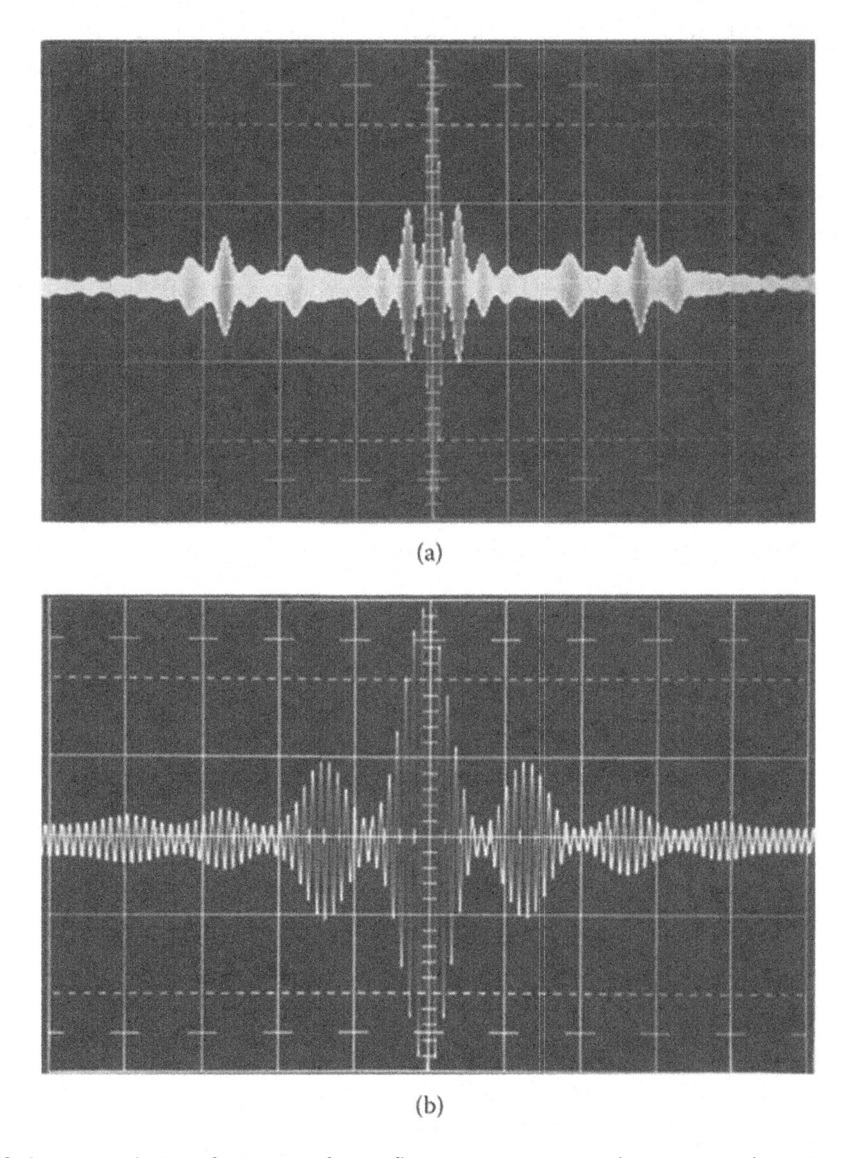


*Figure 8.8* Autocorrelation function of a reflectogram, measured in a reverberation chamber with the playback techniques. The room was excited with a filtered impulse of about 1 ms duration and a centre frequency of 2,000 Hz. Abscissa unit is (a) 20 ms and (b) 5 ms.

techniques. If Equation 8.22 is applied to two different impulse responses, $g(t)$ and $g'(t)$, the result is their cross-correlation function, $\phi_{gg'}(\tau) = g(\tau) * g'(-\tau)$.

In order to decide whether or not a certain side maximum of the autocorrelation function indicates audible colouration, we form a weighted autocorrelation function:

$$\phi'_{gg}(\tau) = b(\tau) \cdot \phi_{gg}(\tau) \tag{8.23}$$

Now, let us denote by $\tau_0$ the value of the argument at which a prominent side maximum appears. According to Bilsen (1968), we have to expect audible colouration if

$$\phi'_{gg}(\tau_0) = 0.06 \cdot \phi'_{gg}(0) \tag{8.24}$$

no matter if this side maximum is caused by an isolated reflection or by a periodic succession of reflections. The weighting function $b(\tau)$ has been calculated from the thresholds represented in Figure 7.10. It is shown in Figure 8.9.

The temporal structure of a room's impulse response determines not only the shape of its autocorrelation function but also its modulation transfer function (MTF), which was introduced in Section 5.6. Indeed, Schroeder (1981) has shown that for white noise as a primary sound signal, the complex MTF is related to the impulse response by

$$m(\Omega) = \int_0^\infty [g(t)]^2 \exp(-i\Omega t)\,dt \Big/ \int_0^\infty [g(t)]^2\,dt \tag{8.25}$$

This formula is readily obtained from Equations 5.36 and 5.38 by replacing $\exp(-2\delta\tau)$ with $[g(\tau)]^2$. It means that the complex modulation transfer is the inverse Fourier transform of the squared room impulse response divided by the integral over the squared response. Of course, this relation applies as well to an impulse response which has been confined to a suitable frequency band by band-pass filtering.

Most of the parameters introduced in Chapter 7 can be evaluated from impulse responses by relatively simple operations using a digital computer. Table 8.2 lists some of them along

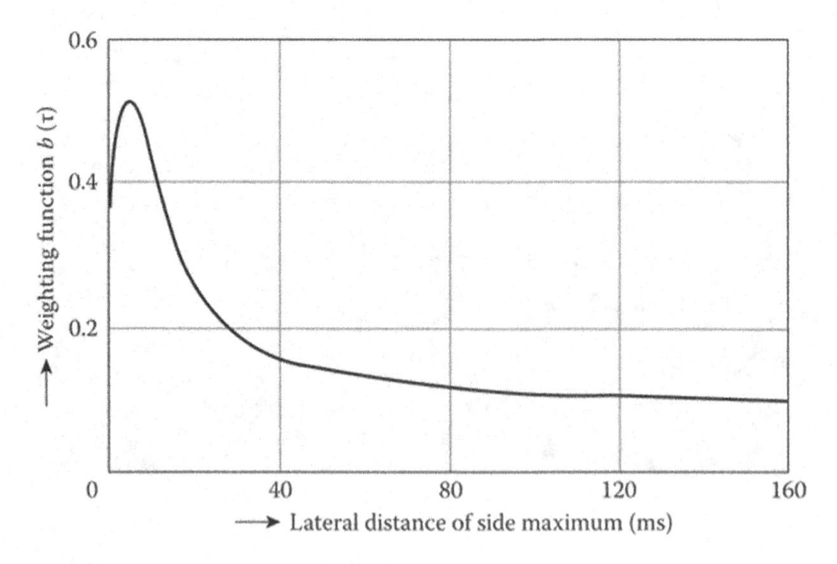


*Figure 8.9* Detecting colouration: weighting function for autocorrelation functions.

Source: Bilsen (1968).

*Table 8.2* Some objective sound field parameters (ISO 3382)

| *Name of parameter* | *Symbol* | *Defined by equation* |
|---|---|---|
| Definition/('Deutlichkeit') | D | 7.9 |
| Clarity index ('Klarheitsmaβ') | C,$C_{80}$ | 7.10 |
| Centre time | $t_c$ | 7.13 |
| Support | STe | 7.18 |
| Echo criterion (Dietsch and Kraak) | EC | 8.20, 8.21 |
| Speech transmission index | STI | 7.17 |
| Lateral energy fraction | LEF | 7.19 |
| Late lateral energy | $LG_{80}^{\infty}$ | 7.22 |
| Interaural cross-correlation | IACC | 7.20, 7.21 |

with the equations by which they are defined. (Only the evaluation of the speech transmission index [STI] is somewhat more involved.) The experimental determination of the 'lateral energy fraction' (LEF) and the 'late lateral energy' $LG_{80}^{\infty}$ requires the use of an additional microphone with gradient characteristics (figure-of-eight microphone) oriented in such a way that the sound source lies in its plane of zero sensitivity. Both microphones are placed at the same position. The 'interaural cross-correlation' (IACC) is obtained by cross-correlating the impulse responses describing the sound transmission from the sound source to both ears of a human head. If such measurements are carried out only occasionally, the responses can be obtained with two small microphones fixed at the entrance of both ear channels of a real person, whose only function is to scatter the sound waves in a realistic way. For routine work, it is certainly more convenient to replace the human head by a dummy head with built-in microphones.

## 8.4 MEASUREMENT OF REVERBERATION

Although both the reverberation time of a room as well as the early decay time (EDT) can be derived from its impulse response, we start by describing the more traditional method, the principle of which agrees with Sabine's measuring procedure, apart from his experimental equipment. Since the reverberation time is usually determined by the evaluation of decay curves, the first step is recording decay curves over a sufficiently wide range of sound levels.

The standard set-up for this measurement, which can be modified in many ways, is schematically depicted in Figure 8.10. A loudspeaker LS driven by a signal generator excites the room to steady-state conditions. The output voltage of the microphone *M* is fed to an amplifier (not shown in Figure 8.10) and filter *F*, and then to a logarithmic recorder *LR*, whose read-out corresponds to the sound pressure level in decibels. At a given moment, the excitation is switched off, and at the same time, the recorder starts to record the decay process.

The signal produced by the generator is either a frequency-modulated sinusoidal signal whose momentary frequency covers a narrow range or it is random noise filtered by an octave or third-octave filter. The range of mid-frequencies, for which such measurements are usually carried out, extends from about 50 to 10,000 Hz; most frequently, however, the range from 100 to 5,000 Hz is considered. As mentioned in Section 8.1, excitation by a pistol shot is often a practical alternative. Pure sinusoidal tones are used only occasionally, for example, to excite individual modes in the range well below the Schroeder frequency.

The loudspeaker, or more generally, the sound source, is placed at the location of the original source when the room is in its normal use. This applies not only to reverberation measurements but also to other measurements. However, because of the reciprocity principle (see Section 3.1), the location of the sound source and the microphone can be interchanged without altering the results, provided the sound source and the microphone have no directionality. In any case, it is important that the distance between the sound source and the microphone is much larger than the critical distance given by Equation 5.44 or 5.48; otherwise, the direct sound would have an undue influence on the shape of the decay curve.

If the sound field were completely diffuse, the decay should be the same throughout the room. Since this ideal condition is hardly ever met in normal rooms, it is advisable to carry out several measurements for each frequency band at different microphone positions. This does not hold even more if the quantity to be evaluated is the EDT, because this varies considerably from one place to another within the same hall.

In Figure 8.10, the microphone signal is passed through a band filter — usually an octave or third-octave filter — mainly to improve the signal-to-noise ratio, that is, to reduce the disturbing influence of noise produced in the hall itself or in the microphone and the electrical amplifier. If the room is excited by a pistol shot or another wide-band signal, it is this filter which defines the frequency band for which the reverberation time is measured and thus yields a rough measure of its frequency dependence.

For recording the decay curves, the conventional electromechanical level recorder has been superseded nowadays by the digital computer, which converts the sound pressure amplitude of the received signal into the instantaneous level. Usually, the level $L(t)$ in the experimental decay curves does not fall in a strictly linear way but contains random fluctuations which are due, as explained in Section 3.8, to complicated interferences between decaying normal modes. If these fluctuations are not too strong, the decay curve can be approximated by a straight line. This can be done manually, that is, by ruler and pencil. If high precision is required, it may be advantageous to carry out a 'least square fit': Let $t_1$ and $t_2$ denote the interval in which the decay curve is to be approximated (see Figure 8.11). Then, the following integrations must be performed:

$$I_1 = \int_{t_1}^{t_2} L(t)\mathrm{d}t, \quad I_2 = \int_{t_1}^{t_2} L(t)t\mathrm{d}t \tag{8.26}$$

With the abbreviations $\Delta_1 = t_2 - t_1$, $\Delta_2 = \left(t_2^2 - t_1^2\right)/2$ and $\Delta_3 = \left(t_2^3 - t_1^3\right)/3$, the mean slope of the straight line in the defined interval is

$$-\frac{\Delta L}{\Delta t} = \frac{\Delta_1 I_2 - \Delta_2 I_1}{\Delta_1 \Delta_3 - \Delta_2^2} \tag{8.27}$$

This procedure is particularly useful when applied to EDT.

In any case, the slope of the decay curve is related to the reverberation time T by

$$T = 60 \cdot \left|\frac{\Delta t}{\Delta L}\right| \tag{8.28}$$

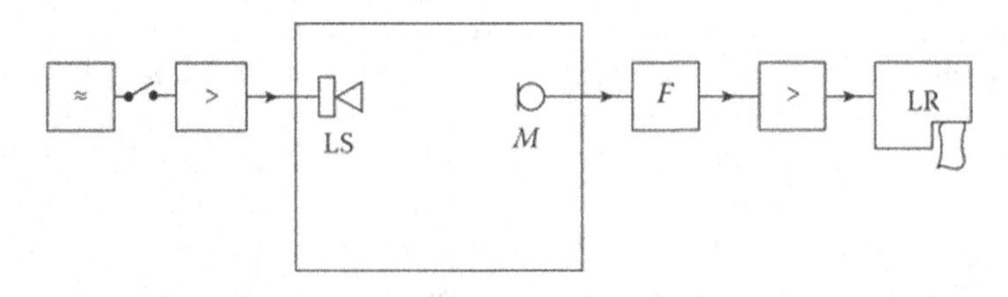


*Figure 8.10* Measurement of sound decay and reverberation time.

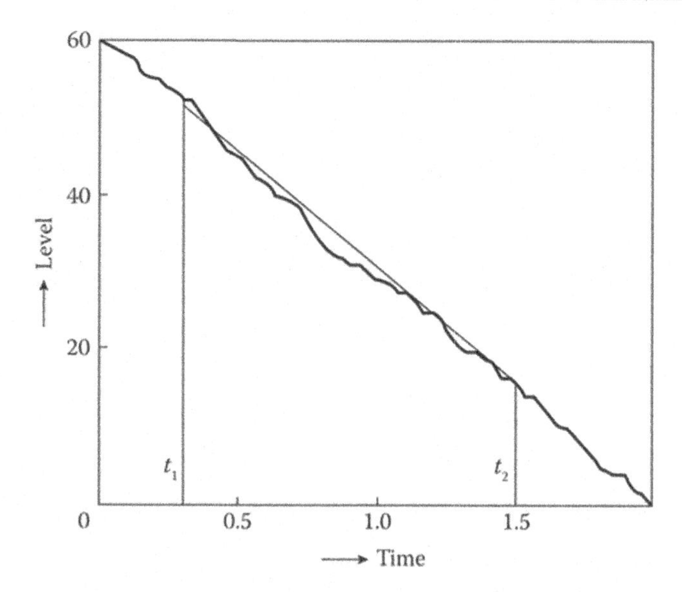


*Figure 8.11* Logarithmic decay curve and its approximation within the limits $t_1$ and $t_2$ by a straight line.

If random noise is used as an excitation signal, each member of a series of repeated decay measurements is slightly different from all others, and none of them is representative for all decay processes. This is an immediate consequence of the random character of the input signal. This uncertainty can be avoided, in principle, by averaging over a great number of individual reverberation curves, taken under otherwise unchanged conditions. Fortunately, this tedious procedure can be circumvented by applying an elegant method, called 'backward integration', which was proposed and first applied by Schroeder (1965). It is based on the following relationship between the impulse response $g(t)$ and the average $\langle h^2(t)\rangle$ over all individual decay curves (for a given measurement configuration and noise band):

$$\langle h^2(t)\rangle = \int_t^\infty [g(x)]^2\,\mathrm{d}x = \int_0^\infty [g(x)]^2\,\mathrm{d}x - \int_0^t [g(x)]^2\,\mathrm{d}x \tag{8.29}$$

To prove this relationship, let us suppose that the room is excited by white noise $n(t)$ switched off at the time $t = 0$. According to Equation 1.63, the decaying sound pressure is given by

$$h(t) = \int_t^\infty g(x)\cdot n(t-x)\,\mathrm{d}x \qquad \text{for } t \geq 0$$

Squaring this latter expression yields a double integral, which, after averaging, reads

$$\langle h^2(t)\rangle = \int g(x)\,\mathrm{d}x \int g(y)\langle n(t-x)\cdot n(t-y)\rangle\,\mathrm{d}y$$

The acute brackets indicate ensemble averaging, strictly speaking, which, however, is equivalent to temporal averaging. Hence,

$$\langle n(t-x)\cdot n(t-y)\rangle = \phi_{\mathrm{nn}}(x-y)$$

is a Dirac function, namely, the autocorrelation function of white noise. Thus, the double integral is reduced to the single integral of Equation 8.29. This derivation is valid no matter whether the impulse response is that measured for the full frequency range or for only a part of it.

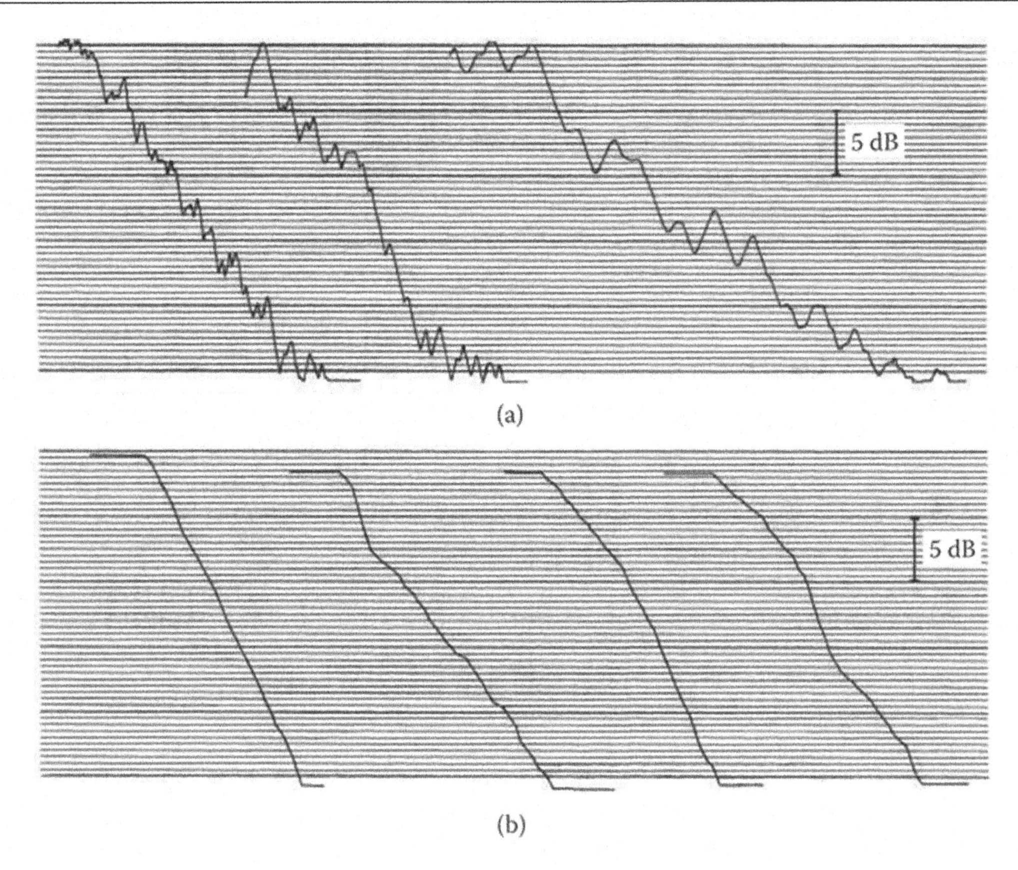


*Figure 8.12* Experimental reverberation curves: (a) conventional procedure according to Figure 8.10 and (b) recorded with backward integration according to Equation 8.29.

The merits of this method are demonstrated by the examples shown in Figure 8.12. The upper decay curves have been measured with the traditional method, that is, with interrupted random noise excitation, according to Figure 8.10. The decaying level shows strong fluctuations, which do not reflect any acoustical properties of the transmission path, and hence of the room, but are mainly due to the random character of the exciting signal; if these measurements were repeated, each new result would differ from the preceding one in many details. In contrast, the lower curves, obtained by backward integration according to Equation 8.29, are free of such confusing fluctuations and hence contain only significant information. Repeated measurements for one situation yield identical results, which is not surprising, since these decay curves are unique functions of impulse responses. It is clear that from such registrations the reverberation time can be determined with much greater accuracy than from decay curves recorded in the traditional way. This holds even more for EDT. Furthermore, any characteristic deviations of the sound decay from exponential behaviour are much more obvious.

For the practical execution, the second version of Equation 8.29 is more useful in that it yields the decaying quantity in real time. This means that the integral of the squared impulse response must be measured and stored beforehand. Of course, the upper limit $\infty$ must be replaced with a finite value, named $t_\infty$, which is not uncritical, because there is always some acoustical or electrical background noise. Its effect on exponential decay curves corresponding to a reverberation time $T = 2$ s is demonstrated by Figure 8.13. If the limit $t_\infty$ is too long, the microphone will pick up too much noise, which causes the characteristic tail at the lower

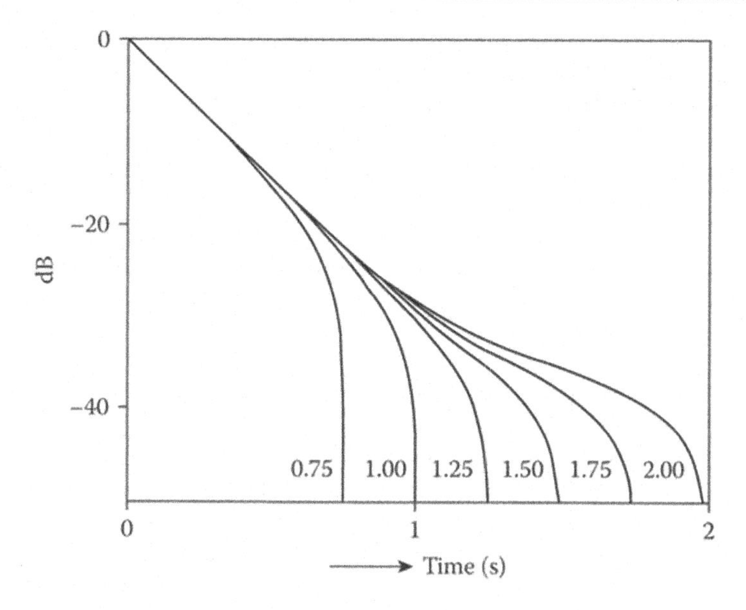


*Figure 8.13* Effect of background noise on decay curves processed with backward integration according to Equation 8.29. The noise level is −41.4 dB. Parameter: upper integration limit $t_\infty$, replacing ∞ (in seconds).

end of the decay curve. Too short an integration time will lead to an early downward bend of the curve, which is also awkward. Obviously, there exists an optimum (about 1.25 s in Figure 8.13) for $t_\infty$ which depends on the relative noise level and the decay time.

To determine the reverberation time after Equation 8.28, the slope of the decay curve is often evaluated in the level range from −5 to −35 dB relative to the initial level (ISO 3382). This procedure is to improve the comparability and reproducibility of results in such cases where the fall in level is not linear. It is doubtful, however, whether the evaluation of an average slope from curves which are noticeably bent is very meaningful or whether the evaluation should rather be restricted to their initial parts, that is, to EDT, which is anyway a more reliable indicator of the subjective impression of reverberance. The same argument applies if the reverberation measurements are carried out with the goal to determine the absorption coefficient of a test material (Section 8.7), since the average damping constant of all excited normal modes is related to the initial slope of a decay curve and not to some average slope (see Equation 3.68).

It should be mentioned that the absorption of a room, and hence its reverberation time, could be determined, at least in principle, from the steady-state sound level or energy density, according to Equation 5.42. Likewise, the modulation transfer function could be used to obtain the reverberation time. In practice, however, these methods do not offer any advantages over those described earlier, since they are more time-consuming and less accurate.

## 8.5 DIRECTIONAL DISTRIBUTION, DIFFUSENESS OF A SOUND FIELD

To this day, it is not clear whether the diffuseness of a sound field is an acoustical quality parameter in its own right or whether it is just the condition which ensures the validity of the simple reverberation formulae which are in common use. This is because the direct measurement of diffuseness is difficult and time-consuming. Accordingly, not many data on this sound field property, collected measured in concert or other large halls, are available. Since sound field 'diffusion' or 'diffusivity' is a kind of magic word in room acoustics, this book would remain incomplete without a section on the measurement of diffuseness.

The straightforward way to determine sound field diffuseness is certainly to measure the directional distribution of sound energy flow (sound intensity). As mentioned in Section 4.3, this distribution is characterized by a function $I(\varphi,\vartheta)$. It can be measured by scanning all directions with a directional microphone with sufficiently high angular resolution. Let $\Gamma(\varphi,\vartheta)$ be the directional factor, that is, the relative sensitivity of the microphone as a function of angles $\varphi$ and $\vartheta$ in a suitably chosen polar coordinate system, then the squared output voltage of the microphone is proportional to

$$I'(\varphi,\vartheta) = \iint_{4\pi} I(\varphi',\vartheta')\cdot\left|\Gamma(\varphi-\varphi',\vartheta-\vartheta')\right|^2 \mathrm{d}\Omega' \tag{8.30}$$

This expression is the two-dimensional convolution of the true directional distribution $I(\varphi,\vartheta)$ with $|\Gamma(\varphi,\vartheta)|^2$, and $\mathrm{d}\Omega' = \sin\vartheta'\mathrm{d}\vartheta'\mathrm{d}\varphi'$ is the solid angle element. Obviously, the agreement of $I'$ with $I$ is better the closer the directivity factor $\Gamma$ comes to a Dirac function.

A measure of the isotropy or diffuseness of the sound field is the *'directional diffuseness'* introduced by Meyer and Thiele (1956). It is based on Equation 8.30, however, with the integration extended over the upper hemisphere only. We denote with $\langle I'\rangle$ the measured intensity averaged over all directions and with

$$m = \frac{1}{4\pi\langle I'\rangle}\iint\left|I' - \langle I'\rangle\right|\mathrm{d}\Omega \tag{8.31}$$

the relative mean absolute deviation from it. Let $m_0$ be the same quantity obtained in an anechoic chamber. Then, the directional diffuseness is defined as

$$d = \left(1 - \frac{m}{m_0}\right)\cdot 100\% \tag{8.32}$$

Dividing $m$ by $m_0$ produces a certain normalization and, consequently, $d = 100\%$ in a perfectly diffuse sound field, whereas in the sound field consisting of one single plane wave, the directional diffusion becomes zero. This simple procedure, however, does not completely eliminate the influence of the microphone characteristics. Therefore, experimental $d$ values are only comparable when they have been determined with similar microphones. (The correct way to 'clean' the experimental data would be to perform a two-dimensional deconvolution.)

Many distributions measured in this way can be found in the cited publication by Meyer and Thiele. As a directional microphone, these authors used a concave metal mirror (see Figure 8.21) with a microphone capsule arranged in its focus. The diameter of the mirror is 1.20 m, which leads to a half-power width of its directional characteristics (angular distance between 3 dB points) of about 8.6° at 2,000 Hz. The directional diffusivity $d$, as determined from the directional distribution, varied between 35% and 75% without showing a clear tendency.

A more manageable device has been employed by Tachibana et al. (1989) in the course of a survey of 20 large hall auditoriums in Europe and Japan. To detect the directional distribution of low order (i.e. of early reflections), they applied a method which was developed by Yamasaki and Itow (1989). The sensor used by these authors consists of four 1/4-inch pressure microphones which define a Cartesian coordinate system, with one of them situated at the origin of the imagined coordinate system while the remaining ones are placed on its three axes at a distance of 50 or 33 mm. The room is excited by an impulse with 5 μs duration. To improve the signal-to-noise ratio, the impulse responses picked up simultaneously by the microphones are averaged over up to 256 shots. Suppose the impulse responses contain just the direct sound and one reflection. The arrival times of both components are slightly different in the four final responses, and from these differences, the direction of the incident reflections can be determined. Together with the delay common to the four microphones, this leads to the location of the image source which caused this reflection. The arrival times are obtained from the peaks of the six short-time correlation coefficients:

$$\psi_{ik}(\tau) = \int_{t_a}^{t_a+\Delta} g_i(t) g_k(t+\tau)\mathrm{d}t \Bigg/ \left\{ \int_{t_a}^{t_a+\Delta} [g_i(t)]^2 \mathrm{d}t \int_{t_a}^{t_a+\Delta} [g_k(t)]^2 \mathrm{d}t \right\}^{1/2}$$

with $g_i$, $g_k$ denoting the averaged impulse responses ($i$, $k$ = 0, 1, 2, 3 and $i \neq k$). $\Delta$ is a suitably chosen time interval. A few of their results, measured in a living room, in the Boston Symphony Hall, and in a cathedral (the Münster in Freiburg, Germany), are presented in Figure 8.14. The first line shows the three impulse responses obtained in these rooms, the second line contains the spatial distributions of image sources indicated by small circles (second line), and in the third line, there are the corresponding directional distributions of the sound. The lengths of the 'rays' in the latter are proportional to the level of the received sound. (Both the pattern of image sources and the directional distributions are projected into the 'floor' plane.)

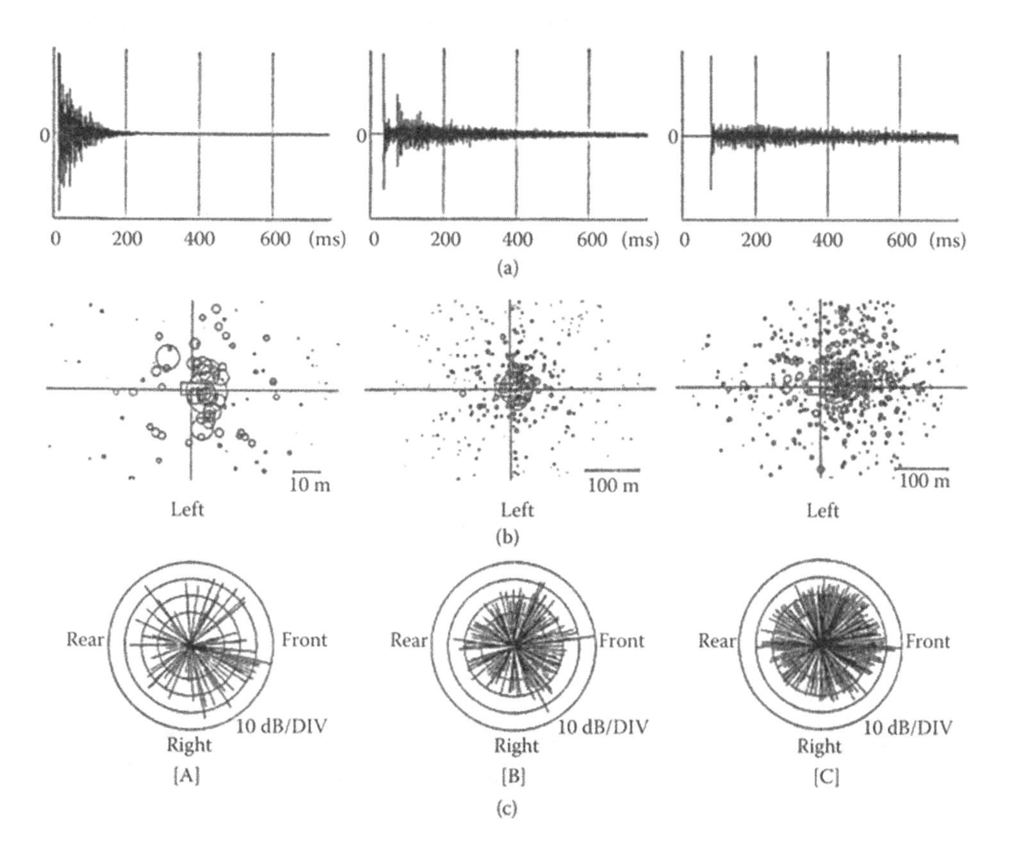


*Figure 8.14* Impulse response: (a) image sources (b) and directional distribution of reflected sound (c) for three different enclosures: living room [A], Boston Symphony Hall [B], and Münster in Freiburg/Germany [C]. The pattern of image sources and the directional distributions are projected into the ground plane.

Source: Yamasaki and Itow (1989), courtesy of *J Acoust Soc Japan*.

As an alternative, Yamasaki and Itow used the aforementioned four-microphone probe for measuring the three components of the particle velocity and the intensity sensor, according to Equation 8.1, with the goal of identifying image sources. To this end, they filtered the impulse responses with a suitable band-pass filter and applied a variable time window with a width of 10 ms to them. From the output of each microphone pair, the corresponding component of the short-time intensity was obtained. These components determine the magnitude and the direction of the intensity vector.

A procedure similar to that of Yamasaki and Itow was developed by Sekiguchi et al. (1992). These authors employed a probe consisting of four 1/4-inch microphones placed at the corners of a regular tetrahedron with a side length of 17 cm. For room excitation, they used 2 kHz tone bursts with a duration of 80 μs; to improve the signal-to-noise ratio, 32 or, alternatively, 64 impulse responses were averaged. Because of the relatively large dimensions of the probe, the authors achieved sufficiently high accuracy without performing the time-consuming correlation operations described earlier.

It should be noted that the application of both methods is restricted to the early part of a room impulse response, which consists of a few well-separated reflections. In the reverberant part of the response, the density of reflections is so high that no individual image sources can be identified.

A very powerful approach for directional sound field analysis starts with solving the wave equation in spherical coordinates (see Section 1.2). One of the fundamentals of angular sound field analysis is plane-wave decomposition. Since about 2000, higher-order spherical microphone arrays can be used to sample the sound field in multi-channel data formats. These were introduced as part of the development of spatial audio technology for sound recording and 3D sound reproduction but can also be used for room-acoustic measurements. 'Sampling' in this context means that the sound field is captured at discrete microphone position on a sphere, where, of course, higher precision is achieved with higher sampling density and thus more microphone channels (Rafaely 2019). In practice, the state of the art allows for 64 channels and more (Meyer and Elko 2002).

Accordingly, spherical microphone arrays are used to measure directional room impulse responses (Gover et al. 2004). The representation of the sound field in the spherical harmonic (SH) domain (Section 1.2) provides an elegant processing methodology with decomposition, filtering, and analysis of the spatial (directional) information in the stationary or transient room response.

We start with a series expression of the room impulse response, *h(t)*, as sum of spherical harmonics for each incidence direction of the direct sound and all reflections (Berzborn and Vorländer 2021):

$$h_{nm}(t) = \sum_{l=1}^{L} s_l(t) Y_n^m(\theta_l, \phi_l). \tag{8.33}$$

$Y_n^m(\theta_l, \phi_l)$ are the real-valued spherical harmonic basis functions (Equation 1.32) of order $n$ and degree $m$ for the angles of incidence, $\theta_l, \phi_l$, of the sound field components on the receiver array. Equation (8.33) is called 'plane-wave composition' or SH expansion of the sound field (Rafaely 2019). In this expression, $s_l(t)$ contains the signal of the wave arriving at the microphone array after travelling in the room from the source via reflection, scattering, and diffraction processes in the room. It can also be considered as a specific sound 'path' arriving at the receiver at the angles $\theta_l, \phi_l$ with a propagation delay $t$ between emission at the source and immission at the receiver.

In this type of expression, the room sound field can be analyzed in various ways in the spatio-temporal domain. One application can be to determine a mixing time, that is, the time after which the sum of reflections forms a uniform (diffuse) sound incidence mainly contained in the low SH orders, rather than a sparse reflection pattern with distinct angles of incidence. Together with other time-domain approaches to measuring mixing times (Jeong et al. 2010), this analysis can provide insight into the gradual process of transition from early reflections to late decay, similar to what is observed in the frequency domain for the modal and the statistical domains (Schroeder 1996).

Another application is the direct measurement of diffuseness, similar to the 3D intensity analysis in Equation 8.32. In the plane-wave decomposition, the wave components can be decomposed into a discrete set of plane waves. Hence, the diffuseness is assessed by comparing the distribution of the energies per angle of incidence with the ideal isotropic case of an omnidirectionally uniform distribution of incoherent waves. This can be done in the stationary case as well as in the impulse response or in the integrated impulse response (decay curve). Nolan et al. (2018, 2020) investigated the isotropy by spatial wave number decomposition, combining the wave number spectrum analysis (Section 3.2) with an SH expansion. The less energy is contained in SH orders other than $n = 0$, the more diffuse is the sound field. Berzborn and Vorländer (2021) used a spherical receiver array to estimate the isotropy of a sound field from the relative differences of the energy incident onto the receiver for a number of steering directions which can be interpreted as the various observation directions in certain beam apertures in $\theta,\phi$. The definition of the time-dependent diffuseness metric is actually very similar to intensity-based Equation 8.3 of Meyer and Thiele (1956) and Gover et al. (2004). It is defined to be observed in the ongoing energy decay process by

$$m_d(t) = 1 - \frac{\sigma_d(t)}{\sigma_{e,0}} \tag{8.34}$$

where $\sigma_d(t)$ is now the absolute difference between the incident directional energy after plane-wave decomposition and its directional mean. After normalization with the same difference calculation with the beamformer pattern, $\sigma_{e,0}$, this decay-related diffuseness metric has values between 0 and 1, where 1 denotes a perfectly isotropic field.

This analysis was applied to the integrated impulse response to investigate the degree of isotropy during the ongoing decay. Three room shapes show quite different patterns in their spatial decays and differences in their diffuseness too (Fig. 8.15). An interesting conclusion is that the definition of a mixing time (or transition time) between the early part and the nominally diffuse part is ambiguous, as the isotropy in a room sound field may reach a maximum before it decreases. However, a decrease of isotropy means a segregation of the sound field, which again leads to dominant angles of incidence. Another finding is that rooms with concentrated absorption tend to show curved decays with first increasing and then decreasing isotropy. Although this observation is not new, having been theoretically investigated by Hunt et al. (1939), it can now be measured experimentally, hopefully stimulating new research to improve the method of measuring diffuse-field absorption in reverberation chambers (see Section 8.7).

If one is not interested in the details of the directional distribution but only in a measure for its uniformity, more indirect methods can be applied, that is, one can measure a quantity whose value depends on the degree of diffuseness. One of these quantities is the spatial correlation coefficient of the steady-state sound pressure at two different points which assumes characteristic values in a diffuse field. Or more precisely, we consider the correlation coefficient $\Psi$ defined in Equation 1.59:

$$\Psi = \frac{\overline{p_1 \cdot p_2}}{\sqrt{\overline{p_1^2} \cdot \overline{p_2^2}}} \tag{8.35}$$

To calculate the correlation coefficient in a diffuse sound field, we assume that the room is excited by random noise with a very small bandwidth. The sound field can be considered

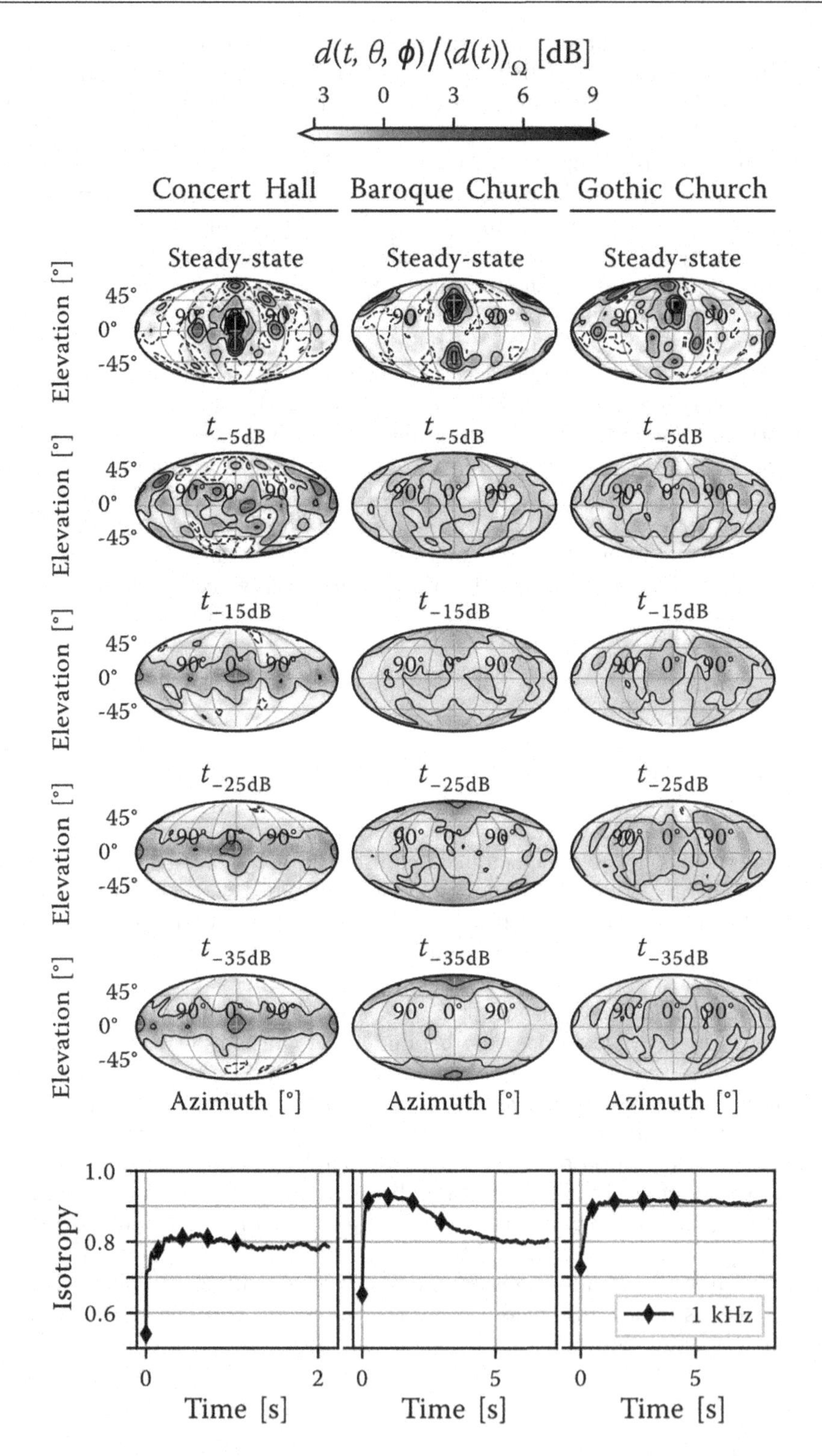


*Figure 8.15* Directional energy decay distributions normalized by their directional mean (top) and estimated diffuseness (bottom) for the 1 kHz octave frequency band at a receiver position (example). Top-down: time steps from steady state to −5 dB, −15 dB, −25 dB, −35 dB after switch-off. The contour lines represent the levels marked in the greyscale bar.

Source: Berzborn and Vorländer (2021).

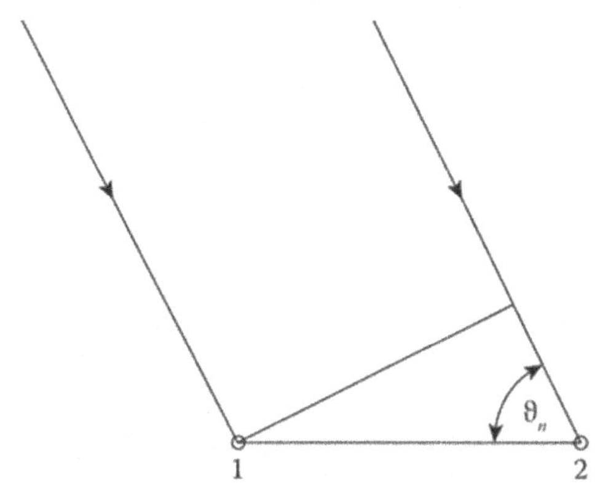


*Figure 8.16* Derivation of Equations 8.36–8.38.1 and 2: microphone positions.

as being composed of plane waves with random amplitudes and phase angles $\psi_n$. Then, the sound pressure due to one such wave at two points 1 and 2 at distance $x$ (see Figure 8.16) is

$$p_1(t) = A\cos(\omega t - \varphi_n) \quad \text{and} \quad p_2(t) = A\cos(\omega t - \varphi_n - kx\cos\vartheta_n)$$

$\omega = kc$ is the centre frequency of the exciting frequency band. The angle $\vartheta_n$ characterizes the direction of the incident sound wave. Time averaging of the squared sound pressures yields

$$\overline{p_1^2} = \overline{p_2^2} = \frac{A^2}{2}$$

The time average of the product of both pressures is

$$\overline{p_1 \cdot p_2} = \frac{A^2}{2}\cos(kx\cos\vartheta_n)$$

Inserting these expressions into Equation 8.35 leads to $\Psi(x,\vartheta_n) = \cos(kx\cos\vartheta_n)$. Finally, this expression is averaged over all possible directions of incidence with equal weight, which means that the sound field is assumed as diffuse. The result is

$$\Psi(x) = \frac{\sin kx}{kx} \tag{8.36}$$

If, however, the directions of incident sound waves are not uniformly distributed over the entire solid angle but only in a plane containing both points 1 and 2, we obtain, instead of Equation 8.36,

$$\psi(x) = J_0(kx) \tag{8.37}$$

where $J_0$ is the Bessel function of order zero. If the line connecting both points is perpendicular to the plane of two-dimensional sound propagation, the result is

$$\psi(x) = 1 \tag{8.38}$$

The functions presented in Equations 8.36–8.38 are plotted in Figure 8.17. The most interesting one is curve *a*; it is expected that any lack of diffuseness manifests itself in poor agreement between the measured and the theoretical correlation curve.

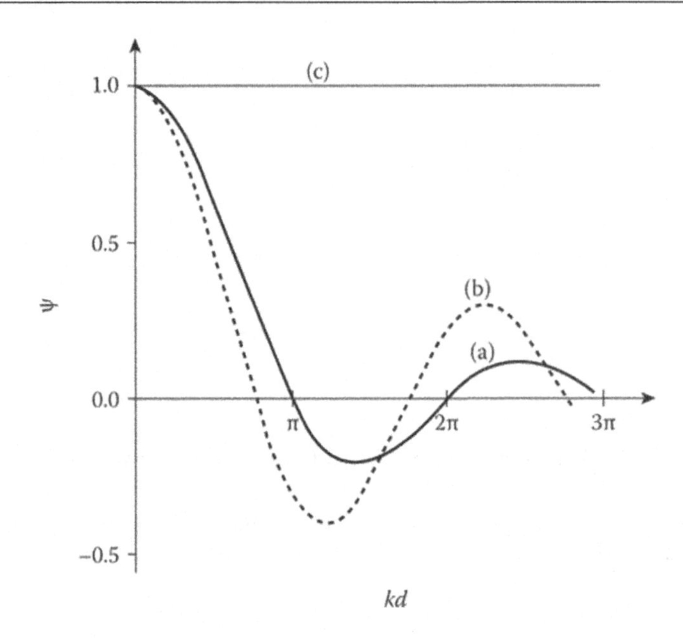


*Figure 8.17* Correlation coefficient $\Psi$ as a function of the distance $x$ between two observation points: (a) in a three-dimensional diffuse field, (b) in a two-dimensional diffuse field with the measuring axis 1–2 (in Figure 8.17) lying within its plane, and (c) same as (b), but with the measuring axis perpendicular to the plane of sound propagation.

A relatively simple way to observe the correlation coefficient is the playback method described in Section 8.3. Its basis is an obvious generalization of Equation 8.22, which is now applied to two different impulse responses, $g(t)$ and $g'(t)$, and yields the cross-correlation function of both responses:

$$\phi_{gg'}(\tau) = \int_{-\infty}^{\infty} g(t)g'(t+\tau)\mathrm{d}t = \int_{-\infty}^{\infty} g(-t)g'(\tau - t)\mathrm{d}t \tag{8.39}$$

The recipe contained in this expression reads as follows: First, the room is excited with a short impulse; its response, g(t), received at point 1, is stored on a magnetic tape or a computer. In a second step, the time-reversed function g(−t) is used as an input signal; the response to it is observed at point 2, for which the impulse response would be g′(t). The signal received during the second step is the cross-correlation function $\phi_{gg}'(\tau)$; here, $\tau$ appears as the real time. The correlation coefficient $\Psi$ is proportional to $\phi_{gg}'(\tau_{\max})$with $\tau_{max}$ denoting the time where $\phi_{gg}'$assumes its absolute maximum. Figure 8.18 presents as an example the correlation coefficient measured in the reverberation chamber. The floor of this room was completely covered with 5 cm Rockwool, while the other walls are virtually free of absorption. The correlation coefficient has been measured as a function of the distance $x$; the line 1–2 connecting both microphones in Figure 8.16 was vertical. As a sound signal, a short impulse has been used filtered by a third-octave filter with a centre frequency of 1,000 Hz. The three series of measurement refer to different configurations of scatterers, namely 0, 16, and 25 diffusers irregularly suspended in the room. The effect of the diffusers is obvious.

The derivation leading to Equations 8.36 to 8.38 is strictly valid only for signals with very small frequency bandwidth. If the bandwidth $\Delta\omega$ is finite but still small compared with the central frequency $\omega$, Equation 8.39 can be averaged with the approximate result (Sekiguchi et al. 1992)

$$\Psi(x) = \frac{1}{\Delta k}\int_{k_1}^{k_2} \frac{\sin(kx)}{kx}\mathrm{d}k \approx \frac{\sin(x\Delta k)}{x\Delta k} \cdot \frac{\sin(kx)}{kx} \quad \text{with } \Delta k = k_2 - k_1 = \Delta\omega / c \tag{8.39}$$

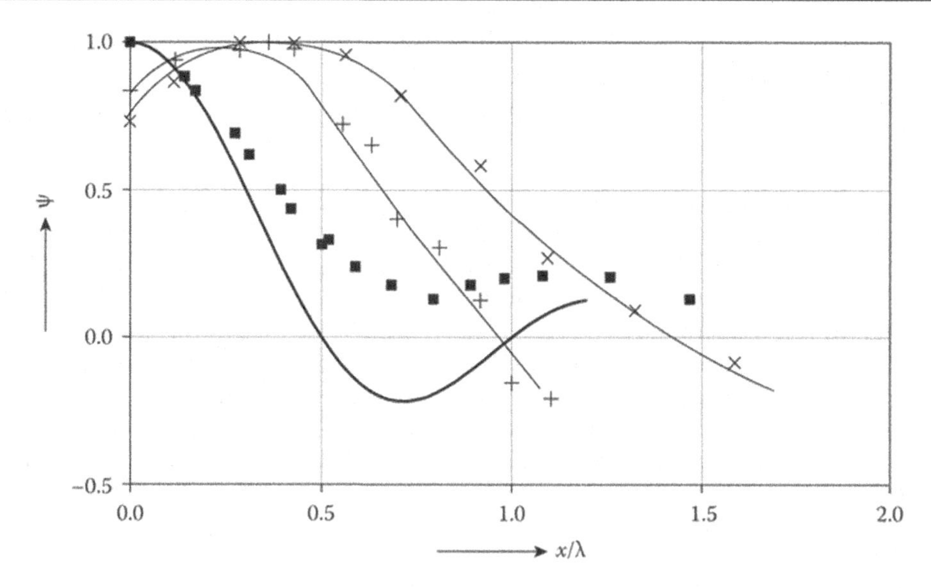


*Figure 8.18* Correlation coefficient measured in a reverberation chamber for various configurations of suspended scattering bodies (Kuttruff 1963). (■ ■ ■, 25 scatterers; + + +, 16 scatterers; × × ×, no scatterers.)

The correlation coefficient is not quite unambiguous, although it is hard to imagine a non-diffuse sound field leading to Equation 8.36 or 8.40. Nevertheless, it is sometimes recommended to amend it by an additional criterion for the diffuseness of a sound field. One possibility is to observe the spatial constancy of the energy density in the room (except in areas next to reflecting walls; see Figure 2.11).

However, we know from Section 3.6 that the energy density in a room, produced by a monofrequent source signal, shows pronounced spatial fluctuations irrespective of whether the sound frequency is below or above the Schroeder limit (Equation 3.44). In the first case, the sound field consists of isolated modes, whereas in the latter one, the sound source excites several or many modes simultaneously which interfere with each other in a complicated way (see Figure 3.8a). From this fact, we conclude that a sound field excited with a pure tone cannot be diffused. However, if the sound source emits filtered random noise with restricted bandwidth, some smoothing of the fluctuations will take place, depending on the bandwidth $B$ and the reverberation time $T$ as described in Section 3.7. As earlier, we denote with $z(f)$ the result of this averaging process:

$$z(f) = \frac{1}{B}\int_{f_1}^{f_2} y(f)\,\mathrm{d}f \quad \text{with}\, f_2 - f_1 = B \tag{8.41}$$

The relative variance of this quantity is given by Equation 3.57, which, for large values of the product $BT$, can be approximated by

$$\frac{\mathrm{Var}(z)}{\langle z^2 \rangle} \approx \frac{6.91}{\mathrm{BT}} \tag{8.42}$$

This variance corresponds to a certain fluctuation of the level $L = 10\log_{10} z = 4.34\ln(z)$:

$$\Delta L \approx \frac{\mathrm{d}L}{\mathrm{d}z}\Delta z = 4.34\frac{\sqrt{\mathrm{Var}(z)}}{\langle z \rangle} \approx \frac{11.4}{\sqrt{\mathrm{BT}}} \tag{8.43}$$

From this relation, we conclude that the level of fluctuation will remain below 1 dB as long as BT > 11.4 2, or roughly

$$\mathrm{BT} > 100 \tag{8.44}$$

Of course, this rule can only be applied if all frequencies contained in $B$ are lying above the Schroeder frequency.

An alternative method of testing the diffuseness is to measure the squared sound pressure amplitude in front of a sufficiently rigid wall as a function of the distance, as discussed in Section 2.5. In fact, Equation 8.36 agrees — apart from a factor 2 in the argument — with the second term of Equation 2.50. This, similarity, is not accidental, since the fluctuations shown in Figure 2.11 are caused by interference of the incident and the reflected waves which become less pronounced with increasing distance from the wall according to the decreasing coherence of those waves. The additional factor of 2 in the argument of Equation 2.41 accounts for the fact that the distance of some point from its mirror image, the rigid wall being considered as a mirror, is equivalent to the distance of both observation points in Figure 8.17.

Finally, the degree of sound field diffuseness can be checked by measuring the sound intensity, which should be zero in a perfectly diffuse field. This method is not restricted to a particular kind of room excitation. A useful measure of the diffuseness is the quantity

$$q_\mathrm{d} = 1 - \frac{c|\mathbf{I}|}{w} \tag{8.45}$$

It varies between 0 (plane wave) and 1 (diffuse field). The magnitude of the intensity vector **I** is calculated from its three Cartesian components,

$$|\mathbf{I}|^2 = I_x^2 + I_y^2 + I_z^2$$

which are determined by using a three-component intensity probe (see Section 8.1).

It should be emphasized that the conditions on which all the described indicators of diffuseness are based are necessary but not sufficient. The only exception is the direct measurement of the directional distribution as described earlier.

## 8.6 SOUND ABSORPTION — TUBE METHODS

The knowledge of sound absorption of typical building materials is indispensable for all tasks related to room acoustical design: for the prediction and control of reverberation times of auditoria and other rooms, for the acoustical computer simulation of environment, for model experiments, and for several other purposes.

Usually, the absorption of a surface is characterized by its absorption coefficient, which is a function of the angle of sound incidence and, of course, of the frequency. It is closely related to the reflection factor and the wall impedance of the material, described in Section 2.1. So all methods to determine the latter quantities can be used to obtain the absorption coefficient (but not vice versa). However, the direct determination of the reflection factor in the free field is laborious and requires large test samples and an anechoic environment. Hence, it is not in common use nowadays. An exception is the measurement of the wall impedance according to its definition in Equation 2.2, since small sensors for both the sound pressure and the particle velocity are available today, the latter in form of small hot-wire anemometers, as already mentioned in Section 8.1. Since there is usually a small distance $d$ between the location of the probe and the surface, a correction may be in place. If it is assumed that the probe and the test surface are separated by an air layer with the thickness $d$, the impedance transformation effected by the latter is, according to Equation 2.24,

$$Z = \rho_0 c \frac{Z' - i\rho_0 c \tan(kd)}{\rho_0 c - iZ' \tan(kd)} \tag{8.46}$$

In this expression, $Z'$ is the impedance observed at the position of the probe, while $Z$ denotes the true impedance of the surface.

Basically, there are two standard methods of measuring the acoustic absorption. They will be described in this and the next section. In the first of them, the test specimen and the probing sound field — a plane wave — are enclosed in a rigid tube. The measurement is restricted to the examination of small samples of locally reacting materials with a plane or nearly plane surface, and also to normal wave incidence onto the test specimen.

A typical set-up, known as 'Kundt's tube' or 'impedance tube', is shown in Figure 8.19. It has smooth and rigid walls and a rectangular or circular cross section. At one of its ends, there is a loudspeaker which generates a sinusoidal plane sound wave travelling toward the test specimen. This specimen terminates the other end of the tube and must be mounted in the same way as it is used in practice (for instance, with or without an air gap between the material and the hard backing). To suppress tube resonances, it may be useful, although not essential for the principle of the method, to place a wedge-like absorber in front of the loudspeaker, as shown in the figure. The test sample reflects the incident wave more or less; the result is a partially standing wave in front of the sample, as described in Section 2.2. Its pressure maxima and minima are measured by a movable microphone probe that must be small enough not to distort the sound field to any great extent. As an alternative, a miniature microphone mounted on the tip of a thin movable rod may be employed as well.

The tube must be long enough to permit the formation of at least one maximum and one minimum of the pressure distribution at the lowest frequency of interest. Its lateral dimensions should be chosen in such a way that, at the highest measuring frequency, they are still smaller than a certain fraction of the wavelength $\lambda_{min}$. Or more exactly, the following requirements must be met:

$$\begin{aligned} &\text{Rectangular tubes: Dimension of the wider side} < 0.5\lambda_{min} \\ &\text{Circular tubes: Diameter} < 0.586\lambda_{min} \end{aligned} \tag{8.47}$$

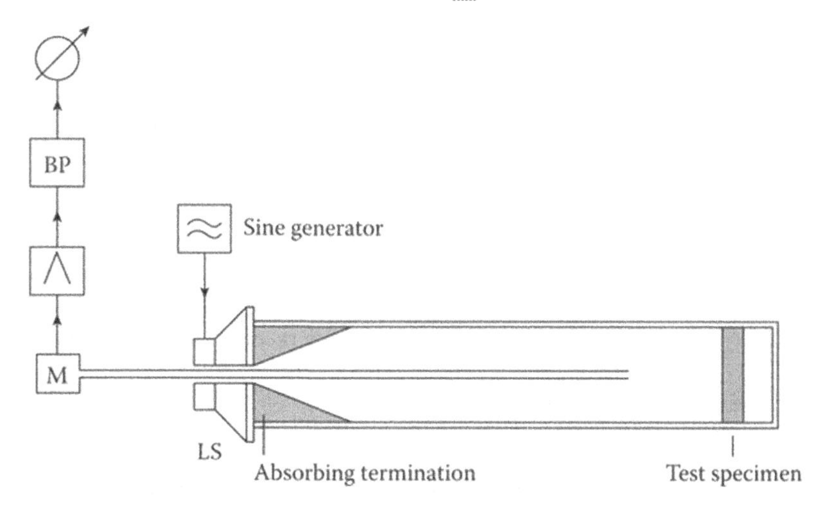


*Figure 8.19* Conventional impedance tube, schematic (*LS* = loudspeaker; *M* = microphone; *BP* = band-pass filter).

Otherwise, higher-order wave types may occur with non-constant lateral pressure distribution and with different and frequency-dependent wave velocities. On the other hand, to avoid unduly high wall losses, the cross section of the tube must not be too small. Generally, at least two tubes of different dimensions are needed in order to cover the frequency range from about 100 to 5,000 Hz.

For the determination of the absorption coefficient, it is sufficient to measure the maximum and the minimum values of the sound pressure amplitudes, that is, the pressures in the nodes and the antinodes of the standing wave. According to Equation 2.17 (with $\theta = 0$), their ratio is

$$\frac{\hat{p}_{\text{max}}}{\hat{p}_{\text{min}}} = \frac{1+|R|}{1-|R|}$$

from which both the magnitude of the reflection factor and the absorption coefficient are easily obtained:

$$|R| = \frac{\hat{p}_{\text{max}} - \hat{p}_{\text{min}}}{\hat{p}_{\text{max}} + \hat{p}_{\text{min}}} \tag{8.48}$$

$$\alpha = 1 - |R|^2 = \frac{4\hat{p}_{\text{min}} \cdot \hat{p}_{\text{min}}}{\left(\hat{p}_{\text{min}} + \hat{p}_{\text{min}}\right)^2} \tag{8.49}$$

If possible, the maxima and minima closest to the test specimen should be used for the evaluation of $|R|$ and $\alpha$, since these values are least affected by the attenuation of the waves. It is possible, however, to eliminate this influence by interpolation or by calculation, but in most cases, it is hardly worthwhile doing this.

If the test sample is of the locally reacting type, its absorption coefficient $\alpha_{\text{uni}}$ for random sound incidence can be calculated from its complex wall impedance, using Equation 2.54 or Figure 2.12. The wall impedance can be obtained, in turn, from the complex reflection factor after Equation 2.9 (with $\theta = 0$). To determine the phase angle $\chi$, of the latter, we observe the location $x_{\text{min}}$ of the pressure minimum (pressure node) next to the test specimen. This is related to the phase angle (see Equation 2.17) by

$$\chi = \pi\left(1 - \frac{4x_{\text{min}}}{\lambda}\right) \tag{8.50}$$

Several attempts have been made to replace the somewhat involved and time-consuming standing wave method by faster and more modern procedures. A typical arrangement is sketched in Figure 8.20. The movable probe is replaced with two fixed microphones which are mounted flush into the wall of the tube. Let $S(f)$ denote the spectrum of the stationary sound signal emitted by the loudspeaker. If we choose the position of microphone 2 as a reference for the phases, the spectra of the sound signals received at both microphones are

$$S_1(f) = S(f)\exp(-ik\Delta)\cdot\left[1 + R(f)\cdot\exp(-i2kd)\right] \tag{8.51}$$

and

$$S_2(f) = S(f)\left[1 + R(f)\exp(-i2kd')\right] \text{ with } d' = d + \Delta \tag{8.52}$$

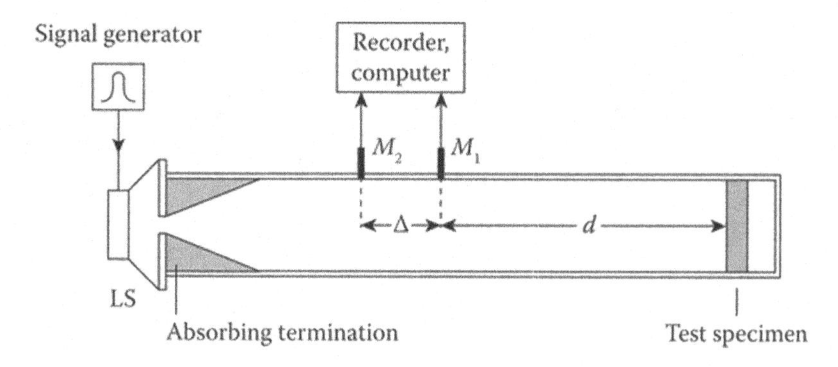


*Figure 8.20* Impedance tube with two fixed microphones.

As shown in Figure 8.20, $d$ is the distance of microphone $M_1$ from the surface of the sample under test, and $\Delta$ denotes the distance between both microphones. From these equations, the complex reflection factor is easily isolated:

$$R(f) = \exp(2ikd)\frac{\exp(ik\Delta) - H_{12}}{H_{12} - \exp(-ik\Delta)} \quad \text{with} \quad H_{12} = S_2(f)/S_1(f) \tag{8.53}$$

Critical are those frequencies for which $\exp(ik\Delta)$ is close to +1 or −1, that is, when the distance $\Delta$ is about an integer multiple of half the wavelength. In such regions, the accuracy of measurement is not satisfactory. This problem can be circumvented by providing for a third microphone position. Of course, the relative sensitivities of all microphones must be taken into account, or the same microphone is used to measure $S_1$ and $S_2$ in succession.

If a short impulse, idealized as a Dirac impulse $\delta(t)$, is used as a test signal, the measurement can be carried out with one microphone, since the signals due to the incident and the reflected wave can be separated by applying proper time windows. The sound signal received by the microphone $M_1$ is

$$s'(t) = \delta(t) + r(t) * \delta(t - 2d/c) = \delta(t) + r(t - 2d/c) \tag{8.54}$$

where $r(t)$ is the 'reflection response' of the test material, defined as the inverse Fourier transform of the reflection factor $R$ (see Section 4.2). If necessary, the signal-to-noise ratio can be improved by replacing the test impulse by a time-stretched test signal with constant amplitude spectrum, for example, by a sine sweep, as described in Section 8.2.

In order to separate safely the reflected signal from the primary one, the latter must be sufficiently short, and the distance $d$ of the microphone from the sample must be large enough. The same holds for any spurious reflections, for instance, from the loudspeaker or some other reflecting object. This may lead to impractically long tubes. An alternative is to omit the tube, as depicted in Figure 8.21. In this form, the one-microphone techniques can be used for *in situ* measurement of acoustical wall and ceiling properties in existing enclosures. In this case, however, the waves are not plane but spherical. For this reason, the results may be not too exact, because the reflection of spherical waves is somewhat different from that of plane waves (see Section 2.4). Of course, the $1/r$ law of spherical wave propagation must be accounted for by proper correction terms in Equation 8.54. Further refinements of this useful method are described in a paper published by Nélisse and Nicolas (1997).

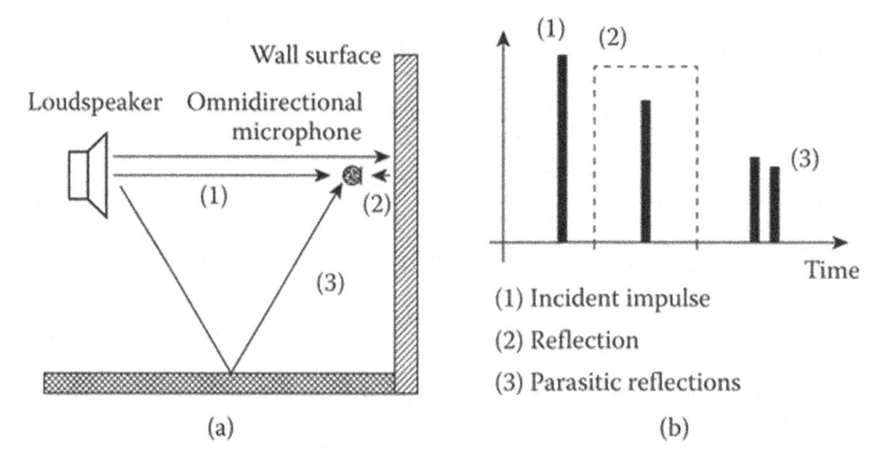


*Figure 8.21 In situ* measurement of acoustical wall properties: (a) experimental set-up and (b) sequence of received signals. Reflection 2 is isolated by applying a suitable time window.

Source: Mommertz and Vorländer (1995).

## 8.7 SOUND ABSORPTION — REVERBERATION CHAMBER

The reverberation method of absorption measurement is superior to the impedance tube method in several respects. First of all, the measurement is performed with a diffuse sound field, that is, under conditions which are much more realistic than those encountered in a one-dimensional waveguide. Secondly, there are no limitations concerning the type and construction of the absorber, and the test specimen can be set up in much the same way in which the material is to be used in the particular practical application. Hence, the reverberation method is well suited for measuring the absorption coefficient of almost any type of wall linings and of ceilings. And finally, the absorption of discrete objects which cannot be characterized by an absorption coefficient can be determined in a reverberation chamber. This concerns, for instance, any kind of chairs, empty or with persons seated on them.

A reverberation chamber is a small room with a volume of at least 100 m$^3$, better still, 200–300 m$^3$, whose walls are as smooth and rigid as possible. The absorption coefficient $a_0$ of the bare walls, which should be uniform in construction and finish, is determined by measuring the reverberation time of the empty chamber:

$$T_0 = 0.161 \frac{V}{S\alpha_0} \tag{8.55}$$

($V$ = volume in m$^3$, $S$ = wall area in m$^2$). For the actual measurement, a certain amount of the material under investigation (or a certain number of absorbers) is brought into the chamber; the arrangement of these materials should correspond to the arrangement in the intended application. The test specimen with the area $S_s$ reduces the reverberation time from $T_0$ to $T$ with

$$T = 0.161 \frac{V}{S_s\alpha + (S - S_s)\alpha_0} \tag{8.56}$$

From this equation, the absorption coefficient $a$ of the test sample is easily calculated. In the case of discrete objects, the product $S_s\alpha$ in the denominator of Equation 8.56 is replaced by the total absorption area of the objects.

As long as there are constant conditions between the two measurements, the air attenuation term $4mV$ introduced in Section 5.4 can be neglected, since it is contained in the absorption of the empty chamber as well as in that of the chamber containing the test material and, therefore, will almost cancel. Its effect is small, anyway, because of the small chamber volume.

The measurement of reverberation has been described in detail in Section 8.4, and therefore, no further discussion on this point is necessary. To increase the accuracy of the decay measurement, it is recommended to repeat the measurement with different source and microphone positions and to average the results. Usually, these measurements are performed with frequency bands of third-octave bandwidth.

As mentioned earlier, the application of the Sabine equation produces systematic errors in that it overrates the absorption coefficient; sometimes, the value determined with Equation 8.45 is larger than unity in contrast to the definition of the absorption coefficient. An example is shown in Figure 6.13 (dashed line). Such inconsistencies could be avoided by using other, more exact decay formulae, for instance, Eyring's formula (5.24) (without the attenuation term $4mV$). Another possible reason for unrealistically high absorption coefficients exceeding unity is edge diffraction, as will be discussed later. Nevertheless, the international standardization of this measuring procedure recommends the use of Equation 8.55 (ISO 354). The coefficient determined in this way is sometimes given the name 'Sabine absorption coefficient' in order to distinguish it from the 'statistical absorption coefficient' or the random incidence absorption coefficient $a_{uni}$ after Equation 2.54.

The advantages of the reverberation method as mentioned at the beginning of this section are offset by considerable uncertainty as regards its reliability and the accuracy of the results obtained with it. In fact, several round robin tests, in which the same specimen of an absorbing material has been tested in different laboratories, have revealed a remarkable disagreement in the results, especially in the low-frequency range. Obviously, these discrepancies must be attributed to different degrees of sound field diffuseness established in the reverberation chambers. Therefore, as already mentioned in Section 8.5, great attention must be paid to this issue. Depending on the placement of the absorber sample, some modes are damped more than others. The modes with least attenuation, such as the tangential modes for an absorber placed on the floor, have longer decay times. All this leads to the fact that the absorption coefficient according to Equation 8.56 depends on where in the decay curve the slope is evaluated. Isotropy analysis and consideration of the best-fit slope estimation for curved decays may lead to more reproducible results of the random-incidence absorption coefficient in the future.

A first step towards a high degree of sound field diffuseness is to avoid pairs of parallel walls in the design of a reverberation chamber. Otherwise, sound waves would be reflected to and fro between such wall pairs without being significantly influenced by other walls. Further improvement is achieved by means of providing for sound scattering either during reflections from the boundary or during the propagation of the sound waves in the free volume. In the former case, the increase in diffuseness is brought about by corrugations of the boundary, for example, by spherical or cylindrical segments or other rigid bodies that are attached to the boundary. It is important that the dimensions of these 'boundary diffusers' are comparable to the acoustic wavelength. A useful alternative to this kind of scattering bodies are 'volume diffusers', as described in Section 5.2. Practically, these scatterers can be realized in the form of bent shells of wood, plastics, or metal of variable size which are hanging from the ceiling in an irregular arrangement (see Figure 8.21). If necessary, bending resonances of such shells can be damped by applying thin layers of some lossy material onto them. Up to now, no clear decision has been arrived at about which kind of diffusers is more efficient (Bradley et al. 2014). However, from a more practical standpoint, it seems that hanging diffusers are preferable, since a given arrangement of scatterers can easily be changed if it does not prove satisfactory. The increase in diffuseness due to the number of diffusers has already been demonstrated in Figure 8.17.

It should be noted that too many volume diffusers may also affect the validity of the usual reverberation formulae, and therefore, the density of scatterers has a certain optimum characterized by

$$0.5 < \langle n \rangle Q_s h < 2 \tag{8.57}$$

with $\langle n \rangle$ and $Q_s$ denoting the average density and scattering cross section of the diffusers, respectively; $h$ is the distance of the test specimen from the opposite wall. This condition has also been proven experimentally. For somewhat-elevated frequencies, the scattering cross section $Q_s$ of a bent shell is roughly half the geometrical area of one of its sides.

Another source of systematic errors is 'edge effect'. If an absorbing area has free edges, that is, edges not adjacent to a perpendicular rigid wall, it will usually absorb more sound

*Figure 8.22* Reverberation chamber fitted out with 25 diffusers of Perspex (volume 324 m$^3$; dimensions of one shell 1.54 m × 1.28 m).

energy per second than in proportion to its geometrical area, the difference being caused by diffraction of sound into the absorbing area. This effect can be reduced (but not completely eliminated) by covering the free edges of a test specimen with a frame made of reflective panels. Such a frame is mandatory for measuring the absorption of chairs or of a seated audience. Formally, the edge effect can be accounted for by introducing an 'effective absorption coefficient' (de Bruijn 1973):

$$\alpha_{\text{eff}} = \beta L' + \alpha_{\infty} \tag{8.58}$$

In this formula, which agrees with Equation 6.39, $\alpha_{\infty}$ denotes the absorption coefficient of the unbounded test material, and $L'$ is the total length of free edges divided by the area of the actual sample. The factor $\beta$ depends on the frequency and the type of material. It may be as high as 0.2 $m^{-1}$ or more and can be determined experimentally by using test pieces of different sizes and shapes. In rare cases, $\beta$ may even turn out slightly negative.

The edge effect is completely absent if one wall of the chamber is entirely covered with the material under test, since then there will be no free edges. However, this arrangement introduces some asymmetry into the sound field and, hence, requires particular efforts to maintain a diffuse sound field. Furthermore, Equation 2.50 and Figure 2.11 tell us that the sound pressure amplitude in front of and near a perpendicularly adjacent rigid wall is different from that at some distance from all reflecting walls. Exactly at the edge, the pressure level is increased by 3 dB, and hence, more sound energy is dissipated per unit area near the edge than at more distant parts of the specimen. This effect, sometimes referred to as 'Waterhouse effect' (Waterhouse 1955), can be corrected by replacing the geometrical area $S_s$ of the test specimen with

$$S_{\text{eff}} = S_s + \frac{1}{8} L'' \lambda \tag{8.59}$$

where $\lambda$ is the wavelength corresponding to the centre frequency of the selected frequency band and $L''$ is the perimeter of the sample.

Finally, a remark may be appropriate on the frequency range in which a given reverberation chamber can be used. If the linear chamber dimensions are in the range of a few acoustical wavelengths, then statistical reverberation theories are no longer applicable to the decay process. Likewise, a diffuse sound field cannot be expected when the number and density of eigenfrequencies is small. Suppose we require that at least ten normal modes be excited by a sound signal of third-octave bandwidth. This is the case if the mid-frequency of the band exceeds a limiting frequency $f_1$, which can be found from Equation 3.26:

$$f_1 \approx \frac{500}{\sqrt[3]{V}} \tag{8.60}$$

Here, the room volume $V$ is expressed in $m^3$, and the frequency in Hz. This may be considered as the lower frequency limit of a given reverberation chamber.

## 8.8 SCATTERING COEFFICIENT

As discussed in Section 2.7, a sound wave reflected from some surface will continue its way either according to the reflection law of geometrical acoustics (specular reflection) or will be diffused more or less in all directions (diffuse reflection), depending on the structure of the wall. Most real walls will produce a mixture of both components. Complete information on the reflecting properties of a given wall can be obtained by direct measurement of the scattering characteristics. For this purpose, a test specimen of the surface is irradiated with a sound wave; at the same time, the sound reflected (or scattered) from the specimen is recorded by swivelling the microphone at fixed distance around the specimen. Usually, this measurement is carried out in the model scale. Its results can be represented as a scattering diagram as the one shown in Figure 2.18 or a collection of such diagrams.

The measurement described previously is a cumbersome procedure which often yields much more information as is really needed. Usually, one is interested just in a single figure which characterizes the diffuse reflectivity of a wall. Such a parameter is the scattering coefficient already introduced in Section 5.2: Let $I_0$ denote the intensity of the incident wave, then $(1 - \alpha)I_0$ is the energy per second reflected from the surface. The scattering coefficient $s$ is defined as that fraction of this energy which travels into non-specular directions, while $I_{\text{spec}}$ is the intensity of the specularly reflected component. Hence,

$$s = 1 - \frac{I_{\text{spec}}}{I_0(1-\alpha)} \tag{8.61}$$

Sometimes, it may prove to be difficult, however, to separate the specular component $I_{spec}$ from the scattered ones in an unambiguous manner. Therefore, more indirect procedures have been developed which have the additional advantage of circumventing the time-consuming measurement of scattering diagrams. We describe here two of them which are due to Vorländer and Mommertz (2000). They have in common that the measurement is carried out with the circular sample placed on a turntable; thus, it may be practical to perform the examination with a scaled-down model of the test object.

One of these methods is carried out in the free field with a source-microphone arrangement, as shown in Figure 8.23. Suppose the loudspeaker emits $n$ band-limited impulses, all of them reaching the sample at different positions. Each of the reflection responses contains a specular part and a diffuse part, and the same holds for their complex Fourier transforms, the reflection factors $R_i(\omega)$:

$$R_i(\omega) = R_{\text{spec}}(\omega) + S_i(\omega) \quad (i = 1,2,\ldots,n)$$

$S_i(\omega)$ is the spectrum of the $i^{\text{th}}$ scattered signal. To separate both components, we form the average

$$\langle R_i \rangle = \frac{1}{n}\sum_{i=0}^{n} R_i \approx R_{\text{spec}} \tag{8.62}$$

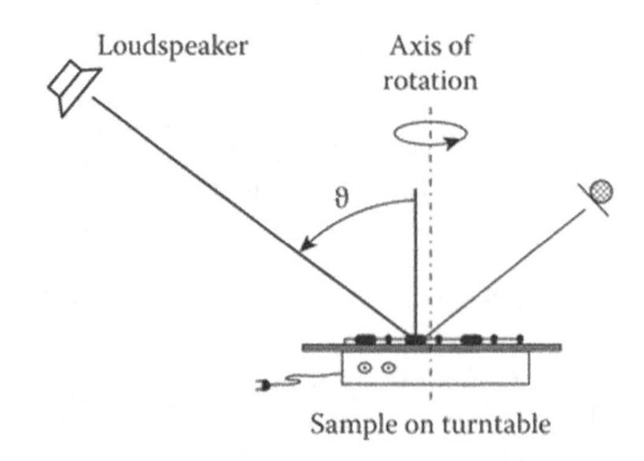


*Figure 8.23* Experimental set-up for measuring the angle-dependent scattering coefficient.

This is because each of the $n$ takes yields the same specular component $R_{spec}$, while the scattered components $S_i$ are different. Under the assumption that the latter are independent, they will cancel each other in the sum of the $S_i$, which can be neglected if $n$ is sufficiently large. Inserting $I_{spec}/I_0 = |R_{spec}|^2$ into Equation 8.59, we finally obtain the angle-dependent scattering coefficient of the test specimen:

$$s(\vartheta) = 1 - \frac{|R_{spec}|^2}{1-\alpha} = 1 - \frac{|\langle R_i \rangle|^2}{1-\alpha} \tag{8.63}$$

The absorption coefficient of the test sample is given by

$$\alpha \approx 1 - \frac{1}{n}\sum_{i=1}^{n} |R_i|^2 \tag{8.64}$$

The second method is based essentially on the same idea. However, the experimental set-up is placed in an otherwise empty reverberation chamber. Accordingly, we expect as a final result the scattering coefficient for random sound incidence. Again, the test signals are $n$ short band-limited impulses, emitted at slightly different positions of the sample. Each of them produces an impulse response $g_1(t)$, $g_2(t)$, $g_n(t)$ of the reverberation room, and each of the latter consists of two parts, $g_i(t) = g_0(t) + g'_i(t)$ with $i = 1, 2, n$. The first part is the specular reflection from the test specimen and does not depend on the position of the turntable. The second part represents the scattered signal and varies from one take to the other. To separate both parts, we add all these impulse responses:

$$h(t) = \sum_{i=1}^{n} g_i(t) = ng_0(t) + \sum_{i=1}^{n} g'_i(t)$$

The square of this sum has the expectation value

$$\langle h^2 \rangle = n^2 g_{spec}^2 + n \cdot \langle g'^2 \rangle \tag{8.65}$$

where the acute brackets indicate ensemble averages. This formula describes the sound decay in the room. Both terms decay with different decay rates: the first term decays faster than the second one since its energy is diminished not only by absorption but also by continuous conversion into 'diffuse energy' by the test sample. Hence, Equation 8.65 can be written as

$$\langle h^2 \rangle = C \cdot \left(n^2 \cdot \exp(-2\delta_1 t) + n \cdot \exp(-2\delta_2 t)\right) \tag{8.66}$$

with the arbitrary constant $C$. The damping constant $\delta_2$ is determined by the sound absorption of the reverberation chamber:

$$\delta_2 = \frac{cA}{8V} = \frac{c}{8V}\left[(S - S_s)\alpha_0 + S_s\alpha\right] \tag{8.67}$$

($V$ is the volume of the chamber; $S$ is the area of its boundary with the absorption coefficient $\alpha_0$; $S_s$ is the area of the sample.) On the other hand, the additional scattering 'losses' to be regarded in the first term of Equation 8.66 lead to an increase in the decay constant:

$$\delta_1 = \delta_2 + s(1-\alpha)\frac{cS_s}{8V} \tag{8.68}$$

Figure 8.24 shows several logarithmic decay curves measured with this method. They are double sloped, as was to be expected. The parameter is $n$, the number of individual decays from which the averaged decay has been formed. With increasing $n$, the initial slope corresponding to the average decay constant (see Equation 3.68),

$$\overline{\delta} = \frac{n\delta_1 + \delta_2}{n+1} \approx \delta_1 \quad \text{if } n \gg 1 \tag{8.69}$$

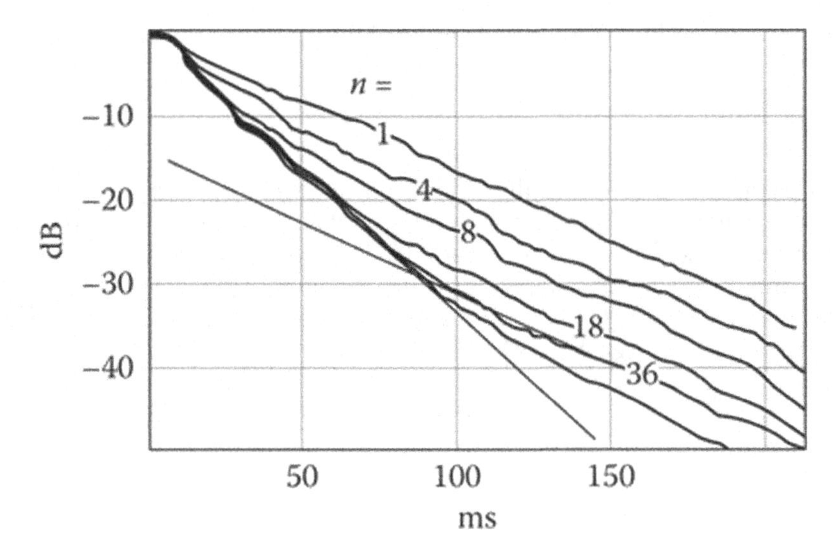


*Figure 8.24* Average decay curves (model measurements). Here, $n$ is the number of individual decays from which the average has been formed.

Source: Vorländer and Mommertz (2000).

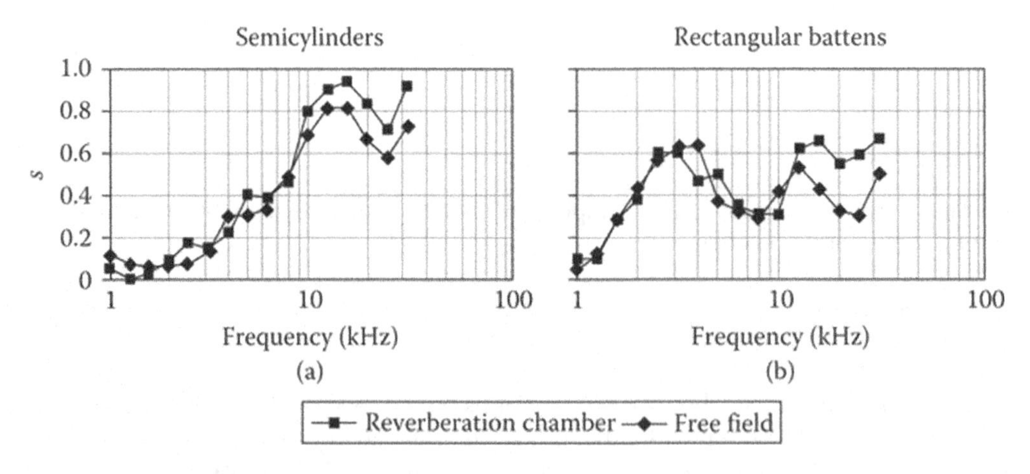


*Figure 8.25* Random incidence scattering coefficients s of irregular arrangements of battens on a plane panel as a function of frequency, ♦——♦ in free field method; ▪——▪ in reverberation method: (a) battens with quadratic cross section, side length = 2 cm; (b) battens with semicircular cross section, diameter = 2 cm.

Source: Vorländer and Mommertz (2000).

becomes more prominent. If the number $n$ of decays is high enough, both decay constants $\delta_1$ and $\delta_2$ can be evaluated with sufficient accuracy, particularly if Schroeder's backward integration technique (see Section 8.4) is applied. From their difference, the scattering coefficient for random sound incidence $s$ is readily obtained, using Equations 8.67 and 8.68.

Figure 8.25 represents the 'random incidence scattering coefficient' of two surfaces obtained with the direct method (averaged over all directions of incidence) and with the reverberation method. The test objects were battens with quadratic or semicylindrical cross section (side length or diameter 2 cm) irregularly mounted on a plane panel. The agreement of both results is obvious. The small differences at high frequencies are probably caused by the quadratic shape of the sample.

More data on scattering surfaces can be found in the books of Vorländer (2008) and of Cox and D'Antonio (2004).

## REFERENCES


Alrutz H., Schroeder M.R. A fast Hadamard transform method for the evaluation of measurements using pseudorandom test signals. Proceedings of the 11th International Congress on Acoustics, Vol. 6, Paris, 1983, p. 235.

Berkhout A.J., de Vries D., Boon M.M. A new method to acquire impulse responses in concert halls. *J Acoust Soc Am* 1980; *68*: 109.

Berzborn M., Vorländer M., Directional sound field decay analysis in performance spaces. *Building Acoustics* 2021; *28*(3): 249.

Bilsen F.A. Thresholds of perception of pitch. Conclusions concerning coloration in room acoustics and correlation in the hearing organ. *Acustica* 1968; *19*: 27.

Borish J., Angell J.B. An efficient algorithm for measuring the impulse response using pseudorandom noise. *J Audio Eng Soc* 1983; *31*: 478.

Bradley D.T., Müller-Trapet M., Adelgren J., Vorländer M. Effect of boundary diffusers in a reverberation chamber: Standardized diffuse field quantifiers. *J Acoust Soc Am* 2014; *135*: 1898.

Cox T.J., D'Antonio P. *Acoustic Absorbers and Diffusers*. London: Spon Press, 2004.

de Bruijn A. A mathematical analysis concerning the edge effect of sound absorbing materials. *Acustica* 1973; *28*: 33.

Dietsch L., Kraak W. An objective criterion for the detection of disturbing echos in the presentation of music and speech (in German). *Acustica* 1986; *60*: 205.

Fahy F. *Sound Intensity*, 2nd edn. London: E & FN Spon, 1995.

Gover B.N., Ryan J., Stinson, M. Measurements of directional properties of reverberant sound fields in rooms using a spherical microphone array. *J Acoust Soc Am* 2004; *116*(4): 2138.

Hunt F.V., Beranek L.L., Maa D.Y. Analysis of sound decay in rectangular rooms. *J Acoust Soc Am* 1939; *11*(1): 80.

ISO 3382-1:2009 International Standardization Organization. Measurement of the reverberation time of rooms with reference to other acoustical parameters.

ISO 354:2000 International Standardization Organization, Geneva, Switzerland 2003. Measurement of sound absorption in a reverberation room.

Jeong, C.-H., Brunskog, J., Jacobsen, F. Room acoustic transition time based on reflection overlap. *J Acoust Soc Am* 2010; *127*(5): 2733.

Kuttruff H. Room acoustical correlation measurements with simple means (in German). *Acustica* 1963; *13*: 120.

Kuttruff H. *Acoustics—An Introduction*. London, New York: Taylor & Francis, 2007.

MacWilliams F.J., Sloane, N.J.A. Pseudo-random sequences and arrays. *Proceedings of IEEE* 1976; *84*: 1715.

Meyer E., Thiele R. Room acoustical investigations in numerous concert halls and broadcasting studios using more recent measurement techniques (in German). *Acustica* 1956; *6*: 425.

Meyer J., Elko G. A highly scalable spherical microphone array based on an orthonormal decomposition of the soundfield. 2002 IEEE International Conference on Acoustics, Speech, and Signal Processing, Vol. II, 1781.

Mommertz E. Angle-dependent in-situ measurements of reflection coefficients using a subtraction technique. *Appl Acoustics* 1995; *46*: 251.
Mommertz E., Vorländer M. Measurement of the scattering coefficient of surfaces in the reverberation chamber and in the free field. Proceedings of the 15th International Congress on Acoustics, Vol. II, Trondheim, 1995, p. 577.
Müller S., Massarani P. Transfer-function measurement with sweeps. *J Audio Eng Soc* 2001; *49*: 443.
Nélisse H., Nicolas J. Characterization of a diffuse sound field in a reverberant room. *J Acoust Soc Am* 1997; *101*: 3517.
Nolan M., Berzborn M., Fernandez-Grande E. Isotropy in decaying reverberant sound fields. *J Acoust Soc Am* 2020; *148*: 1077.
Nolan M., Fernandez-Grande E., Brunskog J., Jeong C-H. A wavenumber approach to quantifying the isotropy of the sound field in reverberant spaces. *J Acoust Soc Am* 2018; *143*(4): 2514.
Rafaely B. *Fundamentals of Spherical Array Processing*. Cham: Springer Nature Switzerland AG, 2019.
Sabine W.C. *Collected Papers on Acoustics*. New York: Dover Press, 1964. First published 1922.
Schroeder M.R. New method of measuring reverberation time. *J Acoust Soc Am* 1965; *37*: 409.
Schroeder M.R. Modulation transfer functions: Definition and measurement. *Acustica* 1981; *49*: 179.
Schroeder M.R. The "Schroeder frequency" revisited. *J Acoust Soc Am* 1996; *99*: 3240.
Sekiguchi K., Kimura S., Hanyu T. Analysis of sound field on spatial information using a four-channel microphone system based on a regular tetrahedron peak point method. *Appl Acoust* 1992; *37*: 305.
Stan G.B., Embrechts J.J., Archambeau D. Comparison of different impulse response measurement techniques. *J Audio Eng Soc* 2002; *50*: 249.
Tachibana H., Yamasaki Y., Morimoto M., et al. Acoustic survey of auditoriums in Europe and Japan. *J Acoust Soc Japan (E)* 1989; *10*: 73.
Vorländer M. *Auralization*. Berlin: Springer-Verlag, 2008.
Vorländer M., Mommertz E. Definition and measurement of random-incidence scattering coefficients. *Appl Acoust* 2000; *60*: 187.
Waterhouse R.V. Interference patterns in reverberant sound fields. *J Acoust Soc Am* 1955; *27*: 247.
Yamasaki Y., Itow T. Measurement of spatial information in sound fields by closely located four point microphone method. *J Acoust Soc Japan (E)* 1989; *10*: 101.

Chapter 9

# Design considerations and design procedures

The purpose of this chapter is to discuss some practical aspects of room acoustics, namely, the acoustical design of auditoria in which some kind of performance (lectures, music, theatre, etc.) is to be presented to an audience, or of spaces in which the reduction of noise levels is the main interest. Its contents are not just an extension of fundamental laws and scientific insights towards the practical world, nor are they a collection of guidelines and rules deduced from them. In fact, the reader should be aware that the art of room acoustical design is only partially based on theoretical considerations and that it cannot be learned from this or any other book, but that successful work in this field requires considerable practical experience. On the other hand, mere experience without at least some insight into the physics of sound and without some knowledge of psychoacoustic facts is of little worth or is even dangerous in that it may lead to unacceptable generalizations.

Usually, the practical work of an acoustic consultant starts with drawings being presented to him which show details of a hall or some other room which is at the planning stage or under construction, or even one which is already in existence and in full use. First of all, he must ascertain the purpose for which the hall is to be used, that is, which type of performances or presentations are to take place in it. This is more difficult than appears at first sight, as the economic necessities sometimes clash with the ambitious ideas of the owner or the architects. Secondly, he or she must gain some idea of the objective structure of the sound field to be expected, for instance, the values of the parameters characterizing the acoustical behaviour of the room. Thirdly, he must decide whether or not the result of his investigations favours the intended use of the room, and finally, if necessary, he must work out proposals for changes or measures which are aimed at improving the acoustics, keeping in mind that these may be very costly or may substantially modify the architect's original ideas and, therefore, have to be given very careful consideration.

In order to solve these tasks, there is so far no generally accepted procedure which would lead with absolute certainty to a good result. Perhaps it is too much to expect there ever to be the possibility of such a 'recipe', since one project is usually different from the next due to the efforts of architects and owners to create something quite new and original in each theatre or concert hall.

Nevertheless, a few standard methods of acoustical design have evolved which have proved useful and which can be applied in virtually every case. The importance which the acoustic consultant will attribute to one or the other, the practical consequences which he will draw from his examination, whether he favours reverberation calculations more than geometrical considerations or vice versa — all this is left to him, to his skill and to his experience. It is a fact, however, that an excellent result requires close and trustful cooperation with the architect — and a certain amount of luck too.

DOI: 10.1201/9781003389873-9

As we have seen in preceding chapters, there are a few objective sound field properties which are beyond question regarding their importance for what we call good or poor acoustics of a hall. One of them is the strength of the direct sound which is responsible for the loudness and for a natural impression of sounds. Another one is the temporal and directional distribution of the 'early' reflected sound energy which supports the direct sound, and finally, we quest for an appropriate reverberance of a hall. These properties depend on constructional data, in particular, on the following:

1. Shape of the room
2. Volume of the room
3. Number of seats and their arrangement
4. Materials of walls, ceiling, floor, seats, and so on

While the reverberation time is determined by factors 2–4 and not significantly by 1, the room shape influences strongly the number, directions, delays, and strengths of the early reflections received at a given position or seat. The strength of the direct sound depends on the distances to be covered and also on the arrangement of the audience.

Before going into some more detail of these items, we shall deal with an ever-important matter, namely, with predicting the noise level to be expected in a room and some methods of reducing it.

## 9.1 PREDICTION OF NOISE LEVEL

There are many spaces that are not intended for any acoustical presentations but where some acoustical treatment is nevertheless desirable or necessary. Although they show wide variations in character and structural details, they all fall into the category of rooms in which people are present and in which noise is produced, for example, by noisy machinery or by the people themselves. Examples of this are staircases, concourses of railway stations and airports, and entrance halls and foyers of concert halls and theatres. Most important, however, are working spaces, such as open-plan offices, workshops, and factories. Here, room acoustics has the relatively prosaic (however, important) task of reducing the noise level.

Traditionally, acoustics does not play any important role in the design of a factory or an open-plan office, to say the least; usually, quite different aspects, as, for instance, those of efficient organization, of the economical use of space, or of safety, are predominant. Therefore, the term 'acoustics' applied to such spaces does not have the meaning it has with respect to a lecture room or a theatre. Nevertheless, the way in which noise propagates in such a room, and hence, the noise level in it, depends highly on its acoustical properties.

A first idea of the steady-state sound pressure level that a non-directional sound source with power output $P$ produces in a room with the equivalent absorption area $A$ is obtained from Equation 5.49. Converting it into a logarithmic scale, with $L_w$ denoting the sound power level (see Equation 1.66), yields for the sound pressure level

$$\mathrm{SPL}^{\infty} = L_w - \log_{10}\left(\frac{A}{A_0}\right) - 4.34\frac{A}{S} + 6\mathrm{dB} \qquad \text{with } A_0 = 1\mathrm{m}^2 \tag{9.1}$$

This relation is valid if the distance from the sound source is significantly larger than the 'diffuse-field distance' $r_c$ as given by Equations 5.44 or 5.48, that is, it describes the sound pressure level in the reverberant field. For observation points at distances $r$ comparable to or smaller than $r_c$, the sound pressure level is, according to the more general equation (5.45),

$$\mathrm{SPL} = \mathrm{SPL}^{\infty} + 10\log_{10}\left(1 + \frac{r_c^2}{r^2}\right) \tag{9.2}$$

Both equations are valid under the assumption that the reverberant sound field is diffuse.

Numerous measurements in real spaces have shown, however, that the reverberant sound pressure level $SPL^{\infty}$ decreases more or less with increasing distance, in contrast to what Equation 9.1 says. Obviously, sound fields in such spaces are not completely diffuse. As discussed in Section 5.2, this lack of diffusion may have several reasons. Often, one dimension of a working space is much larger (very long rooms) or smaller (very flat rooms) than the remaining ones. Another possible reason is non-uniform spatial distribution of absorption. In all these cases, a different approach is needed to calculate the sound pressure level.

For calculating the distribution of sound energy in enclosures bounded by plane walls, the concept of image sources can be employed, which has been discussed at some length in Section 4.1. It must be noted, however, that real working spaces are not empty but contain machines, piles of material, furniture, benches, and so on; in short, numerous obstacles scatter the sound and may also partially absorb it.

One way to simulate the scattering of sound in fitted working spaces is to replace sound propagation in the free space by that in an 'opaque' medium containing many randomly arranged scattering objects, as explained at the end of Section 5.2. Here we restrict the discussion to the steady state. (The transient sound propagation in enclosures containing sound-scattering obstacles is much more complicated than the steady-state case. It has been treated successfully by several authors, for instance, by Hodgson [1983a].) Then, the energy density of the unscattered component, that is, of the direct sound, is

$$w_0(r) = \frac{P}{4\pi c r^2}\exp(-r/\overline{r}_S) \tag{9.3}$$

which is to be compared with Equation 5.43; $\overline{r}_S$ is the 'scattering mean free path length' $\left[\langle n_S\rangle Q_S\right]^{-1}$. The reach of the direct sound characterized by the diffuse-field distance $r_c$ is smaller than it is in an empty room.

To calculate it, we imagine the sound field as being composed of numerous sound particles (with the same meaning as discussed in the introduction to Chapter 5). Furthermore, it is assumed that $\overline{r}_S$ is so small that virtually all sound particles will be scattered at least once before reaching a wall of the enclosure. Then, we need not consider any reflections of the direct sound. Instead, the scattered sound particles will uniformly fill the whole enclosure due to the equalizing effect of multiple scattering. Since the scattered sound particles propagate in all directions, they constitute a diffuse sound field with its well-known properties. In particular, its energy density is constant and given by $w_s = 4P/cA$. Hence, the steady-state level can be calculated from Equations 9.1 and 9.2. In the latter case, however, the 'diffuse-field distance' $r_c = (A/16\pi)^{1/2}$ from Equation 5.44 has to be replaced with a modified value $r_c'$ which is smaller than $r_c$. In fact, equating $w_s$ with $w_0$ from Equation 9.3 leads to a transcendental equation:

$$r_c'/r_c = \exp(-r_c'/2\overline{r}_s) \tag{9.4}$$

Solving it yields $r_c' / r_c = 0.7035$ if $r_c$ equals the scattering mean free path length $\overline{r}$, while this fraction becomes as small as 0.2653 for $\overline{r}_s / r_c = 0.1$, that is, when a sound particle undergoes 10 collisions per distance $r_c$ on average. With such a high density of scatterers, the application of the diffusion equation may be justified (see Section 4.6). In this case, in Equation 4.36, the length $\overline{\ell}$ must be replaced with the scattering mean free path length $\overline{r}$. However, there remains some uncertainty on the scattering cross sections $Q_s$ of machinery or other pieces of equipment, because there is no practical way to calculate them exactly from geometrical data. Several authors (see, for example, work by Ondet and Barbry [1989]) identify $Q_s$ with one-quarter of the scatterer's surface. This procedure agrees with the rule given by the end of Section 5.2.

In a different approach, the scatterers are imagined as being projected onto the walls, so to speak, that is, the walls are assumed to produce diffuse sound reflections rather than purely specular ones. Then the problem can be treated by application of the radiosity integral (4.26). This holds in particular for the calculation of the steady-state sound propagation in certain 'disproportionate' rooms for which the diffuse-field theory is not applicable. One of them is the infinite flat room, that is, the space confined by two parallel planes. As already mentioned in Section 4.5, Equation 4.26 has a closed solution in this case. This is of considerable practical interest since this kind of 'enclosure' may serve as a model for many working spaces in which the ceiling height $h$ is very small compared with the lateral dimensions (factories, or open-plan bureaus). Therefore, sound reflections from the ceiling are absolutely predominant over those from the side walls, and hence, the latter can be neglected unless the source and the observation point are located next to them.

The exact solution is not well suited for the practical application. Therefore, we present here the following approximation (Kuttruff 1985, 1989) for both planes having the same constant absorption coefficient $\alpha$ or 'reflection coefficient' $\rho = 1 - \alpha$, and for both the sound source and the observation point being located in the middle between both planes:

$$w(r) = \frac{P}{4\pi c}\left\{\frac{1}{r^2} + \frac{4\rho}{h^2}\left[\left(1 + \frac{r^2}{h^2}\right)^{-3/2} + \frac{b\rho}{1-\rho}\left(b^2 + \frac{r^2}{n^2}\right)^{-3/2}\right]\right\} \tag{9.5}$$

The constant $b$ depends on the absorption coefficient of the floor and the ceiling. Some of its values are listed in Table 9.1. Equation 9.5 may also be used if both boundaries have different absorption coefficients; in this case, for $\alpha$, the average of both absorption coefficients is inserted.

Figure 9.1b shows how the sound pressure level, calculated with this formula, depends on the distance from an omnidirectional sound source for various values of the (average) absorption coefficient $\alpha = 1 - \rho$ of the walls. For comparison, the corresponding curves for specularly reflecting planes, computed using Equation 4.3, are presented in Figure 9.1a. The

*Table 9.1* Values of the constant $b$ in Equation 9.5

| *Absorption coefficient, $\alpha$* | *Reflection coefficient, $\rho$* | b |
|---|---|---|
| 0.7 | 0.3 | 1.806 |
| 0.6 | 0.4 | 1.840 |
| 0.5 | 0.5 | 1.903 |
| 0.4 | 0.6 | 2.002 |
| 0.3 | 0.7 | 2.154 |
| 0.2 | 0.8 | 2.425 |
| 0.1 | 0.9 | 3.052 |

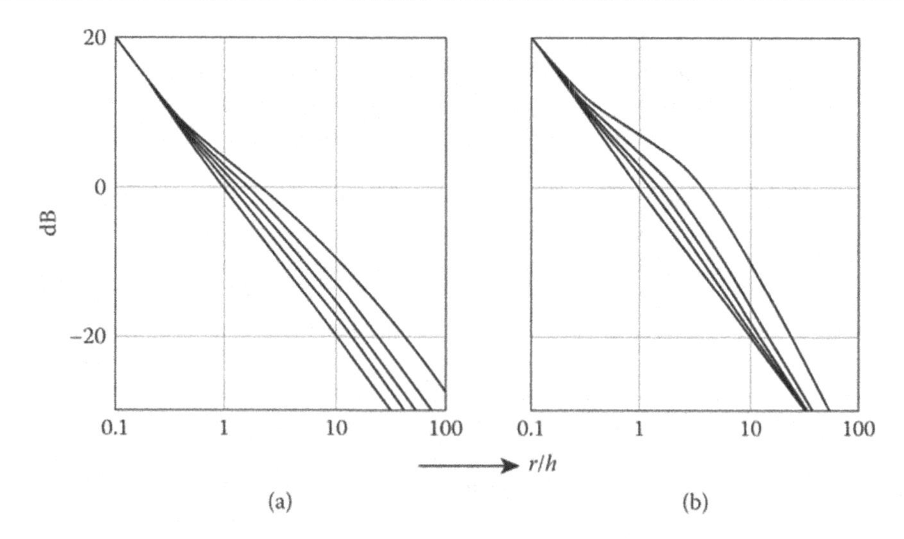


*Figure 9.1* Sound pressure level in an infinite flat room as a function of distance $r$ ($h$ = room height). The absorption coefficient of both walls is (from bottom to top) 1, 0.7, 0.5, 0.3, and 0.1: (a) smooth walls, calculated with Equation 4.3, and (b) diffusely reflecting walls, calculated with Equation 9.5.

plotted quantity is 10 times the logarithm of the energy density divided by $P/4\pi ch^2$. Both diagrams show characteristic differences: smooth boundaries direct all the reflected energy away from the source, and this results in an increased level, with the increment approaching a constant value at large distances. In contrast, diffusely reflecting boundaries reflect some energy back towards the source; accordingly, the level increment caused by the boundary — compared to that of free field propagation ($\alpha = 1$) — reaches a maximum at a certain distance and vanishes at large distances from the source. This behaviour is typical for enclosures containing scattering objects and was experimentally confirmed by numerous measurements carried out by Hodgson (1983a, 1983b), in model spaces as well as in full-scale factories.

Both aforementioned methods are well suited for predicting noise levels in working spaces and estimating the reduction which can be achieved by absorbing treatment of the ceiling, for instance. Other possible methods are measurements in a scale model of the space under investigation or computer simulation as described in Chapter 10.

Another disproportionate room is the long room, that is, a tube with cross-sectional dimensions that are large compared with the acoustical wavelength. Again, we simplify the problem by supposing that the tube is infinitely long. If it has smooth and rigid walls, and if its cross section is rectangular, the sound propagation can be calculated by an obvious extension of Equation 4.3 in two dimensions, leading to a double sum over the full pattern of image sources. If the wall of the tube scatters the impinging sounds, a closed solution of Equation 4.26 is also available, provided the tube is cylindrical, which is not a severe restriction (Kuttruff 1985, 1989). The results of these calculations, both for smooth and scattering walls, are similar to those shown in Figure 9.1b; the deviations of the sound level from the $1/r^2$ law are even somewhat more pronounced than that in Figure 9.1b.

Generally, some moderate absorbing treatment of the walls or the ceiling has a beneficial effect on the noise level as long as the 'diffuse-field distance' $r_c$ is well below the linear dimensions of the room. This is true not only for working spaces, such as factories or large offices, but also for many other rooms where many people gather together, for example, in staircases or in theatre foyers. A noise level reduction by just a few decibels can increase the acoustical comfort to an amazing degree. If the sound level is too high due to insufficient absorption, people will talk more loudly than in a quieter environment. This, in turn, increases the general noise level and so on and so forth until finally people must shout and still do not achieve satisfactory intelligibility. In contrast, an acoustically damped environment usually makes people behave in a 'damped' manner too — for reasons which are not primarily acoustical — and it makes them talk not louder than necessary.

There is still another psychologically favourable effect of an acoustically damped theatre or concert hall foyer: when a visitor leaves the foyer and enters the performance hall, he will suddenly find himself in a more reverberant environment, which gives him the impression of solemnity and raises his expectations.

The extensive use of absorbing materials in a room, however, causes an oppressive atmosphere, an effect which can be observed quite clearly when entering an anechoic room. Furthermore, since the level of the background noise is reduced, too, by the absorbing areas, a conversation held in a low voice can be understood at relatively large distances and can be irritating to unintentional listeners. Since this is more or less the opposite of what should be achieved in an open-plan office, the masking effect by background noise is sometimes increased in a controlled way by feeding loudspeakers with random white or 'coloured' noise, that is, with a 'signal' without any temporal or spectral structure. The level of this noise should not exceed 50 dB(A). Even so, it is still disputed whether the advantages of such measures surpass their disadvantages.

## 9.2 DIRECT SOUND

In a closed room, the direct sound travelling from the sound source to a listener is just one component of the sound field, although the most important one if the room is used as a performance or assembly hall. It is not influenced at all by the walls or the ceiling of a room, since it propagates along straight lines. Nevertheless, its strength depends on the geometry of the hall: on the (average) length of paths which it has to travel and on the height at which it propagates over the audience or other strongly absorbing surfaces until it reaches a listener.

Of course, the direct sound intensity under otherwise-constant conditions is higher the closer the listener is seated to the sound source. Different plans of halls can be compared in this respect by a dimensionless figure of merit, which is the average distance of all listeners from the sound source divided by the square root of the area occupied by an audience. For illustration, Figure 9.2 presents a few types of floor plans; the numbers indicate this

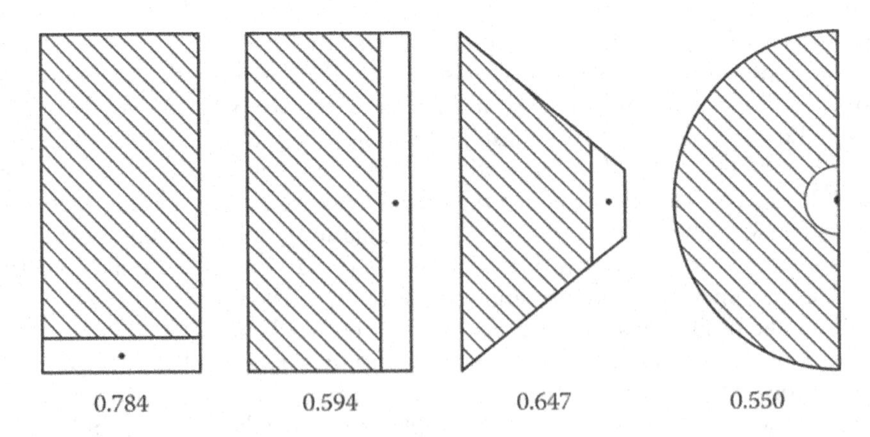


*Figure 9.2* Normalized average distance from listeners to source for various room shapes.

normalized average distance. The audience areas are shaded, and the sound source is marked by a point.

It is seen that a long rectangular room with the sound source on its short side seats many listeners relatively far from the source, whereas a room with a semicircular (or circular) floor plan provides particularly short direct sound paths. For this reason, many large lecture halls, theatres, and session halls of parliaments are of this type. Likewise, most ancient amphitheatres had been given this shape by their builders. However, for a closed room, this shape has severe acoustical risks in that it concentrates the sound reflected from the rear wall toward certain regions. Generally, considerations of this sort should not be given too much weight, since they are only concerned with one aspect of acoustics which may conflict with other ones.

Attenuation of the direct sound due to grazing propagation over the heads of the audience (see Section 6.7) can be reduced or avoided by sloping the audience area upwardly instead of arranging the seats on a horizontal floor. This holds also for the attenuation of side or front wall reflections. As is easily seen by comparing Figure 9.3a and b, a constant slope is less favourable than an increasing ascent of the audience area, because in the former case the angle of incidence, and hence the attenuation, shows a stronger dependence on the distance from the source. The optimum slope (which is optimal as well with respect to the listener's visual contact with the stage) would be reached if all sound rays originating from the sound source *S* strike the audience area at the same incidence angle. The mathematical expression for this condition is

$$r(\varphi) = r_0 \exp(\varphi \cdot \cot \gamma) \tag{9.6}$$

In this formula, which describes what is called a logarithmic spiral, $r(\varphi)$ is the length of the sound ray leaving the source under an elevation angle $\varphi$, and the constant $r_0$ is the length of the sound ray at $\varphi = 0$ (see Figure 9.4). The angle $\gamma$ is named the 'grazing' angle; it is the angle the arriving ray makes with the tangent of the shown curve. For design purposes, Equation 9.6 is not well suited; however, it can be simplified by setting $\cot \gamma \approx 1/\gamma$ and $r \approx x$. With these approximations, the height $y$ of a point can be expressed as a function of its horizontal distance $x$:

$$y(x) = \gamma x \ln\left(\frac{x}{r_0}\right) \tag{9.7}$$

Another important figure is the clearance $h$, that is, the vertical distance between a ray arriving at a particular seating row (a 'line of sight') and the corresponding point of the preceding

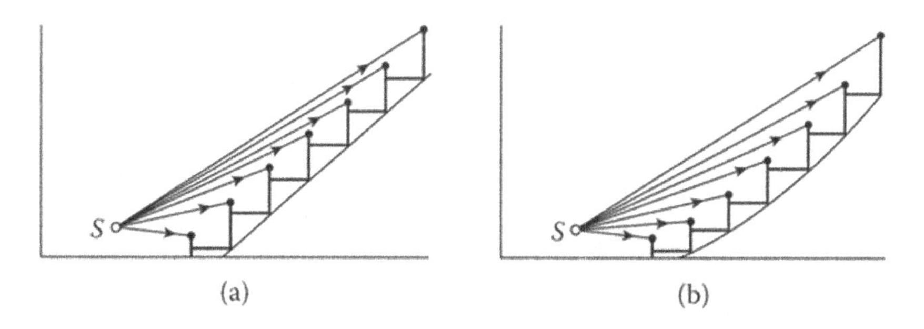


*Figure 9.3* Reducing the attenuation of direct sound by sloping the seating area: (a) constant slope and (b) slope increasing with distance.

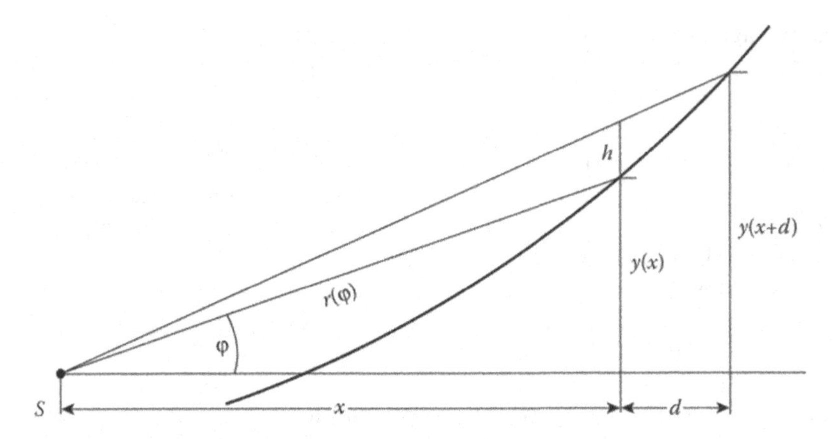


*Figure 9.4* Sloped seating area, schematically. Notations in Equations 9.6, 9.7, and 9.8 ($S$ = sound source; $h$ = clearance of sight lines).

row. Let $d$ denote the distance between the rows, then the clearance for a seating arrangement after Equation 9.7 is

$$h \approx \gamma d\left(1-\frac{d}{2x}\right) \tag{9.8}$$

According to this formula, $h$ does not strongly vary with the distance $x$, except in the region next to the sound source. For very distant seats, it approaches the constant value $\gamma d$. Thus, for $d$ = 1 m and $\gamma$ = 5°, this limiting value is 8.7 cm. To achieve a significant effect, the clearance should not be smaller. Of course, higher values are more favourable. Generally, a clearance of 12 cm is considered satisfactory. However, a gradually increasing slope of the seating area has the consequence that the steps in the upward-going aisles must be of varying height which the user is not accustomed to. This can be circumvented by approximating the sloping function of Equation 9.6 by a few straight lines and thus subdividing the audience area in a few sections with uniform seating rake within each of them.

Front seats on galleries or balconies are generally well supplied with direct sound since they do not suffer at all from sound attenuation due to listeners sitting immediately in front. This is one of the reasons that seats on balconies or in elevated boxes are often appreciated because of the excellent listening conditions met at these places.

## 9.3 EXAMINATION OF ROOM SHAPE

The acoustical power produced by a human speaker or a mechanical musical instrument is rather limited (see Section 1.7). Therefore, a common problem in the acoustical design of any kind of assembly room is the economic use of sound energy in order to provide for sufficient loudness at a listener's ears. Of course, this can be achieved by electroacoustical sound systems, as will be described in some more detail in Chapter 11. However, such systems work better the more the natural properties of the room favour its intended use. Therefore, we should look first for some 'natural' sound reinforcement. This holds for conference rooms and moderately sized classrooms, as well as for large lecture rooms or theatres. This kind of sound reinforcement can be effected by reflections of the original sound signal from the enclosing boundaries, that is, from the walls and the ceiling of the room. Another important precondition of good intelligibility is that a large fraction of the sound energy transported by reflections arrives at a listener's ears shortly after the arrival of the direct sound, say, within the first 50 ms. For then, the reflections subjectively merge with the direct sound, thus increasing the perceived loudness of the latter. The definition of all parameters characterizing the speech intelligibility accounts more or less for this fact (see Section 7.5). Thus, good speech intelligibility requires a high amount of 'early reflected energy'.

The directions, the strengths, and the delays of reflections are determined by the position and the orientation of reflecting areas, that is, by the shape of a room. Thus, it is indispensable to carefully examine the shape of a room in order to get a survey on the reflections produced by the enclosure.

If a room exists just on the paper, that is, if nothing more is known about it other than a plan and a section, one can gain a qualitative picture of the relevant reflection paths by the construction of sound rays with pencil and ruler, assuming that the reflections occur more or less in a specular manner. This picture is certainly not comprehensive but is nevertheless very valuable. If the enclosure is made up of plane boundaries, one can take advantage of the concept of image sound sources described at some length in Chapter 4. This procedure, however, is feasible for first-order or, at best, for second-order reflections only. For the examination of sound reflections from curved walls, the method of image sources cannot be applied. In this case, we have to determine the wall normal in each boundary point of interest and to apply the law of specular reflection as shown in Figure 4.1. The construction of some reflected sound rays in a hypothetical hall is depicted in Figure 9.5. If a sufficiently large portion of the wall or the ceiling appears circular in the sectional drawing or can be approximated by a circle, the location of the focus associated with it may be found from Equation 4.17.

At any event, the construction of sound paths gives us some idea on the distribution of the strongest reflections and on the wall portions which produce them. It tells us whether the reflected sound will be concentrated in a limited region and where a focal point or a caustic will appear which might cause quite annoying acoustical effects, including non-uniformity of the sound field. Furthermore, the directions of sound incidence at various seats can be seen immediately, whereas the delay time between a reflection with respect to the direct sound is easily determined from the difference in path lengths after dividing the latter by the sound velocity.

To find the relative strength of a reflection, the $1/r$ law of spherical wave propagation can usually be applied. In doubt, the reflecting efficiency of a particular wall portion, a balcony face, or a suspended reflector can be checked by application of Equations 2.58 or 2.59. Let $r_0$ and $r_i$ be the path lengths of the direct sound ray and of a particular reflection, measured

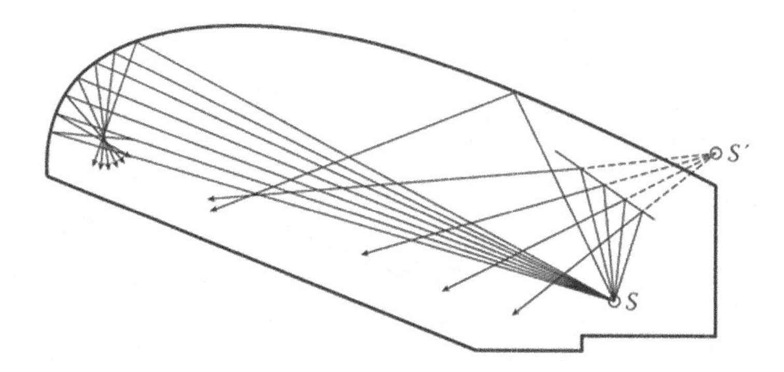


*Figure 9.5* Construction of sound rays paths in the longitudinal section of a hypothetical hall ($S$ = sound source; $S'$ = image source).

from the sound source to the listener via the reflecting wall point, then the level difference between the direct sound and the reflected sound is

$$L_0 - L_i = 20 \cdot \log_{10}(r_i / r_0) \quad \text{dB} \tag{9.9}$$

If the reflecting boundary has an absorption coefficient $\alpha$, the level of the reflected sound portion is reduced by another $-10 \cdot \log_{10}(1 - \alpha)$ dB. Irregularities on walls and ceiling can be neglected as long as their dimensions are small compared to the wavelength. The intensities or pressure levels of reflections from a curved wall section can be estimated by comparing the density of the reflected rays in the observation point with the ray density which would be observed if that wall section were plane. For spherical or cylindrical wall portions, the ratio of the reflected and the incident energy can be calculated from Equation 4.19, which is equivalent to

$$\Delta L = 10 \cdot \log_{10} \left| \frac{1 + x/a}{1 - x/a} \right|^n \tag{9.10}$$

As in Equation 4.19, $a$ and $x$ are the distances of the source and the observation point from the reflecting wall portion, respectively; $n$ is unity if the curved boundary is cylindrical and is 2 for a spherical wall portion.

The method of ray tracing with pencil and ruler applies only to sound paths which are situated in the plane of the drawing to hand. Sound paths in different planes can be constructed by applying the methods of constructive geometry. This, of course, involves considerably more time and labour, and it is questionable whether this effort is worthwhile, considering the rather-qualitative character of the information gained by it. For rooms of more complicated geometry, it is much more practical to apply computerized ray tracing techniques (see Section 9.8).

So far, we have described simple methods to investigate the effects of a given enclosure upon sound reflections. Beyond that, there are some general conclusions which can be drawn from geometrical considerations and also from experiences with existing halls. They are briefly summarized in the following.

If a room is to be used for speech (conference rooms, school classrooms, lecture halls, etc.), it is an advantage to support the direct sound by as many strong reflections as possible. As mentioned earlier, their delays should not exceed about 50 ms. Reflecting areas (wall portions, screens) placed close to the sound source are especially favourable since they collect a great deal of the emitted sound energy and project it towards the audience, provided they are properly orientated and are free of absorption. Any porous curtains, as are often used for decorative purposes, are harmful in acoustical respect if applied in the vicinity of the sound source. The ceiling of the room plays an important role since it is usually low enough to produce reflections which support the direct sound. Thus, absorbent materials as may be needed for the reduction of the reverberation time can only be mounted on more remote ceiling portions (and on the rear wall, of course). If the ceiling is plane, the construction of just one image source allows the designer to decide which part of the ceiling can be used for the supply of a given audience area with reflected sound energy and thus must remain free of absorbent treatment. Sometimes, it turns out that the ceiling is so high that the reflections from it are delayed by more than 50 ms. Then, reshaping the ceiling or the installation of suspended and suitably tilted reflectors should be taken into consideration (see Figure 9.5). Likewise, an unfavourable situation on the stage can be greatly improved by putting up a few portable screens made of reflecting material, for instance, of relatively heavy panels.

Unfortunately, these principles can only be applied to a limited extent to theatres, where such measures could, in fact, be particularly useful. This is because the stage is the realm of the stage designer, of the stage manager, and of the actors — in short, of people who sometimes complain bitterly about the acoustics but who are not ready to sacrifice one iota of their artistic intentions in favour of acoustical requirements. It is all the more important to shape the wall and ceiling portions, which are close to the stage, in such a way as to direct the incident sound immediately onto the audience.

When it comes to the design of concert halls, the economic use of sound energy is not of foremost interest as it is in a lecture hall or a drama theatre, except for very large halls. Here it is advisable to make only moderate use of areas projecting the sound energy immediately towards the audience. This would result in a high fraction of early energy and — in severe cases — to subjective masking of the sound decay in the hall. The effect would be a weak sense of reverberance even if the objective reverberation time is adequate. In a concert hall, a different aspect of early sound reflections is more important, namely, their potential for creating the highly desirable 'spatial impression', which they give to the listener, provided they arrive from lateral directions (see Section 7.7). Whether the boundary of a concert hall produces strong early lateral reflections or not depends critically on the shape of the room, in particular, on the position and orientation of its side walls.

This may be illustrated by Figure 9.6, which shows the spatial distribution of the 'lateral energy fraction' in some two-dimensional enclosures (Vorländer and Kuttruff 1985), computed using Equation 7.19; the area of every hall was assumed to be 600 $m^2$. The position of the sound source is marked by a cross; the densities of shading of the various areas correspond to the following intervals of the early lateral energy fraction (LEF): 0–0.06, 0.06–0.12, 0.12–0.25, 0.25–0.5, and >0.5 (black). In all examples, the LEF is very low at locations next to the sound source, but it is highest in the vicinity of the side walls. Accordingly, in a rectangular hall, the largest areas with high LEF, and hence with satisfactory 'spaciousness', are to be expected if the plan of the hall is long and narrow. Particularly large areas with low early LEF appear in fan-shaped halls opening towards the rear (right side, bottom), a fact which is due to the fact that the sound reflected from the side walls travels almost parallel to the direct sound. If the hall becomes narrower towards the rear (right side, top), the opposite is true. (However, it is not very likely that a hall will ever be used in the latter configuration.) The rectangular shape seems to be a reasonable compromise.

These findings explain — at least partially — why so many concert halls with excellent acoustics have rectangular floor plans with relatively narrow side walls and high ceilings (Vienna Musikvereinssaal, or Boston Symphony Hall, for example). It may be noted, by the way, that the requirement of strong lateral sound reflections favours room shapes which are different from those leading to short distances of the listeners from the sound source (see Section 9.2).

In real, that is, in three-dimensional, halls, additional lateral energy is reflected towards the stalls from the edges formed by a side wall and horizontal surfaces such as the ceiling or the soffits of side balconies (Figure 9.7). These contributions are especially useful since they are less attenuated by the audience than reflections just from the side walls. If no balconies are planned, the beneficial effect of soffits can be achieved as well by properly arranged surfaces or bodies protruding from the side walls.

With regard to the performance of orchestral music, one should remember that most instruments have pronounced directionalities of sound radiation depending on the frequency range. Accordingly, sounds from certain instruments or groups of instruments are predominantly reflected by particular wall or ceiling portions. Since every concert hall is expected to house orchestras of varying composition and arrangement, only some general conclusions can be drawn from this fact. Thus, the high-frequency components, especially from string

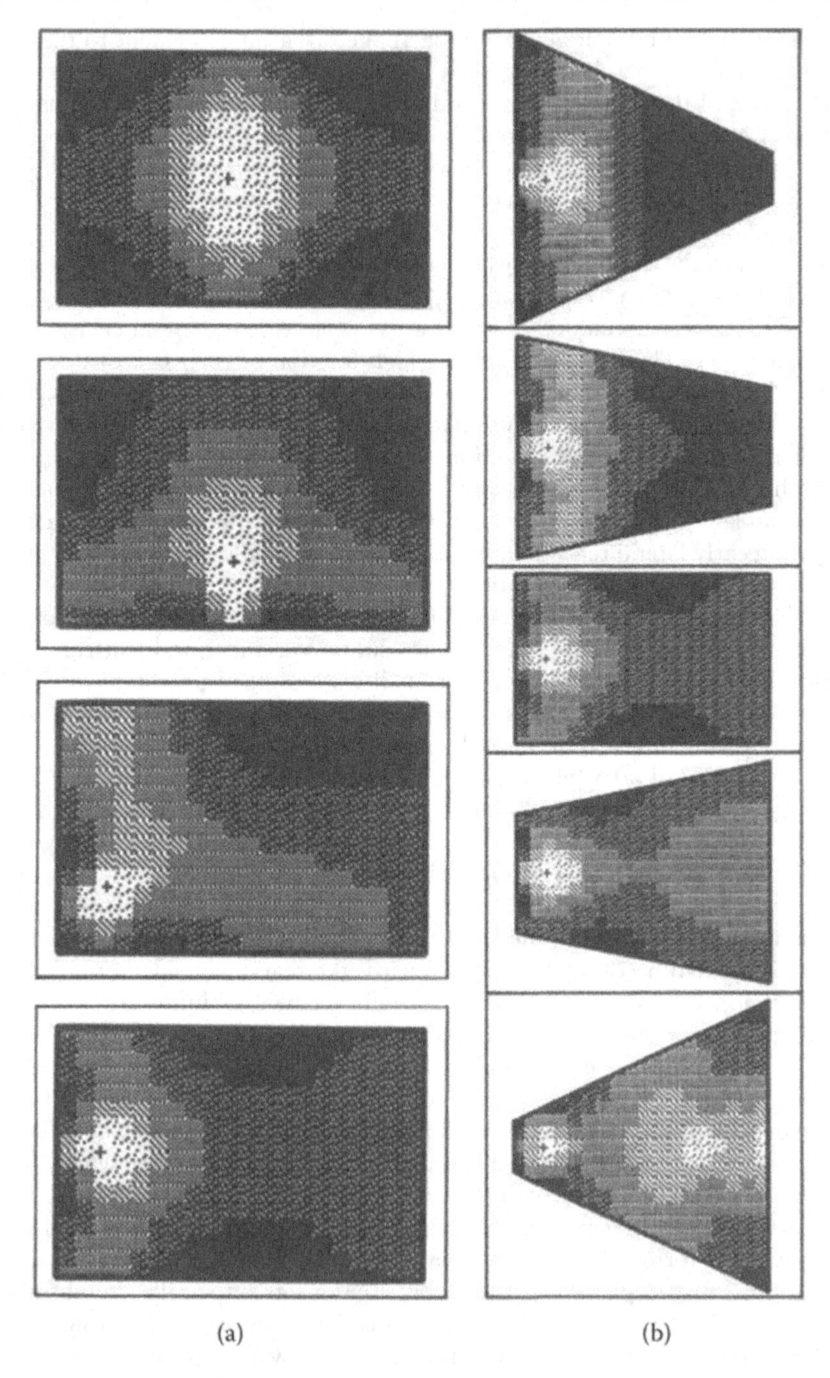


*Figure 9.6* Distribution of early reflected sound energy in two-dimensional enclosures with area 600 m$^2$: (a) rectangular enclosure, various source positions, and (b) various fan-shaped enclosures.

Source: Vorländer and Kuttruff (1985).

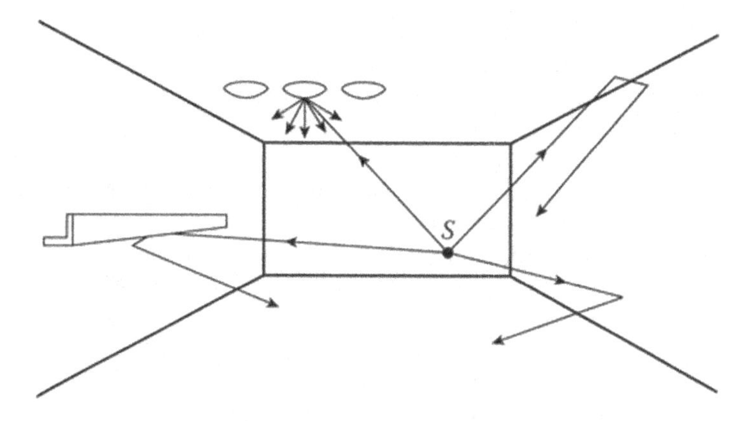


*Figure 9.7* Origin of lateral or partly lateral reflections ($S$ = sound source).

instruments, which are responsible for the brilliance of the sound, are mainly reflected from the ceiling portion next to the stage, whereas the side walls are very important for the reflection of components in the range of about 1,000 Hz and, hence, for the volume and sonority of the orchestral sounds (Meyer 1976). Another consequence of the directionality of instruments and the human voice is that listeners seated behind the orchestra will receive the musical sounds from the 'wrong' direction, which leads to an impaired sense of timbre and ensemble balance.

The examination of room geometry can lead to the result that some wall areas, particularly if they are curved, will give rise to very delayed reflections with relatively high energy, which will neither support the direct sound nor be masked by other reflections but may be heard as echoes. Especially critical in this respect are high ceilings or concavely curved rear walls. The simplest way to avoid echo effects is to line these wall portions with some highly absorbent material. If this measure would cause an intolerable drop in reverberation time or is unfeasible for some other reason, a reorientation of those surfaces could be considered. For example, a curved ceiling of a hall can be split up into plane or convexly curved transversal strips, each of them tilted in a favourable way, a method which still maintains the concave overall shape. Still another possibility is to provide the curved surface with some sound scattering structure (see Section 2.7). If desired, any treatment of these walls — scattering or absorbing — can be concealed behind an acoustically transparent screen consisting of a grid, a net, or highly perforated panels, whose transmission properties were discussed in Section 6.2.

Another undesirable effect is flutter echo, as already described in Section 4.2. It is caused by repeated sound reflections between parallel walls with smooth surface. Whether such a periodic or nearly periodic train of reflections is audible at all depends on its strength relative to that of the of non-periodic components in the impulse response (see also Section 8.3). The methods of controlling a flutter echo are similar to those avoiding the detrimental effects of a curved surface, namely, an absorbing or diffusing treatment of the critical boundary. If the flutter echo is caused by sound reflections from opposite parallel walls, it can be removed by slightly changing the angle between these walls by about 5°, if this is possible. This can also be done section-wise, resulting in a sawtooth-like structure of the wall.

## 9.4 REVERBERATION TIME

Among all significant room acoustical parameters and indices, the reverberation time is the only one which is related to room data by relatively reliable and tractable formulae, despite certain limitations which have been discussed in Chapter 5. Their application helps us decide whether a given room concept has the potential for good acoustics or whether it should be altered or outright discarded.

The room data needed for the application of these formulae are the volume of the room, the materials and the surface treatment of the walls and of the ceiling, and the number, arrangement, and type of seats. Many of these details are not yet fixed in the early phases of planning. For this reason, it makes no sense to carry out a detailed calculation of the reverberation time at this stage; instead, a rough estimate may be sufficient.

An upper limit of the attainable reverberation time can be obtained from Sabine's formula (5.26) by attributing an absorption coefficient of 1 to the areas covered by audience, and an absorption coefficient of 0.05–0.1 to the remaining areas, which need not be known too exactly at this stage.

For halls with a full audience and without any additional sound-absorbing materials, that is, in particular, for concert halls, a few rules of thumb for estimating the reverberation time are in use. The simplest one is

$$T = \frac{V}{4N} \tag{9.11}$$

with $N$ denoting the number of occupied seats. Another estimate is based on the 'effective seating area' $S_a$,

$$T = 0.15\frac{V}{S_a} \tag{9.12}$$

Here, $S_a$ is the area occupied by the audience, the orchestra, and the chorus; furthermore, it includes a strip of 0.5 m around each block of seats. Aisles are added into $S_a$ if they are narrower than 1 m. In order to give an idea of how reliable these formula are, the mid-frequency reverberation times (500–1,000 Hz) of many concert halls are plotted in Figure 9.8 as a

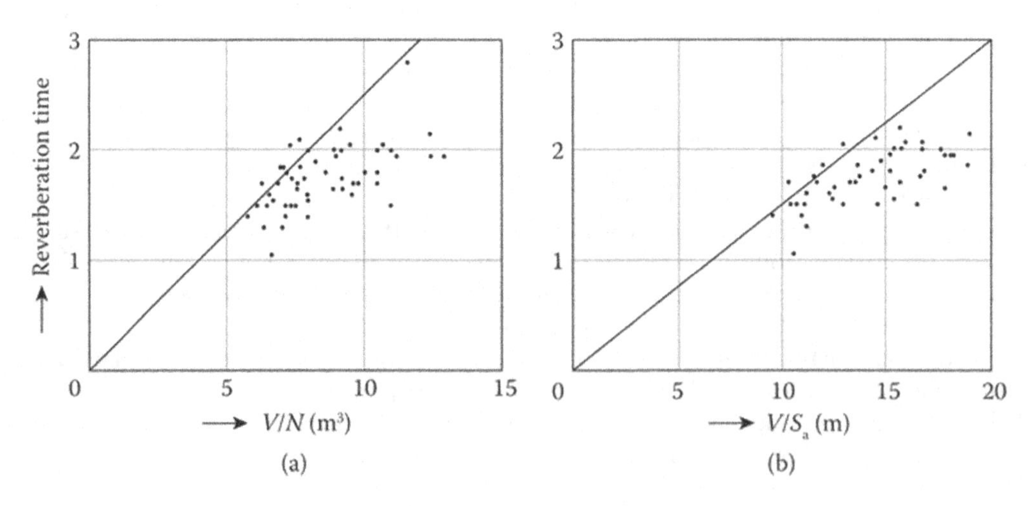


*Figure 9.8* Mid-frequency reverberation times of occupied concert halls: (a) as a function of the volume per seat and (b) as a function of the volume divided by the effective audience area. The straight lines represent Equations 9.11 and 9.12.

Source: Data from Choi et al. (2015).

function of the 'specific volume' $V/N$ (Figure 9.8a) and of volume per square metre of audience $V/S_a$ (Figure 9.8b). Both diagrams are based on data from Choi et al. (2015); each point corresponds to one hall. In both cases, the points show considerable scatter. Furthermore, it seems that Equations 9.11 and 9.12 overestimate the reverberation time.

More reliable is an estimate which involves two sorts of areas, the audience area $S_a$, as before, and the remaining area $S_r$ of the boundary:

$$T \approx 0.161 \frac{V}{S_a \alpha_a + S_r \alpha_r} \tag{9.13}$$

The absorption coefficients $\alpha_a$ and $\alpha_r$ of both types of areas can be looked up in Tables 6.4 and 6.5.

In any case, a more detailed reverberation calculation should definitely be carried out at a more advanced phase of planning when it is still possible to make changes in the interior finish of the hall without incurring extra expense. The most critical aspect is the absorption of the audience. The factors on which it depends have already been discussed in Section 6.7. Regarding the uncertainties caused by audience absorption, it is almost meaningless to debate whether Sabine's formula (5.26) is sufficient or whether the more accurate Eyring equation (5.24) would be more adequate. Therefore, the simpler Sabine formula is preferable, if needed, with an additional term $4mV$ in the denominator accounting for the sound attenuation in air.

As regards the absorption coefficients of the various materials and wall linings, compilations which have been published by several authors can be used (Cox and D'Antonio 2004). The absorption coefficients of some typical materials are listed in Table 9.2; they should be considered as averages. It should be emphasized that the actual absorption, especially of highly absorptive materials, may vary considerably from one sample to the other and that it depends strongly on the particular way in which they are mounted. In case of doubt, it is recommended to test actual materials and the influence of their mounting by measuring the absorption coefficient, either in the impedance tube or, more reliably, in an occupied state reverberation chamber (see Section 8.7). This applies particularly to chairs whose acoustical properties can vary considerably depending on the quantity and quality of the materials used for the upholstery. If possible, the empty chairs should be given about the same absorption as the occupied ones. This has the favourable effect that the reverberation time of the hall will not depend too strongly on the degree of occupation. With tip-up chairs, this can be

*Table 9.2* Typical absorption coefficients of various types of wall materials

| | *Centre frequency of octave band (Hz)* | | | | | |
|---|---|---|---|---|---|---|
| *Material* | *125* | *250* | *500* | *1,000* | *2,000* | *4,000* |
| Hard surfaces (brick walls, plaster, hard floors, etc.) | 0.02 | 0.02 | 0.03 | 0.03 | 0.04 | 0.05 |
| Slightly vibrating walls (suspended ceilings, etc.) | 0.10 | 0.07 | 0.05 | 0.04 | 0.04 | 0.05 |
| Strongly vibrating surfaces (wooden panelling over air space, etc.) | 0.40 | 0.20 | 0.12 | 0.07 | 0.05 | 0.05 |
| Carpet, 5 mm thick, on hard floor | 0.02 | 0.03 | 0.05 | 0.10 | 0.30 | 0.50 |
| Plush curtain, flow resistance 450 Ns/m$^3$, deeply folded, in front of a solid wall | 0.15 | 0.45 | 0.90 | 0.92 | 0.95 | 0.95 |
| Polyurethane foam, 27 kg/m$^3$, 15 mm thick on solid wall | 0.08 | 0.22 | 0.55 | 0.70 | 0.85 | 0.75 |
| Acoustic plaster, 10 mm thick, sprayed on solid wall | 0.08 | 0.15 | 0.30 | 0.50 | 0.60 | 0.70 |

accomplished by perforating the underside of the plywood or hardboard seats and backing them with Rockwool. Likewise, an absorbent treatment of the rear of the backrests can be advantageous.

According to Table 9.2, light partitions such as suspended ceilings or wall linings have their maximum absorption at low frequencies. Therefore, such constructions can make up for the low absorptivity of an audience and thus reduce the frequency dependence of the reverberation time.

If the auditorium is to be equipped with pseudorandom diffusers, their absorption should be taken into account. The same holds for the absorption of an organ (see Section 2.7).

In practice, it is not uncommon to find that a room actually consists of several subspaces which are acoustically coupled to each other by some opening. Examples of coupled rooms are theatres with boxes which communicate with the main room through relatively small openings, or the stage (including the stage house, which may be quite voluminous) of a theatre or opera house which is coupled to the auditorium by the proscenium, or churches with several naves or chapels. Cremer (1961) was probably the first author to point out the necessity of considering coupling effects when calculating the reverberation time of such a room. This necessity arises if the area of the coupling aperture is substantially smaller than the equivalent absorption area of the partial rooms involved.

The following discussion is restricted to a qualitative description of a system consisting of two coupled subspaces, called Room 1 and Room 2. They are coupled to each other by an aperture with the area $S'$ (see Figure 9.9a), which is small compared to both $A_1$ and $A_2$. Their damping constants and reverberation times are

$$\delta_i = \frac{c}{8V_i}(A_i + S'), \mathrm{T}_i = \frac{6.91}{\delta_i} \quad (i = 1,2) \tag{9.14}$$

$A_i$ is the equivalent absorption area in room $i$. It is assumed that the reverberation time of Room 1 is longer than that of Room 2 ($\delta_1 < \delta_2$).

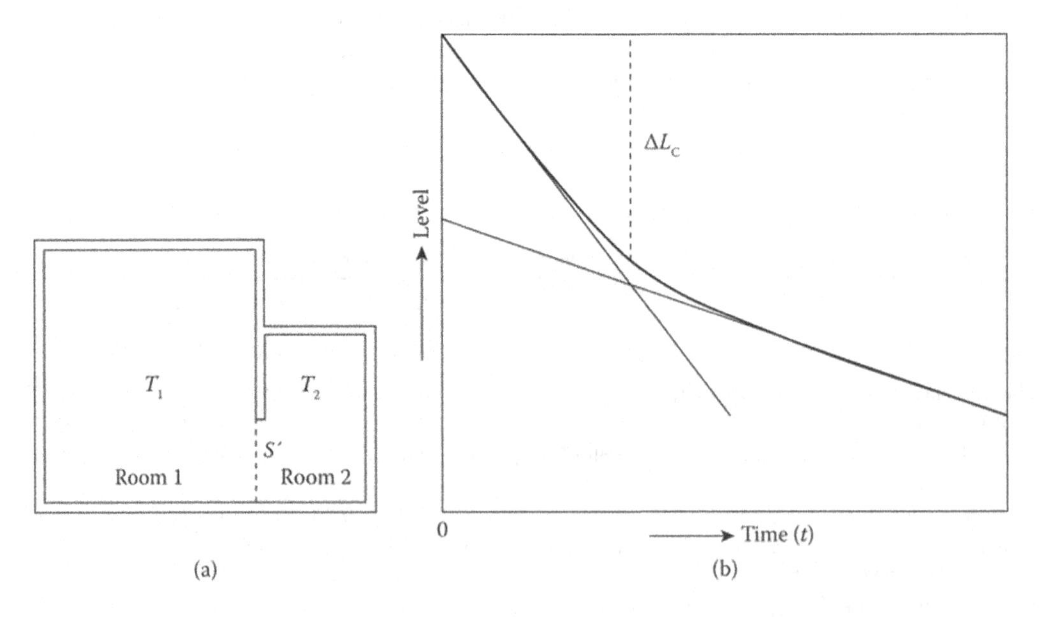


*Figure 9.9* Coupled rooms: (a) geometric situation and (b) bent decay curve made up of two components. $\Delta L_C$ characterizes the position of the bent (see Equation 9.15).

If the listener finds himself in the more reverberant room, the only effect of the coupling to another space is a slight increase of the total absorption, that is, a slight reduction of reverberation, since the coupling area acts merely as an 'open window'. The listener will hardly notice any difference, even if there is a substantial difference between reverberation times $T_1$ and $T_2$. This situation is typical for an auditorium (Room 1) with large seating areas under balconies (Rooms 2) when the listener is located in the main room. Nevertheless, whenever an auditorium has deep balcony overhangs, it is advisable to carry out an alternative calculation of its decay time by treating the 'mouths' of the overhangs as completely absorbing boundaries. Likewise, instead of including the whole stage of a concert hall with all its uncertainties in the calculation, its opening can be treated as an area having an absorption coefficient rising from about 0.4 at 125 Hz to about 0.8 at 4 kHz.

Matters are different if the listener is in the less-reverberant partial room (Room 2). Whether he will become aware of the longer reverberation in Room 1 or not depends mainly on the way the system is excited and on the size of the coupling aperture. If the sound source excites mainly Room 2, the (logarithmic) sound decay in that room will be composed of two straight parts with a bend in between, as depicted in Figure 9.9b. It has been calculated from a simple energy balance like that of Equation 5.4, however adapted to the present problem. For Figure 9.9b, it was assumed that $\delta_1 = \delta_2/3$. This shape of decay curve appears for the impulse response of the system as well as for the decay process subsequent to interrupted steady-state excitation. In the former case, the bent occurs at

$$\Delta L_C = \frac{10}{1-(\delta_1/\delta_2)} \cdot \log_{10}(1/C) - 3 \text{ dB} \tag{9.15}$$

below the beginning of the curve at $t = 0$. The constant $C$ is given by

$$C = \frac{1}{V_1 V_2 (\delta_1 - \delta_2)^2} \cdot \left(\frac{cS'}{8}\right)^2 \tag{9.16}$$

If the level at which the bend occurs is high enough — or $C$ small enough — the listener will hear the reverberant 'tail', at least when the exciting signal contains impulsive components (loud cries, isolated musical chords, drumming, etc.). In no case will he or she experience this tail as natural, because it is not a property of the listener's room but originates from the coupling aperture. However, if the location of the sound source is such that it excites both rooms, as may be the case with actors performing on the stage of a theatre, then the longer reverberation from Room 1 will be heard continually, or it may even be the only reverberation to appear. In any event, it is useful to calculate the reverberation times of both subspaces separately using Equation 9.14.

Coupling phenomena can also occur in enclosures lacking sound field diffusion. This is sometimes observed in rooms with regular geometry, with smooth wall surfaces and non-uniform distribution of the boundary absorption. Thus, a fully occupied hall with a relatively high ceiling and smooth and reflecting side walls will often build up a two-dimensional, highly reverberant sound field in its upper part. It is caused by horizontal or nearly horizontal sound paths and is influenced only slightly by the absorption of the audience. Whether a listener will perceive this particular kind of reverberation depends again on the strength of its excitation and other factors.

For lecture halls, theatre foyers, and so on, such an extra reverberation is, of course, undesirable. In a concert hall, however, this particular lack of diffusion can sometimes lead to a badly needed increase in reverberation time, namely, when the volume per seat is too small to yield a sufficiently long decay time under diffuse conditions. An example of this is the City Hall at Göttingen, a multipurpose hall with a flat floor which has a hexagonal ground plan and, in fact, has a reverberation time of about 2 s, although calculations had predicted a value of 1.6–1.7 s only (both for medium frequencies and for the fully occupied hall). Model experiments carried out afterwards demonstrated very clearly that a sound field of the described type was responsible for this unexpected increase in reverberation time.

## 9.5 MORE ON CONCERT HALLS

So far, much has been said about the acoustical design of concert halls and the underlying principles: according to Table 7.4, typical reverberation times of halls which are known for their excellent concert acoustics are in the range of 1.7–2.2 s. Furthermore, in Section 9.3 the eminent role of early reflections has been discussed. These reflections are important for two reasons: firstly, they support the direct sound and, hence, contribute to the loudness of the received sound, and secondly, if they are incident from lateral directions, they create the subjective impression of space.

Many traditional concert halls are of the 'shoebox' type, originally for non-acoustic reasons. Famous examples are the 'New' Gewandhaus in Leipzig (opened in 1886,

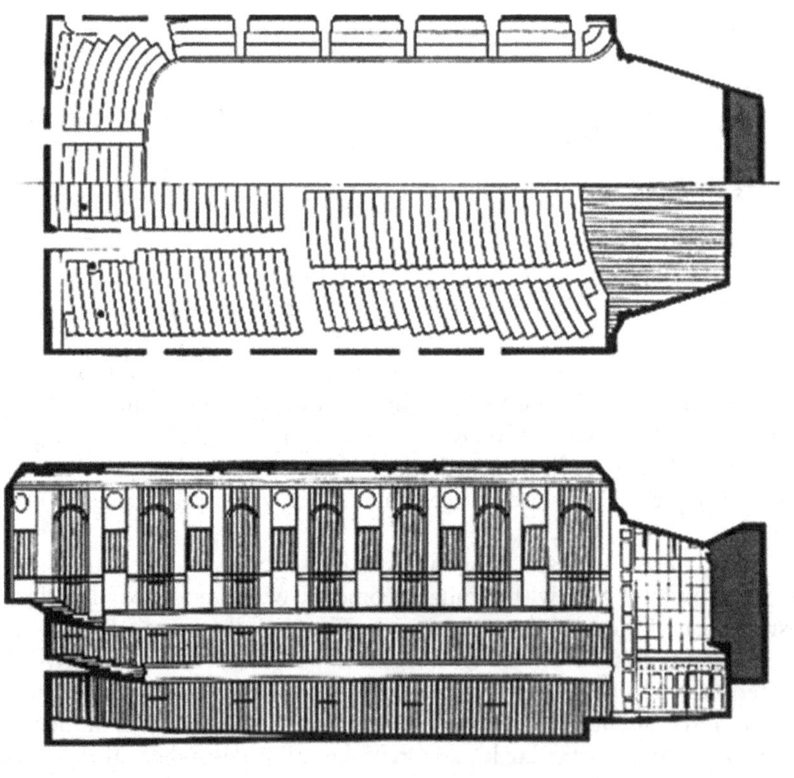

*Figure 9.10* Plan and longitudinal section of a typical 'shoebox' hall.

Source: Reproduced with permission from Beranek L. L., *Concert and Opera Halls: How They Sound.* Copyright (1996), Woodbury, NY: Acoustical Society of America.

reconstructed in 1981), the Grosser Musikvereinssaal in Vienna (1870), and the Boston Symphony Hall (1895). The plan of these and many similar concert halls has basically the shape of a relatively long and narrow rectangle, with the orchestra platform at one end (see Figure 9.10); both their floor and relatively high ceiling are basically flat, apart from plastic decorations of the latter. It is kind of a lucky coincidence that this design, including many subsequent copies, turned out to lead to acoustically great concert halls, not to the least because of the numerous and strong lateral reflections originating from the close the side walls of the room. As we learned in the preceding section, these reflections not only increase the intimacy and clarity of the sounds but also create what we call the spatial impression the sound field conveys. The high ceiling has the potential of producing the long reverberation time which is favourable for concerts of large ensembles (Figure 9.11).

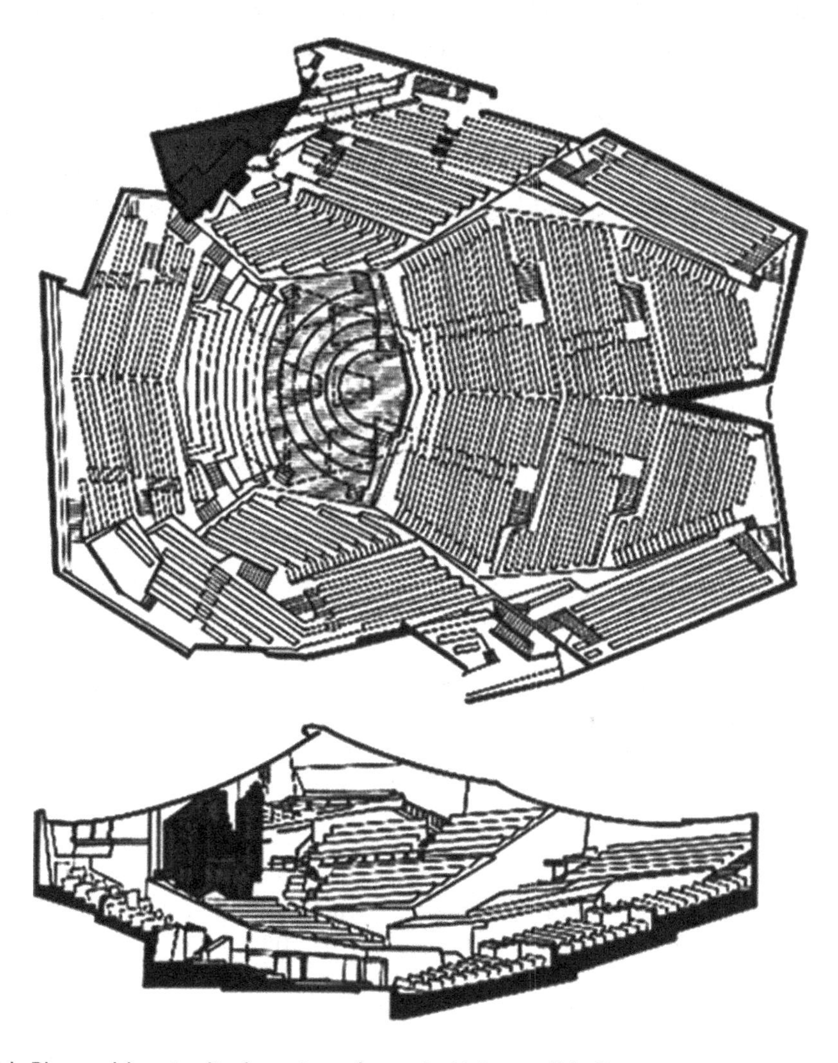

*Figure 9.11* Plan and longitudinal section of a typical 'vineyard' hall.

Source: Reproduced with permission from Beranek L. L., *Concert and Opera Halls: How They Sound.* Copyright (1996), Woodbury, NY: Acoustical Society of America.

In the course of time, the demand for larger halls emerged in order to increase the seating capacity. Furthermore, the listeners should be allotted more space than in the 19th-century halls. This demand cannot be satisfied just by blowing up the proven concept of a shoebox hall, since this would increase the number of listeners which are seated too far from the stage. And also, some of the 'early' lateral reflections would lose their ability to contribute significantly to the spatial impression and also to the intimacy and clarity of the sounds because of their increased delay times. Therefore, many architects gave up the concept of the shoebox hall in favour of shapes which are closer to a fan or even to an arena, which means that the stage is shifted towards the centre of the hall, thus bringing the audience closer to the stage. In principle, such shapes are afflicted with the lack of early lateral reflections, since nearly all seats are quite distant from the next side wall, so we cannot expect much spaciousness in them. This problem has been solved by Cremer (1964), who invented the concept of the 'vineyard' hall, which was first applied in the Berlin Philharmony (opened in 1963). Here, the audience area is subdivided into terraces called 'vineyards' which are located at different heights. They are designed in such a way that the fronts and sides of each terrace reflect early sound energy laterally to the audiences seated in adjacent terraces. The listeners in the first row of each vineyard receive strong and unimpeded direct sound. According to its tent-like shape, the ceiling is too high in the stage region to produce early reflections; therefore, large suspended sound reflectors are arranged over the stage. As mentioned before (see Section 9.3), the listeners seated to the rear of the orchestra hear the music with an incorrect timbre and balance, since nearly all instruments have the 'wrong' for these listeners, directivity especially at elevated frequencies. The same holds for singers. Despite this disadvantage, the hall was, and still is, highly appreciated, not to the least because of its brilliant acoustics.

It is a widespread opinion in room acoustics that the diffuseness of the sound field is an indispensable ingredient of fine concert hall acoustics. This belief stems probably from the observation that many of the famous 19th-century halls have walls which are not smooth surfaces but show — according to the taste of that epoch — many decorative 'irregularities', such as columns, niches, statues, and so on, at the side walls and coffers or deep beams at the ceiling. It is evident that these irregularities disperse the impinging sound waves more or less, depending on their shapes and sizes and, of course, on the sound frequency (see Section 2.7), and it is believed that this effect is responsible for the smooth and pleasant character of the sounds. Although acoustical measurements have been taken in many halls, no quantitative guidelines or recommendations concerning the necessary or desirable degree of diffuseness in a concert hall have emerged so far.

It should be noted that there are also voices which question this general opinion on diffuseness. As early as in 1967, Damaske (1967) found, by experimenting with synthetic sound fields, that a high degree of diffuseness is not necessarily a prerequisite of the spatial impression but that the latter can be created with just a few distinct lateral reflections (see Section 7.7). More recently, T. Lokki and his co-workers (Lokki et al. 2011a) concluded from listening tests in synthetic 'concert halls' that large and smooth side wall portions contribute to clear and open acoustics of a hall because they produce replica of the direct sound, which means they preserve the temporal envelope of the original sound signal. In contrast, diffusing wall areas disperse the sound in space and time and thus distort the envelope of the sound, which will be rendered weak and muddy. This is in clear contradiction to the opinion that the side walls and ceilings of a fine concert hall should be more or less structured, for instance, by applying Schroeder diffusers to them. It is quite possible that the desirable amount of surface scattering, and hence of sound field diffuseness, is — a least within certain limits — a matter of listeners' personal taste and hearing habits, that is, that some listeners prefer clearly defined reflections while others are more in favour of a smooth sound which is usually ascribed to diffusely reflecting boundaries (Lokki et al. 2011b).

Some further comments may be appropriate on the acoustical design of the stage or the orchestra's platform of concert halls. From the acoustical point of view, the stage enclosure of a concert hall has the purpose of collecting sounds produced by the musicians, to blend them and, finally, to project them towards the auditorium, but also to reflect part of the sound energy back to the performers. This latter effect is necessary to establish the mutual auditory contact they need to maintain ensemble playing, that is, proper intonation and synchronism.

At first glance, a separate stage house arranged at one end of the hall seems to serve these purposes better in that the design of its walls can be optimized in acoustical respect. As a matter of fact, however, several famous concert halls have more exposed stages without a separate enclosure. It is clear that, in such halls, reflections from the ceiling play an important role. This holds in particular if the hall is arena-shaped, as the Berlin Philharmony.

Concerning the size of the stage, it should be noted that one musician occupies about 1.5 $m^2$ on the average. This means a large orchestra with 100 musicians needs a stage area of about 150 $m^2$ or more. Stages with an area in excess of, say, 200 $m^2$ bear the risk of impaired ensemble due to the finite sound velocity which causes noticeable delays between the contributions of different musicians.

Meyer (2008) has collected stage data in several concert halls, along with the experience and the outcome of systematic investigations. According to his publication, the ceiling of the stage area should be at least 6 m over the stage floor and should not exceed about 10 m. This holds also for the height of suspended reflectors over the stage of an arena-like hall. However, in many well-known concert halls, the ceiling in the performing area is noticeably higher than that. With regard to the side walls, no clear guideline is available; a large orchestra will usually occupy the whole stage. If symphonic music is to be performed in an opera theatre, it may be useful to install a demountable orchestra shell to provide for the necessary reflections. It goes without saying that the elements of this shell must consist of sufficiently heavy elements with non-porous surface and that their orientation is critical.

Another important aspect of stage design is raking of the platform (Allen 1980), which is often achieved with adjustable or movable risers. It has, of course, the effect of improving the sight-lines between listeners and performers. From the acoustical standpoint, it increases the strength of the direct sound and reduces the obstruction of sound propagation by intervening players. This is absolutely necessary if the audience is seated on a horizontal floor, as is the case in halls which are also used for social events. It seems, however, that this kind of exposure can be carried too far; probably, the optimum rake has to be determined by some experimentation.

The successful implementation of new findings in concert hall acoustics was demonstrated in the concert hall of the famous Sydney Opera House, for example. Between 2015 and 2022, a rather complex renovation took place that included measures to improve the acoustics on the stage and in the stalls, as well as on the seating in the circle audience area, while also taking advantage of new opportunities in construction and building equipment technology. Despite the excellent reputation of the building and the Main Hall in its unique architectural design (Clements 2017), three measures for improving concert hall acoustics were identified. First, the large volume above the stage and stalls resulted in localisable reverberation perception from the partial volume high above the orchestra. Secondly, the seats in the circle, which were a large distance from the stage, suffered from not being well connected to the performance on the stage. In addition, the musicians onstage reported that they could not hear their fellow musicians and even their own instrument.

*Figure 9.12* Sydney Opera House before (a) and after (b) renovation.

Source: Courtesy of Müller-BBM.

The design process included usage of 3D computer models, an acoustic scale model, and thorough data collection with listening tests. It was finally decided to change the hall in three main characteristics (Engel and Reinhold 2023):

1. Installing new stage reflectors
2. Lowering the stage
3. Placing special movable wall reflectors

A new stage reflector cloud was designed to improve the support of the musicians onstage and to reduce the sound energy radiated into the upper part of the room. The four frontal reflectors provide strong early reflections to the stalls and audience area at the circles (see Figure 9.12). It is worth noting that the construction of the new reflector cloud was only possible with state-of-the-art materials and equipment and innovative solutions. The stage was lowered from 1.3 m to 0.9 m in order to improve the perceived orchestra balance in the stalls. In addition to this major interventions, the previously flat vertical wall surfaces around the auditorium were structured, achieving scattering for frequencies of 1 kHz and above. The renovation took place between 2020 and 2022. The construction phase was followed by validation measurements and various rehearsal sessions. After the reopening in July 2022, so far, the comments from musicians, audience, and press have been positive.

## 9.6 MULTIPURPOSE HALLS

Very often, a hall has to serve various types of purposes, including not only the performance of music of different styles or the presentation of theatre plays but also conferences, fashion shows, and many other events.

First of all, the stage should offer some possibility to adapt its size and its surrounding to the various uses of the hall. The simplest way to achieve this is to provide for some portable screens, as already mentioned earlier. Of course, their proper use requires some experience. If, for instance, the room is basically a theatre which is to be used occasionally for orchestra concerts, a carefully designed demountable orchestra shell which takes account of the principles outlined in Section 9.3 would be useful, including a stage machinery to install or to remove it. About the same holds for the orchestra pit (if there is any), which should offer the option to be raised to the stage level for orchestra concerts.

Of equal importance as the stage is the auditorium itself and its acoustical properties. This concerns, in the first place, its reverberation time. Since optimum values of the reverberation time for speech and for classical orchestra music are quite different, one can try to arrive at some compromise which would be in the range of 1.3–1.5 s in the mid-frequency range. This value would be all right for recitals and presentations of chamber music. The achieved speech intelligibility would be still sufficient, especially if the reverberation time at low frequencies does not exceed that at medium frequencies, and if the voice of a speaker is supported by a carefully designed public address system. For orchestra music or for performances with a choir, the acoustics of the hall would be perceived as somewhat too 'dry'.

However, if both orchestral music and speech are to be presented under optimum acoustical conditions, some variability of the room is indispensable, by which the conflicting requirements can be reconciled. Variations of the reverberation time can be achieved by installing variable wall or ceiling elements. Very often, these elements are panels which can be turned back to front and exhibit reflecting surfaces in one position and absorbent ones in the other. Another way is to cover certain reflecting wall portions with an absorbing curtain or, conversely, to cover absorbing wall portions with a reflecting one, that is, with a sufficiently heavy and non-porous curtain. These methods have in common that the hall volume must be high enough to reach the long reverberation required for orchestral presentations. The achieved variation in reverberation time depends on the fraction of boundary area treated in this way. Installations of this type are usually quite costly and require regular maintenance, and they are subject to the risk of inappropriate operation.

As an example, Figure 9.13 shows the effect of variable wall absorption in a broadcasting studio with a volume of 726 m$^3$. Its walls are fitted with strips of glass wool tissue which can be electrically rolled up and unrolled. Behind the porous curtain, there is an air space with an average depth of 20 cm, subdivided laterally in 'boxes' of 0.5 m × 0.6 m. The reverberation time of the studio measured for the two extreme situations (curtain rolled up and curtain completely unrolled) can be changed between 0.6 and 1.25 s.

There are also multipurpose halls with variable volume. In this case, the change of reverberation time is rather a side effect, while the main goal is to adapt the seating capacity to different uses. One example of this kind is the Stadthalle in Braunschweig, which has a maximum volume of 18,000 m$^3$ and accommodates nearly 2,200 persons. Basically, its ground plan has the shape of a regular triangle with truncated corners. In one of these corners, the stage is situated, while the other ones form elevated parts of the seating area. These latter parts can be separated from the main auditorium by folding walls. This measure reduces the volume and the seating capacity of the auditorium by 1,800 m$^3$ and 650 seats, respectively. At the same time, the reverberation time is increased from 1.3 to 1.6 s. Thus, the reduced configuration is more suited for symphonic concerts, while for more popular events or for large meetings, the full capacity of the hall can be employed (Figure 9.14).

Quite a different way to vary the acoustics of a hall is by using special electroacoustic systems by which the reverberation time can be increased. Such methods will be described in Chapter 11.

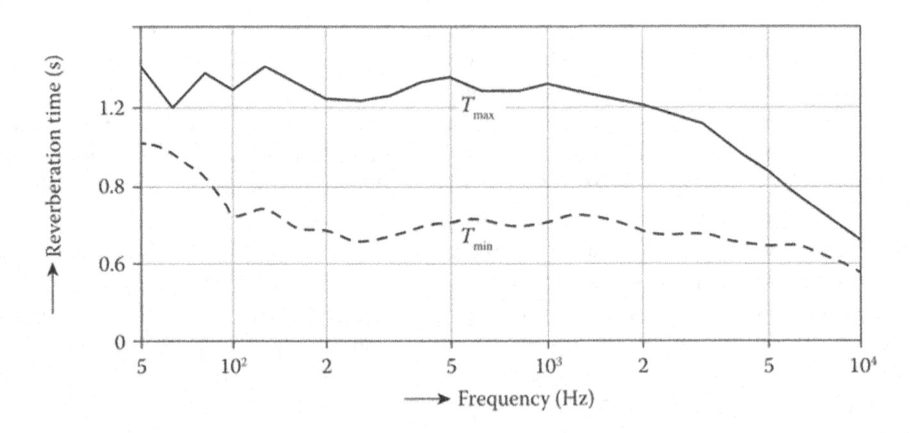


*Figure 9.13* Maximum and minimum reverberation time of a broadcasting studio with variable absorption.

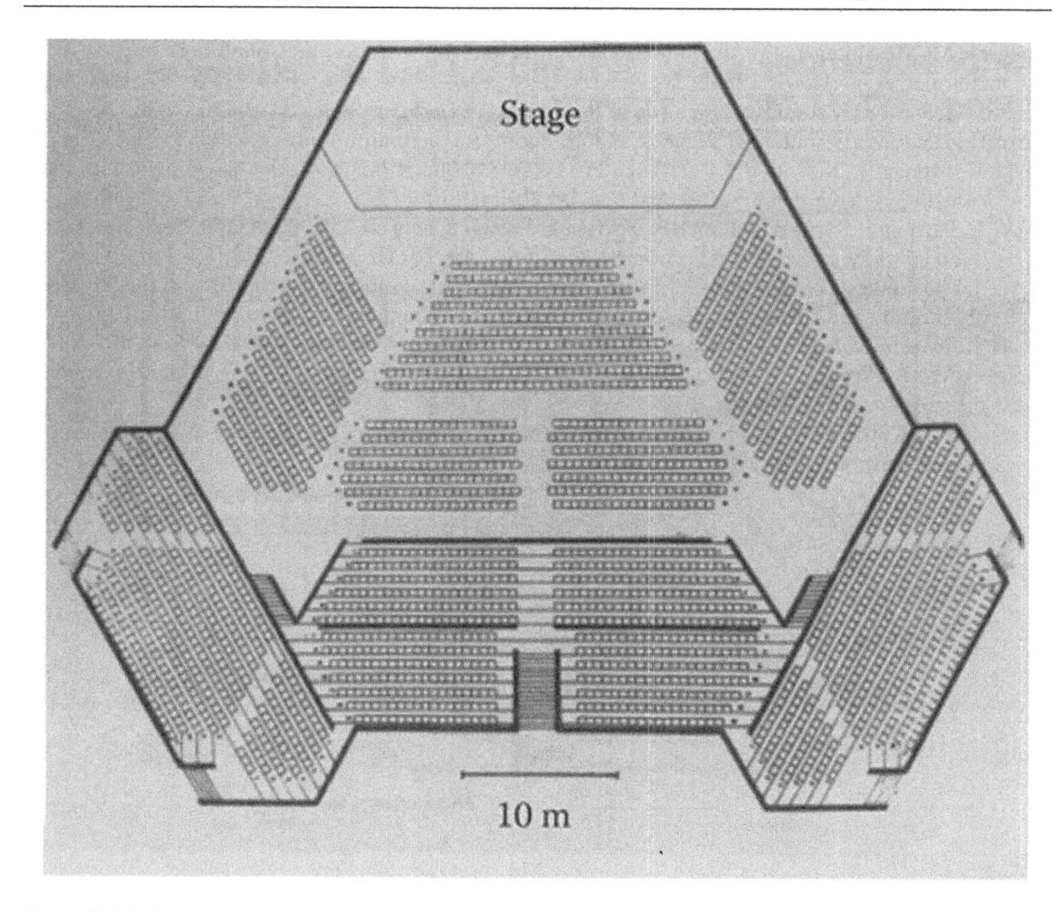


*Figure 9.14* Stadthalle, Braunschweig, plan.

## REFERENCES

Allen W.A. Music stage design. *J Sound Vibr* 1980; *69*: 143.

Choi Y.J., Bradley J.S., Jeong D.U. Experimental investigation of chair type, row spacing occupants and carpet on theatre chair absorption. *J Acoust Soc Am* 2015; *137*: 105.

Clements P.V.L. Jordan and Jørn Utzon: Acoustic and architectural interactions in the early design of the Major Hall at the Sydney Opera House, 1957–1962. *J Acoust Soc Am* 2017; *141(5)*: 3498–3498.

Cox T.J., D'Antonio P. *Acoustic Absorbers and Diffusers*. London: Spon Press, 2004.

Cremer L. *The Scientific Foundations of Room Acoustics*, Vol. II (in German). Stuttgart: S Hirzel Verlag, 1961.

Cremer L. The room acoustical design steps in the reconstruction of the Berlin Philharmony (in German). *Die Schalltechnik* 1964; *57*: 1–11.

Damaske P. Subjective investigations of sound fields (in German). *Acustica* 1967/68; *19*: 199.

Engel G., Reinhold J. Acoustic upgrade for the concert hall of the Sydney Opera House. Proceedings Institute of Acoustics, Auditorium Acoustics, Vol. 45 Pt. 2, Athens, Greece, 2023.

Hodgson M. *Theoretical and physical models as tools for the study of factory sound fields*. PhD thesis, University of Southampton, 1983a.

Hodgson M. Measurements of the influence of fitting and roof pitch on the sound field in panel-roof factories. *Appl Acoust* 1983b; *16*: 369.
Kuttruff H. Stationary sound propagation in flat rooms (in German). *Acustica* 1985; *57*: 62.
Kuttruff H. Stationary sound propagation in long rooms (in German). *Acustica* 1989; *69*: 53.
Lokki T., Patynen J., Kuusinen A., Vertanen H., Tervo S. Concert hall acoustics assessment with individually elicited attributes. *J Acoust Soc Am* 2011a; *130*: 835.
Lokki T., Pätynen J., Tervo S., Siltanen S., Savioja L. Engaging concert hall acoustics is made up of temporal envelope preserving reflections. *J Acoust Soc Am* 2011b; *129*: EL 223.
Meyer J. The influence of the angle-dependent sound radiation of musical instruments on the efficiency of reflecting and absorbing areas near the orchestra (in German). *Acustica* 1976; *36*: 147.
Meyer J. *Acoustics and the Performance of Music*. Berlin: Springer-Verlag, 2008.
Ondet A.M., Barbry J.L. Modelling of sound propagation in fitted workshops using ray tracing. *J Acoust Soc Am* 1989; *85*: 787.
Vorländer M., Kuttruff H. Lateral sound ratio as a function of the shape and the type of walls of a room (in German). *Acustica* 1985; *58*: 118.

Chapter 10

# Prediction models

## 10.1 ACOUSTICAL SCALE MODELS

For the acoustical designer, it would often be very helpful to get an idea of the acoustics of a room which is still in its planning phase, that is, in a state where details still can be altered in order to avoid possible risks or even mistakes.

A well-tried method of gaining information on the acoustics of a non-existing enclosure is to study the propagation of waves in a smaller model that is similar to the original room, at least geometrically. This method has the advantage that, once the model is at hand, many variations can be tried out with relatively little expenditure: from the choice of various wall materials to changes in the shape of the room or the arrangement of seats.

Since some basic laws are common to the propagation of all sorts of waves, it is not absolutely necessary to use sound waves for such model tests. This was an important point in earlier times, when acoustical measuring techniques were not yet at the advanced stage they have reached nowadays. So the propagation of waves on water surfaces was sometimes studied in 'ripple tanks'. However, the use of this method is restricted to two dimensions, that is, to the examination of plane sections of the hall. More profitable is to use light as a substitute of sound. In this case, the energy distribution can be measured with photocells or by photography. Absorbent areas are painted black or covered with black paper or fabric, whereas reflecting areas are made of polished sheet metal. Likewise, diffusely reflecting areas can be quite well simulated by white matt paper. However, because of the high speed of light, this method is restricted to the steady-state energy distribution. Another limitation is the absence of realistic diffraction phenomena, since the optical wavelengths are very small compared to the dimensions of all those objects which would diffract the sound waves in a real hall.

Nowadays, the techniques of electroacoustical transducers have reached a sufficiently high state of the art to generate and to receive sound waves under the condition of a model and, hence, to use sound waves for examining sound propagation in the model scale. For this purpose, a few geometrical and acoustical modelling rules have to be observed. They are based on the fact that for the propagation of sound — including diffraction effects — the physically significant length unit is not the metre but the acoustical wavelength. Hence, the most important modelling rule is

$$\frac{\ell'}{\lambda'} = \frac{\ell}{\lambda} \tag{10.1}$$

where $\lambda$ and $\lambda'$ are the sound wavelengths in the original room and its model, respectively, and $\ell$ and $\ell'$ are corresponding lengths in both rooms. Their ratio $\sigma = \ell/\ell'$ is the scale factor of

DOI: 10.1201/9781003389873-10

the model. Inserting $\lambda = c/f$ and $\lambda' = c'/f'$ with $c'$ denoting the sound velocity of the medium within the model leads us to the scale factor of frequencies:

$$f' = \sigma \frac{c'}{c} \cdot f \tag{10.2}$$

According to this relationship, the sound frequencies applied in a scale model may well reach into the ultrasonic range. Suppose the model is filled with air ($c' = c$) and the model is scaled down by 1:10 ($\sigma = 10$), then a frequency range from 100 to 5,000 Hz in the original room would correspond to 1–50 kHz in its model.

If a model is to yield realistic impulse responses, it must be more than a geometrical replica of the original hall, because the wall absorption, including its frequency dependence, must be modelled as well. This means that any surface in the model should have the same absorption coefficient at frequency $f'$ as the corresponding surface in the original at frequency $f$:

$$\alpha_i'(f') = \alpha_i(f) \tag{10.3}$$

Even more problematic is the modelling of the attenuation, $m$, which the sound waves undergo in the medium. According to Equation 1.21, the pressure amplitude of a plane sound wave travelling a distance $x = 2/m$ within a lossy medium is attenuated by a factor $e = 2.718$. From this we conclude for the attenuation constant $m'$ of the medium in the model

$$m'(f') = \sigma \cdot m(f) \tag{10.4}$$

The accuracy with which these requirements must be met depends on the kind of information we wish to obtain from the model experiments. If only the initial part of the impulse response or 'reflectogram' is of interest (over, say, the first 100 or 200 ms in the original), the air absorption can be neglected at all, or its effect can be numerically compensated, at least if the test signal is nearly monofrequent. It may even be sufficient in this case to provide for only two different kinds of surfaces in the model, namely, reflecting ones (made of metal, glass, gypsum, etc.) and absorbing ones (e.g. felt or plastic foam).

Matters are much more difficult if longer reflectograms are desired, for instance, for creating realistic listening impressions from the model auditorium, as was first proposed by Spandöck (1934). With this method, tape-recorded music or speech signals are replayed in the model at a tape speed elevated by a factor of $\sigma$. At some point within the model, the sound signal is picked up with a small microphone. After transforming the re-recorded signals back into the original time and frequency scale, it can be presented by earphones to a listener who can judge subjectively the 'acoustics' of the hall and the effects of any modifications. Nowadays, this technique is known as 'auralization'. More will be said about it in Section 10.4.

Concerning the instrumentation for measuring the impulse response in scale models, omnidirectional excitation of the model is more difficult to achieve the higher the scale factor and, hence, the frequency range to be covered. Small spark gaps can be used as sound sources, but in any case, it is advisable to check their directivity and frequency spectrum beforehand. Furthermore, electrostatic or piezoelectric transducers have been developed for this purpose; they have the advantage that they can be fed with any desired electrical signal and, therefore, allow the application of the more sophisticated methods described in Section 8.2. The microphone should also be omnidirectional. Sufficiently small condenser microphones are commercially available. Any further processing, including the evaluation of the various sound field parameters, as discussed in Chapters 7 and 8, is carried out, as with full-scale measurements, by means of a computer software (Katz et al. 2023).

*Figure 10.1* 1:10 scale model of the Philharmonie de Paris: (a) view from the stage into the hall, (b) view across the choir balcony to the organ.

Source: Courtesy of Eckhard Kahle.

Nowadays, the use of a physical scale model as a tool for the acoustical design has lost a good deal of its original appeal. The main reason for this is that this method has to compete with the mathematical methods of sound field simulation as described in the next section. One should realize that the construction of a scale model is relatively costly and requires special skills. It is nearly impossible to fulfil the conditions (10.3) and (10.4) correctly in the model despite great efforts to solve this problem. On the other hand, the sound field produced in a physical room model includes all diffraction and other wave effects which will take place in the original room. For this reason, the acoustical model techniques are still employed, provided the project under design is sufficiently expansive and prestigious, as the example of the construction of the concert hall in the Philharmonie de Paris shows (Katz et al. 2015; see Figure 10.1).

## 10.2 COMPUTER SIMULATION

As in other scientific and technical disciplines, the rapid development of digital technology, the tremendous increase in processor speed and memory size, has opened new possibilities of research in acoustics and also in room acoustic modelling. It is state of the art since the beginning of the 1990s.

The introduction of the digital computer into room acoustics is due to Schroeder (1962) and his co-workers. Since then, the computer has become an indispensable research tool for everybody who is active in simulating the propagation of sound in rooms. The first authors who applied digital simulation to concert hall acoustics were Krokstad et al. (1968), who evaluated a variety of acoustical room parameters from impulse responses obtained by a technique nowadays known as ray tracing. Meanwhile, digital computer simulation has been applied not only to all kinds of auditoria but also to factories and other working spaces.

In this section, only a brief description of the principles underlying computer simulation of sound fields in rooms can be given. A more detailed account is found in Vorländer (2020).

Basically, there are two methods of sound field simulation in use nowadays: ray tracing and the method of image sources. Both are based on geometrical acoustics, that is, they rely on the validity and application of the laws of specular or diffuse reflection.

The principle of digital ray tracing is illustrated in Figure 10.2. We imagine that, at some instant $t = 0$, a sound source releases numerous sound particles of equal energy in all directions. Each of them travels on a straight path until it hits a wall which is assumed as plane. At the intersection *PI* of the initial path with the boundary, the particle will be reflected, either specularly or diffusely. In the first case, its new direction is calculated by applying the law of geometrical reflection: Let $\varphi$ and $\vartheta$ denote the azimuth angle and the polar angle of

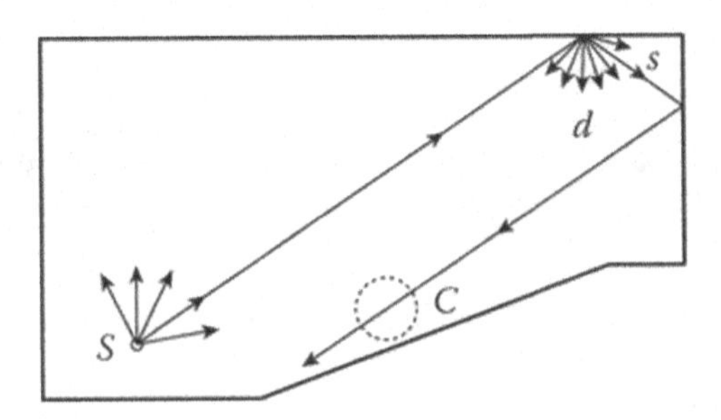


*Figure 10.2* Principle of digital ray tracing (*S* = sound source; *C* = counting sphere; s = specular reflection; *d* = diffuse reflection).

a local spherical coordinates system (see Figure 2.10 with $\theta = \vartheta$ and $\phi = \varphi$). The direction of the incoming particle is given by the angles $\varphi_i$ and $\vartheta_i$, while $\varphi_r$ and $\vartheta_r$ indicate the direction in which the particle travels after its reflection. The latter angles are given by

$$\vartheta_r = \vartheta_i \text{ and } \varphi_r = \varphi_i + \pi$$

In the second case, the stochastic character of diffuse reflection is accounted for by generating two uniformly distributed random numbers $z_1$ and $z_2$ with the computer. These are used to determine the polar angle $\vartheta_r$ and the azimuth angle $\varphi_r$ of the new direction, both in accordance with Lambert's cosine law as expressed in Equation 4.24:

$$\vartheta_r = \arccos\sqrt{z_1} \quad \text{with } 0 \le z_1 < 1 \tag{10.5}$$

$$\varphi_r = 2\pi z_2 \quad \text{with } 0 \le z_2 < 1 \tag{10.6}$$

If partially diffuse reflection is to occur, we can assign a scattering coefficient $s$ to the reflection surface. It is defined as the fraction of the reflected sound which will be scattered according to Lambert's law (see Section 8.8). We apply it by generating a third random number $z_3$ with $0 \le z_3 \le 1$. If $z_3 \le s$, the considered particle will be scattered; otherwise, its reflection will be specular. The absorption of the reflecting wall can be accounted for in two ways: either by reducing the energy of the particle by a factor of $1 - \alpha_i$ after each reflection ($\alpha_i$ = absorption coefficient) or by interpreting $\alpha_i$ as an 'absorption probability'. In the latter case, we again generate another random number $z_4$ between 0 and 1: if it exceeds $\alpha_i$, the particle will proceed carrying the same energy as before; otherwise, it has been annihilated.

After its reflection, the particle will continue on its way through the enclosure until it hits the next wall, where the same procedure takes place, and so on. The air attenuation can be accounted for by reducing the particle energy by another factor $\exp(-md_i)$ whenever the particle has travelled a straight section $d_i$ of its zigzag path. The particle's 'life' ends when its energy has fallen below a prescribed value (or when the particle has been absorbed). Then the path of the next particle will be 'traced'. The whole procedure is repeated until all the particles emitted by the sound source have been followed up.

The results are collected by means of 'counters', that is, of previously assigned counting areas or counting volumes. Whenever a particle hits such a counter, its energy and arrival time are stored, and, if needed, also the direction from which it arrived. After the process has finished, that is, the path of the last particle has been determined, all registered particle energies are classified with respect to the arrival times of the particles; the result is a histogram as shown in Figure 10.3. This can be considered as a short-time-averaged energetic impulse response. The choice of the class width $\Delta t$ is not uncritical: if it is too long, the histogram will be only a crude approximation to the true impulse response, since significant details are lost by averaging. Too short time intervals, on the other hand, will afflict the results by strong random fluctuations. Typical values for the class width $\Delta t$ are of the order of 1 ms.

The class width, and hence the achieved time resolution, is less critical if the main interest is not the true shape of the energetic impulse response of a decay curve but a parameter which involves integrations over the impulse response, or parts of it, anyway. As can be seen from Table 8.1, this holds for most of the parameters mentioned in this book, in particular, for the strength factor $G$, the 'definition' $D$, the 'centre time' $t_s$, or for the lateral energy parameters $LEF$ and $LG_{80}^{\infty}$. As an example, Figure 10.4 depicts the distribution of the strength factor (the stationary sound pressure level) and of the 'definition' obtained with ray tracing applied to a lecture hall with a volume of 3,750 m$^3$ and 775 seats (Vorländer 1988).

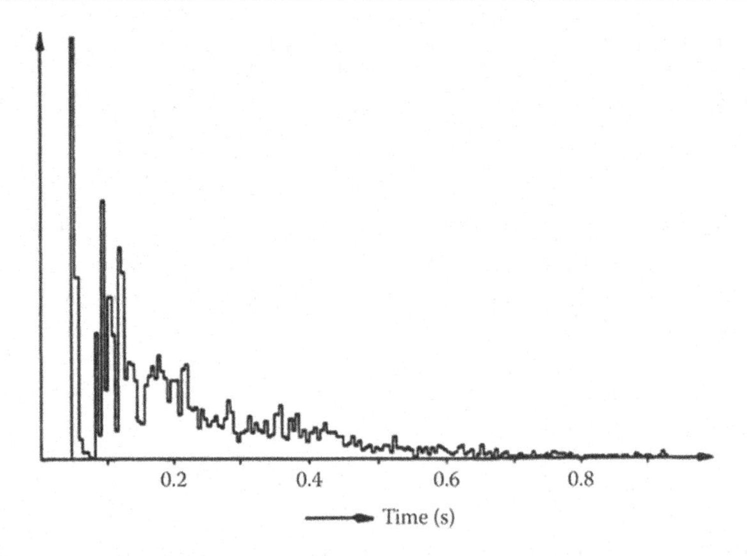


*Figure 10.3* Temporal distribution of received particle energies (energetic impulse response). The class width is 5 ms.

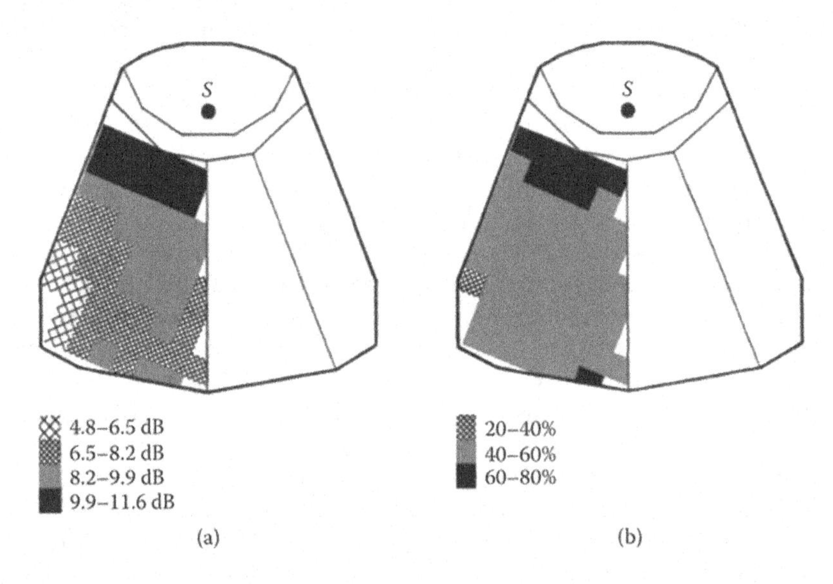


*Figure 10.4* Distribution of (a) the stationary sound level and (b) 'definition' in a large lecture hall.
Source: Vorländer (1988).

In any case, the achieved accuracy of the results depends on the number of sound particles counted with a particular counter. For this reason, the counting area or volume must not be too small; furthermore, the total number of particles contained in one 'shot' of sound impulse must be sufficiently large. As a practical guideline, a total of $10^5$–$10^6$ sound particles will yield sufficiently precise results if the dimensions of the counters are of the order of 1 m.

The most tedious and time-consuming part of the whole process is the collection and input of room data such as the positions and orientations of the walls and their acoustic properties. If the boundary of the room contains curved portions, these may be approximated by planes, unless their shape is very simple, for instance, spherical or cylindrical. The degree of approximation is left to the intuition and experience of the operator. It should be noted, however, that this approximation may cause systematic and sometimes intolerably large errors (Kuttruff 1992). These are avoided by calculating the path of the reflected particle directly from the curved wall by applying Equation 4.1, which is relatively easy if the wall section is spherical or cylindrical.

The ray tracing process can be modified and refined in many ways. Thus, the sound radiation need not necessarily be omnidirectional; instead, the sound source can be given any desired directionality. Likewise, one can study the combined effect of more than one sound source, for instance, of a real speaker and several loudspeakers with specified directional characteristics, amplifications, and delays. This permits the designer to optimize the configuration of an electroacoustic system in a hall. Moreover, any mixture of purely specular or ideally diffuse wall reflections can be taken into consideration; the same holds for the dependence of absorption coefficients on the direction of sound incidence. Further modifications replace sound rays by ray bundles with circular or polygonal cross section, as in cone tracing or pyramid tracing (Stephenson 1996).

The second method to be discussed here is based on the concept of image sources — also known as mirror sources — as has been described at some length in Section 4.1. In principle, this method is very old, but its practical application started only with the advent of the digital computer, by which constructing numerous image sources and collecting their contributions to the sound field has become very easy, at least in principle.

It should be noted that only reflections from the inside of a wall are relevant, as already mentioned in Section 4.1. Even more severe is the problem of valid and invalid image sources addressed in Section 4.1. It is illustrated in Figure 10.5, which shows two plane walls adjacent at an obtuse angle, along with a sound source $A$, both its first-order images $A_1$ and $A_2$, and the second-order images $A_{12}$ and $A_{21}$. It is easily seen that a path running from the source $A$ to the receiver $R$ can be found which involves the images $A_1$ and $A_{12}$. But there is no path reaching $R$ via the image $A_{21}$, since the intersection of the line $A_{21}$–$R$ with the plane (1) is outside the physical wall; hence, $A_{21}$ is 'inaudible' from the point $R$, and $A_{21}$ is an invalid image source.

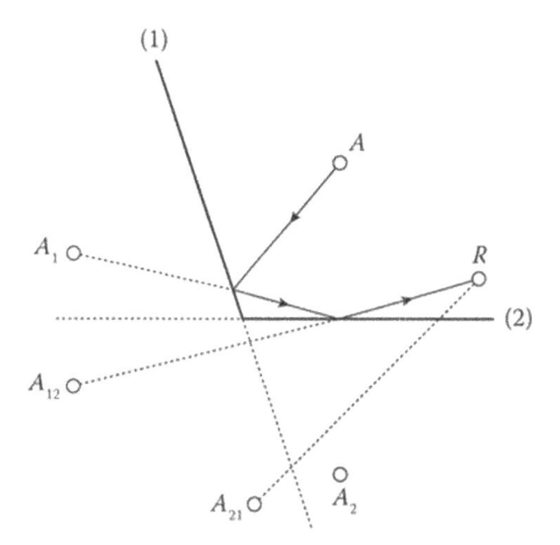


*Figure 10.5* Valid and invalid image sources. Image source $A_2$ is invalid, that is, 'invisible' from receiver position $R$.

Unfortunately, most higher-order source images are inaudible. Consider, for example, an enclosure made up of six plane walls with a total area of 3,600 m² and a volume of 12,000 m³. According to Equation 4.9, a sound ray or sound particle would undergo 25.5 reflections per second on average. To compute only the first 400 ms of the impulse response, image sources of up to the 10th order must be considered. With this figure and $N = 6$, Equation 4.2 tells us that about $1.17 \times 10^7$ (!) image sources must be constructed. However, if the considered enclosure were rectangular, there were only

$$v_r(i_0) = \frac{2}{3}\left(2i_0^3 + 3i_0^2 + 4i_0\right) \tag{10.7}$$

image sources of order $\leq i_0$ (neglecting their multiplicity), and all of them are audible. For $i_0 = 10$, this formula yields $N_r(i_0) = 1{,}560$. This consideration shows that the fraction of audible image sources is very small, unless the room is highly symmetric. And the set of audible image sources differs from one receiving point to another, of course.

In principle, the audibility of image sources is checked, as has been explained in Figure 10.5. Several authors have developed algorithms by which these tests can be facilitated. One of them is by Vorländer (1989), who performs an abbreviated ray-tracing process preceding the actual simulation. Each sound path detected in this way is associated with a particular sequence of valid image sources, for instance, $A \to A_1 \to A_{12} \to R$ in Figure 10.5, which is identified by back-tracing the path of the sound particle, starting from the receiver $R$. Hence, $A_1$ and $A_{12}$ are identified as audible. The next particle which happens to hit the counting volume at the same time can be omitted since it would yield no new image sources. After running the ray tracing for a certain period, one can be sure that all significant image sources — up to a certain maximum order — have been found, including their relative strengths, which depend on the absorption coefficients of the walls involved in the mirroring process. It should be noted that this simulation model is still based on the image source concept and that the only purpose of this ray tracing procedure is checking the audibility of image sources.

Another useful criterion of audibility is due to Mechel (2002). It employs the concept of the 'field angle' of an image source, that is, the solid angle subtended by the physical wall polygon which a source irradiates. To create an audible image source, the reflecting wall polygon must be inside the field angle of the 'mother' image source. With each new generation of image sources, the field angles are diminished. Besides a careful discussion of criteria for interrupting the process of image source construction, this publication offers many valuable computational details.

Having identified the relevant image sources, we are ready to form the energy impulse response by adding their contributions. Suppose the original sound source produces a short power impulse at time $t = 0$, represented by a Dirac function $\delta(t)$, then the contribution of a particular image source of order $m$ to the energetic impulse response at a given receiving point is

$$\frac{E_0}{4\pi c d_m^2} \cdot \rho_1 \cdot \rho_2 \cdots \rho_m \cdot \sigma\left(t - \frac{d_m}{c}\right) \tag{10.8}$$

where $E_0$ is the total energy released by the original sound source, the subscripts 1, 2, . . . , $m$ indicate the walls involved in the particular sequence of a particle's reflections, the factors $\rho_i = 1 - \alpha_i$ are their reflection coefficients, and $d_m$ is the length of the path connecting the considered image with the receiving point, which corresponds to the time delay $\tau_m = d_m/c$.

In Figure 10.6, an impulse response obtained in this way is depicted. Note that no random processes whatsoever are involved in its generation. Since the absorption coefficients are

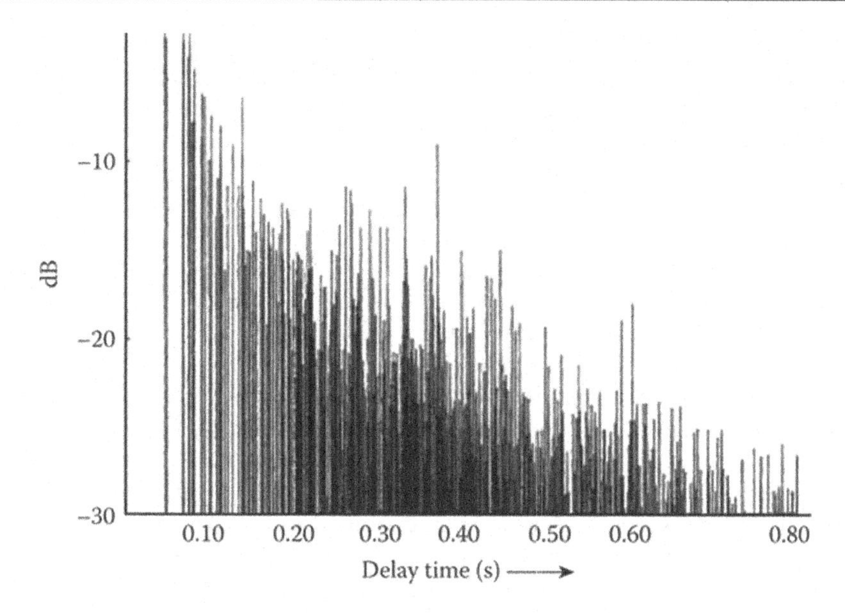


*Figure 10.6* Room impulse response computed from image sources.

usually frequency-dependent, this computation must be repeated for a sufficient number of frequency bands.

The construction of the impulse response by addition of its energetic components may be permissible if we can assume that the contributions of all image sources are mutually incoherent, that is, they cannot interfere with each other. However, Suh and Nelson (1999) have shown experimentally that energetic impulse responses calculated after the expression (10.8) are less accurate than the pressure-related impulse responses which contain all phase effects. To obtain such a response, we have to replace in Equation 10.8 the reflection coefficients $\rho_i$, of the boundary by the 'reflection responses' $r_i(t)$ as already introduced in Section 4.1 as the Fourier transform of the complex reflection factor $R_i(f)$. Accordingly, each multiplication must be replaced by a convolution indicated by asterisks. Assuming non-directional sound radiation from the original sound source and a calibrated free-field source signal $Q$, recorded, for example, at a not too close but fixed distance, we obtain (see also Equation 1.27)

$$\frac{\rho_0 Q}{4\pi d_m} \cdot r_1(t) * r_2(t) * \cdots * r_m(t) * \delta\left(t - \frac{d_m}{c}\right) \qquad (10.9)$$

For the practical computation, it is more convenient to use the Fourier transform of Equation 10.9. Hence, the contribution of an image source of order $m$ to the frequency transfer function of the room reads

$$S_m(f) = \frac{\rho_0 Q}{4\pi d_m} \cdot R_1(f) \cdot R_2(f) \cdots R_m(f) \cdot \exp(-2\pi i f d_m / c) \qquad (10.10)$$

The factors $R_1$, $R_2$, . . . , $R_m$ are the complex reflection factors of all walls from which the sound ray has been reflected. If needed, the air attenuation can be taken into account by an additional factor $\exp(-md_m)$ in Equation 10.8 or $\exp(-m\, d_m/2)$ to Equations 10.9 and 10.10.

In practical applications, it may turn out to be difficult to find the correct reflection factor $R_i$ of a given surface, as this is a complex function of the frequency and of the angle of sound incidence. However, it may be permissible to determine the absolute values of the reflection factors from the absorption coefficients

$$|R_i| = \sqrt{\rho_i} = \sqrt{(1-\alpha_i)} \tag{10.11}$$

and to combine them with some arbitrary phase function, which must be an uneven function of the frequency. To eliminate the angle dependence of the absorption coefficient, its Paris average according to Equation 2.53 can be used.

Since the image model yields the complex transfer function of a room (or, more exactly, of a particular transmission path within the room), which is equivalent to its pressure-related impulse response, it is, in a way, more 'powerful' than ray tracing. Furthermore, its results are strictly deterministic. It fails, however, if the boundary of the room is not piecewise plane or if it is so complicated that it cannot be modelled in every detail. This holds especially for acoustically 'rough' surfaces. And its application is restricted to polyhedral enclosures. Moreover, for small rooms and at low frequencies, it may happen that the assumptions for assuming the 'plane-wave condition' are violated (see Section 2.4). The ray-tracing method, on the other hand, is not subject to any limitations of room shape, and it permits us to treat the reflection from structured surfaces, for instance, a coffered ceiling, by allowing for sound scattering. Therefore, it suggests itself to combine both methods. In fact, several authors have developed hybrid procedures which, in principle, are based on the image model, thus preserving the high temporal and directional resolution of that model. At the same time, every surface element is treated as a source of diffusely reflected sound energy, the amount of which depends on the scattering coefficient of the surface. This latter part of the reflected energy can be processed with a stochastic method, such as low resolution ray tracing (Heinz 1993), or with the radiosity method (Koutsouris et al. 2013) described in Section 4.5. The specularly reflected components are used to construct the early part of the impulse response, which can be considered as the acoustical fingerprint of a room, whereas the later parts of the response, which are perceptually less characteristic for a room's acoustics, are composed of the scattered energy.

The reader may have noted that both the ray-tracing process and the method of image sources completely disregard any diffraction effects. This is a serious shortcoming since diffraction is a very common phenomenon in acoustics. In a real hall, every protruding edge, every column or pillar, and every niche or balcony is the origin of secondary waves produced by diffraction. The same holds for all abrupt changes of the impedance of an otherwise-smooth boundary. Some of these effects can be accounted for in ray tracing, but others cannot. The diffraction by the steps of a stair, for instance, can be taken into regard by treating the stair as an inclined plane and assigning a certain scattering coefficient to it. Obviously, the diffraction by the edges of a balcony face cannot be treated in this way.

Several authors have tried to fill this gap. The basic idea of all these attempts is to apply a simplified version of the diffraction pattern shown in Figure 2.14 to every sound ray which passes a rigid edge within a certain distance. It is obvious, however, that each diffraction process creates many secondary rays or particles travelling in different directions (Stephenson and Svensson 2007). This increases the complexity of the process drastically, and also the processing time. To reduce the computational load, Stephenson (1996) has developed an algorithm which unifies energy portions occurring at neighbouring locations and about equal time in space-time cells of finite size once in a while. In this model, the carriers of sound energy are pyramids which include all rays connecting a particular image source with the points of a wall polygon. During each mirroring at a wall, the wall's edges clip the pyramid, generating a narrower 'daughter' pyramid. Hence, the problem of invisible image sources will not occur at all.

To include diffraction in the modelling of image source paths, the edges can be identified in the polygon model and associated with secondary sources, as introduced by Torres and Svensson (2001). If we can assume that the diffraction edge is in the far field of the source or the previous reflection or diffraction point, the subsequent processing of first-order diffractions is a good approximation. Diffraction can thus be implemented as an extension of the image source model. The edges are first mirrored at the reflection planes in a similar way to the sources. The next task is to find valid paths between source and receiver for an arbitrary sequence of reflections and diffractions by applying Svensson's formulation of unified diffraction theory (Svensson et al. 1999; Tsingos et al. 2001). The final task in this approach is to derive filters that represent the frequency-dependent diffraction effects along the paths (Erraji et al. 2021; Kirsch and Ewert 2021). An alternative is the leading and trailing edge concept (Rozynova and Xiang 2019). However, higher-order diffraction in more complex situations requires specific near-field solutions for the actual geometries.

## 10.3 HYBRID METHODS

Due to the extremely large frequency range in room acoustics, which basically covers the entire range of human hearing, the range of wavelengths is between a few metres and a few centimetres. However, the simulation methods listed earlier are only applicable for specific cases. For geometrical acoustics, this means that the wavelengths are considered sufficiently small so that no wave effects other than diffraction occur. Accordingly, no resonances (modes) can be simulated. However, these effects are especially important at lower frequencies, where the modal response has specific features that are not hidden in the statistical modal superposition, so to speak. Which frequency is 'low' and which is 'high' cannot be decided in general terms, but two aspects are relevant here. The Schroeder frequency determines the transition between the modal regime and the stochastic regime. The relevant object sizes in the room polygon model, for example, the ceiling height for balconies or the dimensions of seats, can serve as a rule of thumb for determining frequencies and wavelengths where pronounced wave effects should be considered. Finally, a crossover frequency separates the low-frequency range for wave-based simulations and a high-frequency range for simulation with energy models (image sources, ray tracing, radiosity, etc.).

The separation into two ranges also makes sense with regard to the reliability of the input data (boundary and medium conditions) and their influence on the physical and psychoacoustic significance of the results. Above the Schroeder frequency, the stochastic properties of the room transfer function dominate. In particular, the phase response with uniform probability distribution determines the modal superposition and the resulting fine structure (see Figure 3.7). Thus, as long as the exact phases of the boundary conditions and the medium (temperature-dependent sound velocity) are not known, a nominally exact wave-based simulation result is an illusion. In this case, wave-based simulations can rather provide a plausible room transfer function with the same depth of information as energy-based methods. However, the latter are computationally more efficient by orders of magnitude.

However, for significant wave effects, wave-based models are an indispensable tool for simulating the room impulse response or room transfer function. Since the computation time of wave-based methods such as the finite element method strongly depends on the upper frequency limit, a cut-off frequency has to be chosen, and the results are only valid for the frequency range below this limit. The limit usually corresponds to the spatial resolution of the mesh; typically, mesh elements of 1/6 of the wavelength of the upper frequency limit are used.

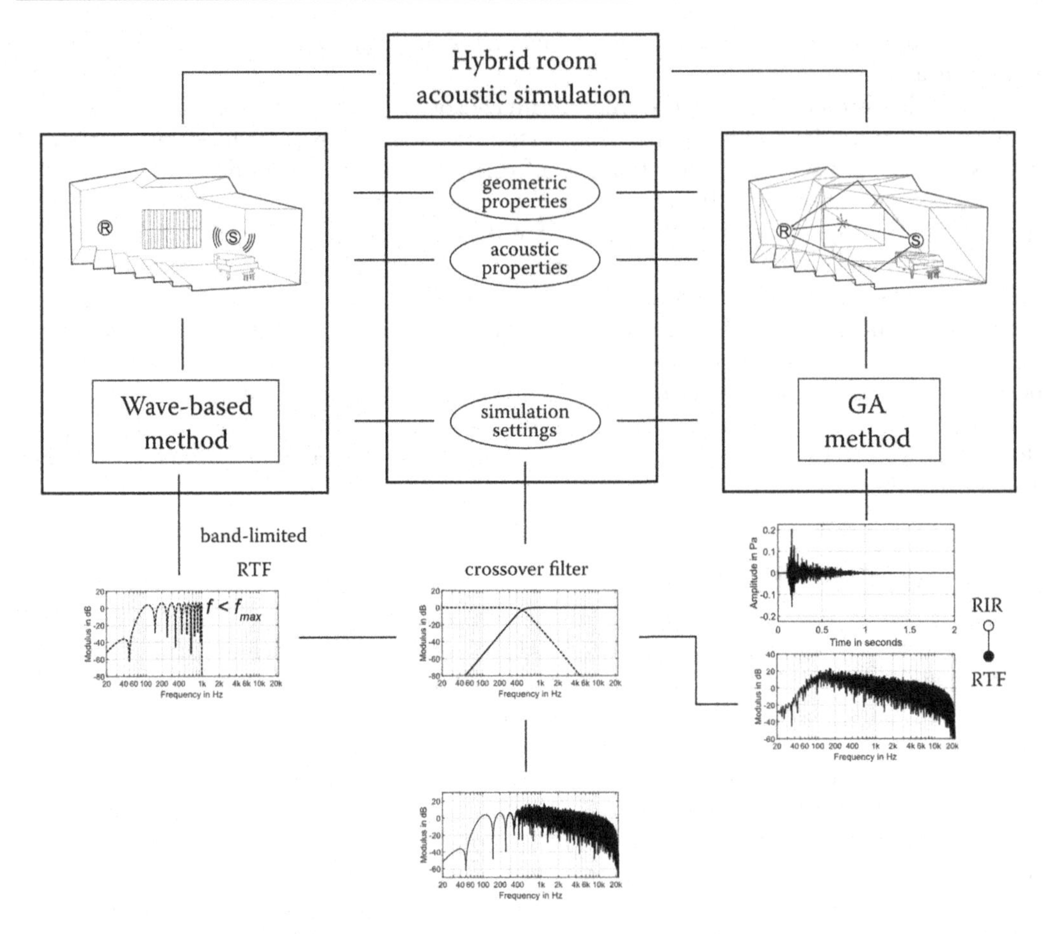


*Figure 10.7* Hybrid wave-based and geometrical acoustics simulation: principle and application example.

Higher frequency limits thus lead to smaller mesh sizes, and for a given volume or surfaces, the number of mesh elements increases rapidly (see also Section 3.5). Finally, the crossover frequency has to be defined by the user. It can be the Schroeder frequency, for which the volume and the average reverberation time must be known, or at least estimated. It can also be any higher frequency to find a compromise between computational effort and consideration of wave effects. The two simulations in their combination, a wave-based simulation for the low frequencies and for the higher frequencies, are shown in Figure 10.7.

## 10.4 AURALIZATION, VIRTUAL REALITY

The term *auralization* was coined to signify all techniques aimed at the creation of audible impressions from enclosures which do not exist in reality but in the form of design data only. Its principles are outlined in Figure 10.8. Music or speech signals ('dry' signals) originally recorded in an anechoic environment are fed to a transmission system which modifies the input signal in the same way as its propagation in a real room would modify it. This system is either a physical-scale model of a room, equipped with a suitable sound source and receiver and performing a twofold frequency transformation (see Section 10.1), or it is a digital filter

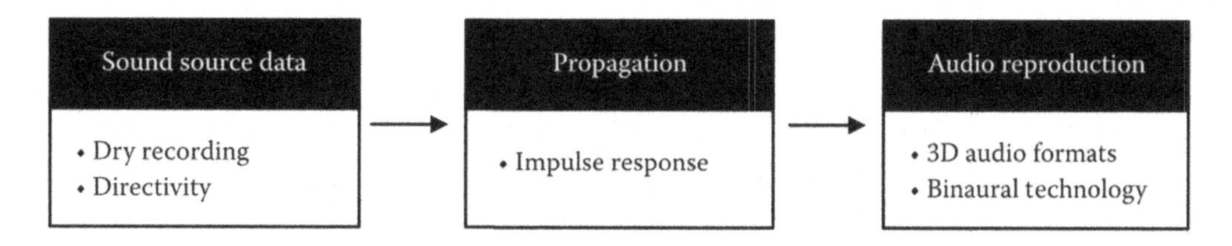


*Figure 10.8* Principle of auralization. The input signal is 'dry' music or speech.

which has the same impulse response as the considered room. The impulse response may have been measured beforehand in a real room or in its scale model, or it is obtained by simulation as described in the preceding section.

In any case, the room simulator must produce a binaural output signal; otherwise, no realistic, that is, spatial, impressions can be conveyed to the listener. The output signal is presented to the listener by headphones or, preferably, by two loudspeakers combined with a cross-talk cancellation system, operated in an anechoic room. The merits and shortcomings of both methods are discussed in Section 7.8.

If the auralization filter consists of a physical scale model, it should cover a wide frequency range without any noticeable linear distortions. Hence, the model transducers have to meet very high standards; otherwise, the listener will not get a realistic impression of a room's acoustics. For the same reason, the requirements concerning the acoustical similarity between an original room and its model are very stringent. The absorption of the various wall materials and of the audience must be modelled quite correctly, including their frequency dependence (see Equation 10.3), a condition which must be carefully checked by separate measurements. Even more difficult is modelling the sound attenuation by the medium according to Equation 10.4. Several research groups have tried to meet this requirement by filling the model either with air of very low humidity or with nitrogen. If at all, the frequency dependence of air attenuation can be modelled only approximately by such measures, and only within a limited frequency range and at a particular scale factor $\sigma$. The first experiments of auralization with scale models were carried out by Spandöck's group in Karlsruhe (Germany) in the early 1950s; a comprehensive report may be found in a publication by Brebeck et al. (1967).

Auralization based on a purely digital room model was first carried out by Allen and Berkley (1979). In this case, most of the mentioned problems do not exist, as both the acoustical data of the medium and all data of the enclosure are fed into the computer from the keyboard or from a database. Generally, digital models are much more flexible, as the shape and the acoustical properties of the room under investigation can be changed quite rapidly. Therefore, nearly all auralization experiments are based on computer models nowadays.

Pressure-related impulse responses or transfer functions computed with the image source model according to Equation 10.9 or Equation 10.10 can immediately be used for auralization, provided they are binaural. This is achieved by multiplying each sum term of expression (10.10) with $L(f, \varphi, \vartheta)$ and $R(f, \varphi, \vartheta)$, the head-related transfer functions for the left and the right ears, respectively (see Section 1.6). The angles $\varphi$ and $\vartheta$ characterize the direction of incidence in a head-related coordinate system. Then, the contributions to the binaural transfer functions are

$$\left[S_m(f)\right]_l = S_m(f) \cdot L(f,\varphi,\vartheta) \quad \text{(left channel)} \tag{10.12}$$

and

$$\left[S_m(f)\right]_r = S_m(f)\cdot R(f,\varphi,\vartheta) \quad \text{(right channel)} \tag{10.13}$$

$L$ and $R$ are either a particular listener's individual head-related transfer functions or they are averages which can be regarded as representative of many individuals.

In contrast to image source results, energy histograms as obtained with the ray tracing method (see, for instance, Figure 10.3) need some post-processing. Since the absorption coefficients of the boundary are usually frequency-dependent, the complete result of a ray tracing process consists of a set of such histograms, namely, one for each frequency band (octave band or third-octave band). These histograms can be summarized in a function, $E(f_i, k)$, where $f_i$ is the mid-frequency of the $i$th frequency band, and $k$ denotes the number of the time intervals. Considered as a function of the frequency, $E(f_i, k)$ approximates a short-time energy spectrum valid for the time interval around $t_k = k\Delta t$. The typical class width $\Delta t$ is of the order of one or a few milliseconds.

In the following, we describe a procedure by which these spectra can be converted into a pressure-related impulse response of the room, as developed by Heinz (1993). By properly smoothing, one obtains from $E(f_i, k)$ a continuous and positive function $E_k(f)$ of the frequency. The pressure-related transfer function $G_k(f)$ is calculated by equating its absolute value $|G_k|$ with the square root of $E_k(f)$ and 'inventing' a suitable phase spectrum $\psi_k(f)$. Then,

$$G_k(f) = \sqrt{E_k(f)}\cdot \exp\left[i\psi_k(f)\right] \tag{10.14}$$

As we know from Section 3.4, the phase spectrum is not critical at all, since the propagation in a room randomizes anyway all phases. Therefore, any odd phase function $\psi_k(f) = -\psi_k(-f)$ could be used for this purpose, provided it corresponds to a system with causal behaviour, that is, to a system the impulse response of which vanishes for $t < 0$. One of several possibilities is to derive $\psi_k(f)$ from $E_k(f)$ as the minimum phase function. This is achieved by applying the Hilbert transform, Equation 8.19, to the natural logarithm of $E_k(f)$:

$$\psi_k(f) = \frac{1}{\pi}\int_{-\infty}^{\infty} \frac{\ln\left[E_k(f-f')\right]}{f'}df' \tag{10.15}$$

What is still missing is the information on the direction from which the particular sound component arrives at the receiver. To take this point into account, the full solid angle is subdivided in a number of — say, 37 — directional groups, each of them characterized by two representative angles $\varphi_j$ and $\vartheta_j$, and each of them associated with a pair of head-related transfer functions $L(f, \varphi_j, \vartheta_j)$ and $R(f, \varphi_j, \vartheta_j)$, namely, one for the left and one for the right channel.

From the directional information collected in the counter $C$ during the ray tracing process (see Figure 10.2), we can derive the probability of a sound particle which is counted in a particular interval $\Delta t$ appearing in one of the directional groups, taking into account the different sizes $\Delta\Omega_j$ of these groups. The directional group with the highest probability will be assigned to all sound particles observed in $\Delta t$. For this purpose, we consider a random succession of equal Dirac impulses with a typical mean density of 10,000 per second. Each of the impulses collected in $\Delta t$ is convolved with:

1. The inverse Fourier transform of $G_k(f)$ after Equation 10.14
2. The inverse Fourier transforms of both head-related transfer functions $L$ and $M$ (see Equations 10.12 and 10.13)

Superposing all impulses which are modified by the described operations yields the binaural impulse response of the room — and hence of the auralization filter in Figure 10.8, which can be convolves with the dry input signal.

If all steps are carefully carried out, the listener to whom the ultimate result is presented will experience an excellent and quite realistic impression. To what degree this impression is identical with what the listener would have in the real room was investigated by Brinkmann (2018). It depends on not only the quality of the simulation but also the quality of the electroacoustic transducers, that is, of the headphones or the loudspeakers used for the presentation. In fact, there are many details which are open to improvement and further development. Nevertheless, an old dream of acousticians has become true through the techniques of auralization. Many new insights into room acoustics are expected from its application. Thus, auralization has also become an important research tool. At the same time, it is an indispensable for the acoustical consultant as it permits to convince the architect, the user of a hall, and — last but not the least — himself on the efficiency of the measures he proposes in order to reach the original design goal.

Meanwhile, several commercial software package for the simulation of sound fields in rooms, computation of characteristic parameters, and presentation of results by auralization have become available (CATT-Acoustic, ODEON, EASE, and others). Each of them exists in several versions, and they all are hybrid image-source cone/ray-tracing algorithms, including phases. EASE (Ahnert and Feistel 1991) has originally been developed as a tool for the design and the test of electroacoustic sound reinforcement systems; nowadays, it is much similar to the aforementioned process described by Heinz (1993). The core of the CATT-Acoustic algorithm (Dalenbäck 1996) is cone tracing. Each cone arriving at a diffusely reflecting surface is imagined to create a secondary source which emits secondary cones according to the absorption and the scattering coefficient of the surface. For auralization, the reverberant tail is simulated by considering an equivalent rectangular room which is derived from the actual room. ODEON is a hybrid image source–ray tracing algorithm (Naylor 1993). To account for diffuse reflections, each diffusing surface is subdivided in square patches that act as receivers for incident energy and later serve as secondary sources for the scattered energy. Comparisons of room acoustic simulation and auralization software have been carried out by Vorländer (1995), Bork (2000, 2005), and Brinkmann et al. (2019).

The process of auralization as described so far suffers from a serious shortcoming: it assumes that the listener keeps his head fixed during the whole session. Even a slight turn of the head has the consequence that the installed head-related transmission functions (HRTFs) in Equations 10.12 and 10.13 are no longer valid and must be replaced with modified ones as soon as these changes exceed certain limits. The same holds when the listener is going around in the room. This concerns mainly the original sound source and the low-order image sources. Here, the rate of updating the auralization filter should typically be about 50 per second in order to achieve sufficient smoothness of the perceived audio signal. This choice leaves a processing time of just 20 ms. This may give an idea about the tremendous challenges connected with real-time auralization, which cannot be met without applying sophisticated procedures of signal processing, such as real-time convolution and the use of a computer with sufficient memory space and processor speed. The simplest way to present the auditory signal to the user of a binaural VR system is by means of high-quality headphones. More convenient for the user, and providing a more realistic presentation, is the use of 'dynamical crosstalk cancellation', CTC. This is an extension of the CTC system described in Section 7.8 (Lentz and Behler 2004); it permits the listener to move his head without distorting the auditory impression. A further refinement is the implementation of the user's interaction with the environment.

One solution to the challenge of quickly adapting binaural signals to the head orientation is to use a 3D spatial sound format, such as Ambisonics or Higher-Order Ambisonics, HOA. In this case, spatial coding at the receiver in the simulation does not require multiplication with the corresponding angle-dependent HRTF but with spherical harmonic (SH) functions, as presented in Section 1.2. For sound reproduction, the multi-channel spatial impulse response can then be processed with the 'dry' sound signal into a multi-channel audio stream that can be fed into an HOA loudspeaker array (Zotter 2009).

The optical counterpart to auralization is visualization, that is, the simulation of the spatial distribution of light energy in a virtual environment with the intention to convey a realistic impression of the scenery a spectator would see in a particular hypothetical ('virtual') environment. Of course, for visualization, the auralization filter in Figure 10.8 has to be replaced with a visualization filter, which has a 'binocular' output, in order to produce a stereoscopic image of the scene. The process starts by defining the positions of light sources with prescribed spectral and directional characteristics, and also the positions and spectral reflectivities of boundaries, including their scattering properties. In the visualization filter, similar concepts as in auralization are implemented, namely, the image source model, statistical ray tracing, or the radiosity integral. One way to present the result of this process to a spectator is either by means of a head-mounted display, fitted out with two small monitors placed immediately in front of the spectator's eyes. An acoustical signal can be presented by built-in earphones. An alternative (and preferable) method employs four or five projection screens surrounding the spectator. Each of them is associated with two projectors which are directed towards the screen in order to transmit the result of the simulation in the form of two images with differently polarized light. The user looking at the scenery wears polarization filters in front of his eyes which separate the two images and thus create a stereoscopic, that is, spatial image of the scenery. It goes without saying that the actual position and orientation of the spectator's head must be continuously tracked. This can be achieved by applying electromagnetic, ultrasonic, or optical means.

Both processes, the visualization and the auralization of an environment, can be combined in what is called virtual reality (VR). If needed (for instance, in driving or flight simulators), haptic sensations such as caused by vibrations can also be included. The output of the auralization filter and the result of visualization are presented simultaneously to the spectator — apart from some unavoidable delay caused by the finite processing speed. As long as the total delay is smaller than about 30 ms, this is still tolerable and does not disturb the impression.

Although there are many parallels between processing the acoustical and the visual branch of VR, there are also important differences between both processes. One of them is that, in visualization, no phases at all must be considered because the light is assumed as incoherent. For this reason, and because optical wavelengths are much smaller than the dimensions of typical obstacles, any diffraction effects can be neglected. Another, very significant difference is due to the high speed of light propagation (about 300,000 km/s compared with the sound velocity of about 340 m/s). This has the consequence that, in visualization, all transient effects can be completely neglected; the optical field is always stationary, even if there are rapid movements of the spectator's head or changes of the scenery. Thus, some of the difficult problems which one faces in auralization, such as real-time convolution, for instance, do not appear when it comes to visualization.

This may be the reason that visualization is much more advanced nowadays than auralization. On the other hand, virtual reality cannot yield realistic impressions of some spatial scenery without the correct transmission of auditory signals, including the acoustic information on source positions, on the effects of the boundary and the medium, and generally on the spatial dimension.

Concerning the practical use of VR systems, much can be repeated with what previously has been said about the use of auralization. In the first place, the designers of theatres, concert halls, large lecture or congress halls, including the electroacoustic systems to be operated in them (architects, acoustical consultants, electrical engineers), will benefit from the possibility to obtain immediate listening and seeing impressions from their project and from the practical and aesthetic effect of modifications. Regarding the rapid progresses of computer technology along with falling costs, it can be observed that VR appears on the consumer market too.

## REFERENCES

Ahnert W., Feistel R. Binaural auralization from a sound system simulation programme. Proceedings of the 91th AES Convection, New York, 1991, Preprint 3127.

Allen J.B., Berkley D.A. Image method for efficiently simulating small room acoustics. *J Acoust Soc Am* 1979; *65*: 943.

Bork I. A comparison of room simulation software — the 2nd round robin on room acoustical computer simulation. *Acta Acust United Acust* 2000; *84*: 943.

Bork I. Report on the 3rd round robin on room acoustical computer simulation, part II: Calculations. *Acta Acust United Acust* 2005; *91*: 75.

Brebeck P., Bücklein R., Krauth E., Spandöck F. Akustisch ähnliche Modelle als Hilfsmittel für die Raumakustik. *Acustica* 1967; *18*.

Brinkmann F. *Binaural processing for the evaluation of acoustical environments*. Doctoral dissertation, Technical University Berlin, 2018.

Brinkmann F., Aspöck L., Ackermann D., Lepa S., Vorländer M., Weinzierl S. Round robin on room acoustical simulation and auralization. *J Acoust Soc Am* 2019; *145*(4): 2746.

Dalenbäck B.-I. Room acoustic prediction based on a unified treatment of diffuse and specular reflection. *J Acoust Soc Am* 1996; *100*: 899.

Erraji A., Stienen J., Vorländer M. The image edge model. *Acta Acustica* 2021; *5*: 17.

Heinz R. Binaural room simulation based on the image source model with addition of statistical methods to include the diffuse sound scattering of walls and to predict the reverberant tail. *Appl Acoust* 1993; *38*: 145.

Katz B., Jurkiewicz Y., Wulfrank T., Parseihian G., Scélo T., Marshall H. La Philharmonie de Paris — acoustic scale model study. Proceedings of Institute of Acoustics, Auditorium Acoustics, Vol. 37 Pt. 2; Paris, France, 2015.

Katz B., Stitt B., Size matters: New ways of working with scale models. Proceedings of Institute of Acoustics, Auditorium Acoustics, Vol. 45 Pt. 2; Athens, Greece, 2023.

Kirsch C, Ewert S. Low-order filter approximation of diffraction for virtual acoustics. 2021 IEEE Workshop on Applications of Signal Processing to Audio and Acoustics, New Paltz, 2021, p. 341.

Koutsouris G.I., Brunskog J., Jeong C.H., Jacobsen F. Combination of acoustical radiosity and the image source method. *J Acoust Soc Am* 2013; *133*: 3963.

Krokstad A., Strem S., Sørsdal S. Calculating the acoustical room response by the use of a ray-tracing technique. *J Sound Vibr* 1968; *8*: 118.

Kuttruff H. Some remarks on the simulation of sound reflection from curved walls. *Acustica* 1992; *77*: 176.

Lentz T., Behler G. Dynamic cross-talk cancellation for binaural synthesis in virtual reality environments. Proceedings of the 117 Convention Audio Engineering Society, San Francisco, CA, 2004.

Mechel F. *Formulas of Acoustics*. Berlin: Springer-Verlag, 2002.

Naylor G.M. ODEON—another hybrid room acoustical model. *Appl Acoust* 1993; *38*: 131.

Rozynova A., Xiang N. Sound diffraction prediction of a rectangular rigid plate using the physical theory of diffraction. *J Acoust Soc Am* 2019; *145*(4): 2677.

Schroeder M.R. Digital computers in room acoustics. Proceedings of the Fourth International Congress on Acoustics, Copenhagen, 1962, Paper M21.

Spandöck F. Raumakustische Modellversuche. *Ann d Physik V* 1934; *20*: 345.

Stephenson U.M. Quantized beam tracing — a new algorithm for room acoustics and noise immission prognosis. *Acta Acust United Acust* 1996; *82*: 517.

Stephenson U.M., Svensson P. Can also sound be handled as stream of particles? An improved approach to diffraction-based uncertainty principle — from ray to beam tracing. Proceedings of DAGA, Stuttgart, 2007.

Suh J.S., Nelson P.A. Measurement of the transient response of rooms and comparison with geometrical acoustic models. *J Acoust Soc Am* 1999; *105*: 230.

Svensson U.P., Fred R.I., Vanderkooy J. Analytic secondary source model of edge diffraction impulse responses. *J Acoust Soc Am* 1999; *106*: 2331.

Torres R., Svensson U.P., Kleiner M., Computation of edge diffraction for more accurate room acoustics auralization. *J Acoust Soc Am* 2001; *109*(2): 600.

Tsingos N., Funkhouser T., Ngan A., Carlbom I. Modeling acoustics in virtual environments using the uniform theory of diffraction. Proceedings of the 28th Conference of Computer Graphics and Interactive Techniques, Los Angeles, CA, 2001, p. 545.

Vorländer M. Ein Strahlverfolgungsverfahren zur Berechnung von Schallfeldern in Räumen. *Acustica* 1988; *65*: 138.

Vorländer M. Simulation of the transient and the steady state sound propagation in rooms using a new combined sound particle-image source algorithm. *J Acoust Soc Am* 1989; *86*: 172.

Vorländer M. International round robin on room acoustical computer simulations. Proceedings of the 15th ICA Trondheim, Norway, 1995, p. 689.

Vorländer M. *Auralization*, 2nd edn. Cham: Springer Nature Switzerland, 2020.

Zotter F. *Analysis and synthesis of sound-radiation with spherical arrays*. Doctoral thesis, Institute of Electronic Music and Acoustics, University of Music and Performing Arts Graz, 2009.

Chapter 11

# Electroacoustical systems in rooms

There currently are many points of contact between room acoustics and electroacoustics, even if we neglect the fact that modern measuring techniques in room acoustics could not exist without the aid of electroacoustics. Thus, we shall hardly ever find a meeting room of medium or large size which is not provided with a public address system for speech amplification; it matters not whether such a room is a church, a council chamber, or a multipurpose hall. We could dispute whether such an acoustical 'prothesis' is really necessary for all these cases or whether sometimes they are rather a misuse of technical aids; it is a fact that many speakers and singers are not only unable but also unwilling to exert themselves to such an extent and to articulate so distinctly that they can make themselves clearly heard even in a lecture or meeting room of moderate size. Instead, they prefer to rely on the microphone which is readily offered to them. But the listeners are also demanding, to an increasing extent, a loudness which will make listening as effortless as it is in broadcasting, television, or cinemas. Acousticians have to come to terms with this trend, and they are well advised to try to make the best of it and to contribute to an optimum design of such installations.

But electroacoustical systems in rooms are by far more than a necessary evil. They open acoustical design possibilities which would be inconceivable with traditional means of room acoustical treatment. For one thing, there is a trend to build halls and performance spaces of increasing size, thus giving large audiences the opportunity to witness personally important cultural, entertainment or sports events. This would be impossible without electroacoustical sound reinforcement, since the human voice or a musical instrument alone would be unable to produce an adequate loudness at most listeners' ears. Furthermore, large halls are often used — largely for economic reasons — for very different kinds of presentations (see also Section 9.6).

In this situation, it is a great advantage that electroacoustical systems permit the adaptation of the acoustical conditions in a hall to different kinds of presentations, at least within certain limits. Imagine a hall with relative long reverberation, well suited for musical performances. Nevertheless, a carefully designed electrical sound system can provide for good speech intelligibility by directing the sounds towards the audience and, hence, avoiding the excitation of long reverberation.

The reverse way is more versatile and also more difficult technically: to render the natural reverberation of the hall short enough in order to match the needs of optimum speech transmission. For the performance of music, the reverberation can be enhanced by a sophisticated electroacoustical system to a suitable and adjustable amount. The particular circumstances will decide which of the two possibilities is more favourable.

Electroacoustical systems for reverberation enhancement simulate acoustical conditions that are not encountered in the given hall as it is. At the same time, they can be considered as a first step towards producing new artificial effects that are not encountered in halls without

DOI: 10.1201/9781003389873-11

an electroacoustical system. In the latter respect, we are just at the beginning of a development whose progress cannot yet be predicted.

Whatever the type and purpose of an electroacoustical system, there is a close interaction between the system and the room where it operates in that its performance depends, to a high degree, on the acoustical properties of the enclosure itself. Therefore, the installation and use of such a system does not dispense with careful acoustical planning. Furthermore, without the knowledge of the acoustical factors responsible for speech intelligibility and of the way in which these factors are influenced by sound reflections, reverberation, and other acoustical effects, it would hardly be possible to plan, install, and operate electroacoustical systems with excellent performance. It is the goal of this chapter to deal particularly with this interaction. More information on electroacoustical sound systems can be found in books on this subject — see that by Ahnert and Steffen (1999).

## 11.1 LOUDSPEAKERS

In the simplest case, an electroacoustic sound reinforcement system consists of a microphone, an electrical amplifier, and one or several loudspeakers. The microphone and the loudspeaker are the electroacoustic links to the world of sound. Another important component of the system is the room in which the system is operated.

The frequency range which the system must cover depends on its intended use. If it is to be used for speech reinforcement only — maybe apart from occasional musical interludes — a range from 100 Hz to about 6,000 Hz may be sufficient, as spectral components with higher frequencies will not appear in speech. When it comes to music transmission, the required quality standards are much higher. Accordingly, the system should transmit frequency components reaching from 20 Hz to 20,000 Hz or even higher, which implies a significant increase of expenditure and costs.

The most critical part of a sound reinforcement system is the loudspeaker because it must generate high acoustical power in a very limited space (compare the size of a loudspeaker or loudspeaker cluster with that of a symphonic orchestra!) without producing noticeable nonlinear distortions. At the same time, it should distribute the sound it generates as uniformly as possible over the audience.

For the construction of a loudspeaker, nearly every known transduction principle can be employed. We mention here the most successful ones, namely, the piezoelectric, the electrostatic, and the electrodynamic principle. Among these, the last one is definitely the most popular, due to its flexibility and its power handling capacity. Figure 11.1 presents a cross-sectional view of an electrodynamic (or just dynamic) loudspeaker system. It consists of a small coil arranged in the cylindrical gap of a strong permanent magnet which produces a radial magnetic field in the gap. An electrical current flowing through the coil produces an axially directed force acting onto a conical (or nearly conical) diaphragm, from which the sound wave is radiated. Both the diaphragm and the coil are kept in correct position by highly compliant spring elements. These elements together with the mass of the moving parts form a mechanical resonator, the resonance frequency of which determines the lower frequency limit of the loudspeaker.

A loudspeaker system according to Figure 11.1, for example, without any additional measures would be a rather inefficient sound source, since both sides of the diaphragm emit sound waves of opposite phase which partially cancel each other especially at low frequencies ('acoustical short-circuit'). This unwanted effect can be reduced or avoided by mounting the chassis system based on Figure 11.1 into a panel or into an open or a closed box. The useful frequency range of the loudspeaker can be extended towards lower frequencies by using more sophisticated boxes.

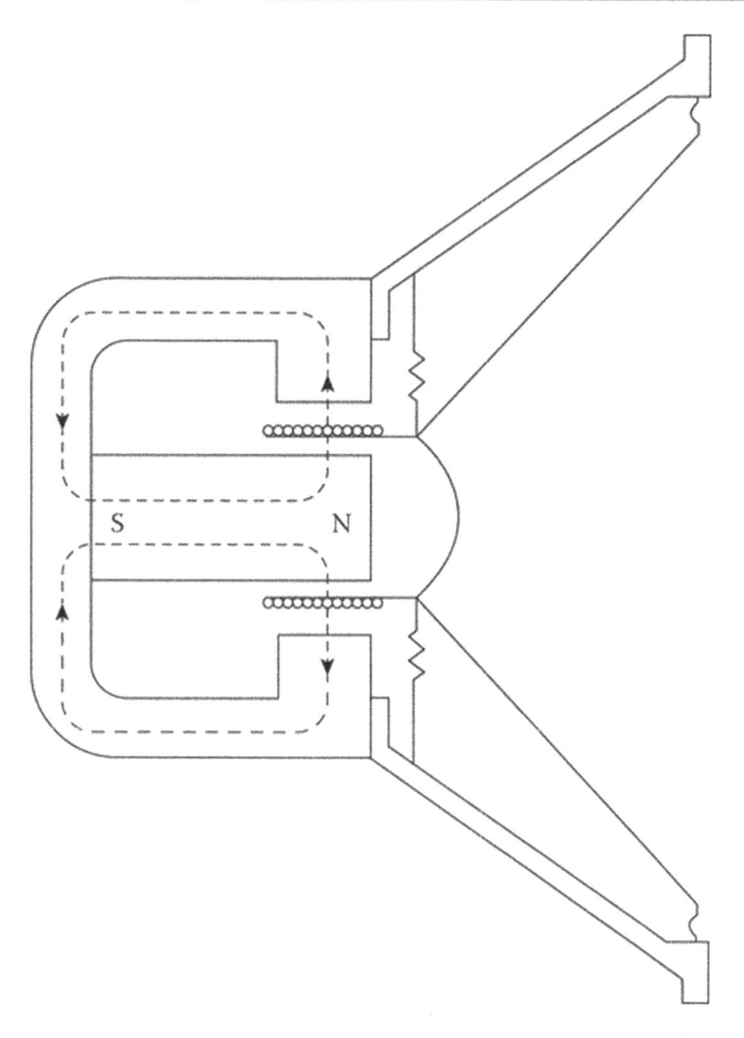


*Figure 11.1* Electrodynamic cone loudspeaker system (schematic).

Source: Reproduced with permission from H. Kuttruff, Akustik — Eine Einführung. Copyright (2004) S. Hirzel Verlag Stuttgart.

As an abstract model of a loudspeaker, we now consider a plane circular piston mounted flush in an infinite baffle which oscillates in the direction of its normal (see Figure 11.2). Each of its surface elements contributes a spherical wave to the sound pressure at some point. At higher frequencies, when the radius *a* of the piston is not small compared with the wavelength of the radiated sound signal, noticeable phase differences between these contributions may occur, resulting in an interference pattern of the sound pressure amplitude. In the far field of the loudspeaker, that is, in the region where the distance from the source is considerably larger the $S/\lambda$ ($S$ = area of the piston), the directional structure of the sound field can be described by the directional factor (see Equation 1.29), which in the present case reads

$$\Gamma(\vartheta) = \frac{2J_1(ka \sin\vartheta)}{ka \sin\vartheta} \tag{11.1}$$

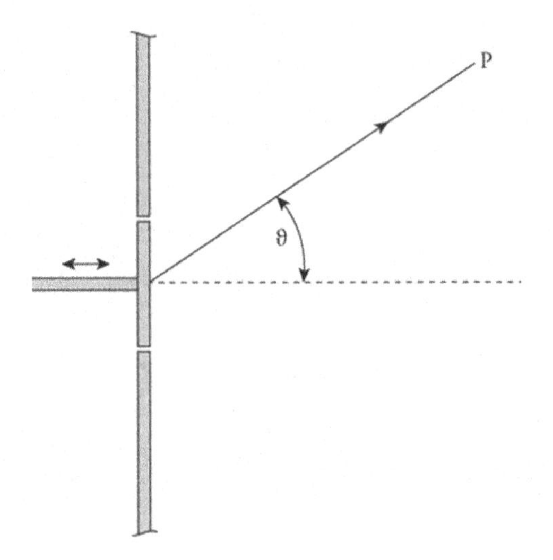


*Figure 11.2* Piston radiator (schematic).

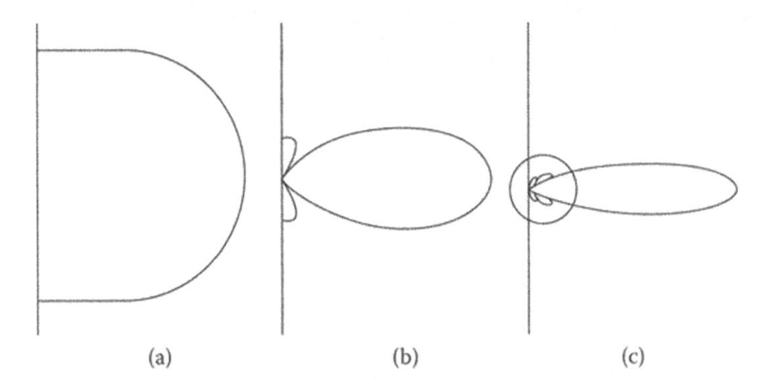


*Figure 11.3* Directional factor (magnitude $|\Gamma(\vartheta)|$) of the circular piston: (a) $ka = 2$, (b) $ka = 5$, and (c) $ka = 10$.

Here, $\vartheta$ denotes the angle between the normal to the piston area and the direction of radiation, $J_1$ is the Bessel function of first order, and $ka = 2\pi a/\lambda$ is the so-called Helmholtz number, which is the circumference of the piston divided by the acoustical wavelength. Figure 11.3 depicts a few directional patterns of the circular piston, that is, polar representations of the directivity factor $|\Gamma|$ for various values of $ka$. (The three-dimensional directivity factors are obtained by rotating these diagrams around their horizontal axes.) For $ka = 2$, the radiation is nearly uniform, but with increasing $ka$, the sound is increasingly concentrated toward the middle axis of the piston. For $ka > 3.83$, additional smaller lobes appear in the diagrams.

The shape of the main lobe of a radiator's directional diagram can be characterized by its half-power bandwidth, that is, the angular distance $2\Delta\vartheta$ of the points for which $|\Gamma|^2 = 0.5$.* For the circular piston with $ka >> 1$, this quantity is approximately

$$2\Delta\vartheta \approx \frac{\lambda}{a}\cdot 30^\circ \tag{11.2}$$

* As shown in Figure 11.6.

Another important characteristic figure is the directivity factor $\gamma$, defined as the ratio of the maximum and the average intensity, both at the same distance from the source (see Equation 5.46). For the circular piston, it is given by

$$\gamma = \frac{(ka)^2}{2} \cdot \left(1 - \frac{2J_1(2ka)}{2ka}\right)^{-1} \tag{11.3}$$

The dependence of this function on $ka$ is shown in Figure 11.4.

The membrane of a real loudspeaker is neither plane nor rigid, and in most cases, it is not mounted flush in a plane infinite baffle but in the front side of a box. Hence, its directivity differs more or less from that described by Equations 11.1 through 11.3. Nevertheless, these relationships yield at least a guideline for the directional properties of real loudspeakers.

Another type of loudspeaker which is used in many sound reinforcement systems is the horn loudspeaker. It consists of a tube with continuously increasing cross-sectional area, called a horn, and an electrodynamic driver attached to the horn at its narrow end. The main advantage of this loudspeaker is its high-power efficiency, because the horn improves the acoustical match between the driver's diaphragm and the free field. This effect is usually augmented by a pressure chamber incorporated in the driver acting as a mechanical impedance transformer (see Figure 11.5). Another advantage of the horn loudspeaker is that it has a certain directivity even at low frequencies. The simplest horn types are the conical and the exponential horn. Unfortunately, no closed formulae are available for calculating or estimating the directional characteristics of a horn loudspeaker, which depends on the shape and the length of the horn as well as the size and the shape of its 'mouth'. Thus, the directional pattern of a horn has to be determined experimentally or with numerical methods. The same holds for its frequency characteristics. The most successful way to do this is by application of the boundary element method (BEM), which has been mentioned in Section 3.5. By combining several horns, a wide variety of directional patterns can be achieved. The most straightforward solution of this kind are the multicellular horns consisting of many single horns, the openings of which approximate a portion of a sphere and yield nearly uniform radiation into the solid angle subtended by the opening of the horn.

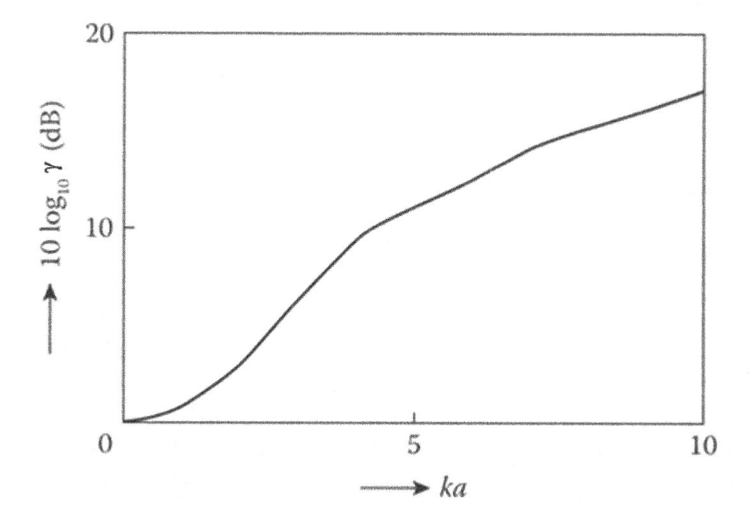


*Figure 11.4* Directivity $\gamma$ (in dB) of the circular piston as a function of *ka*.

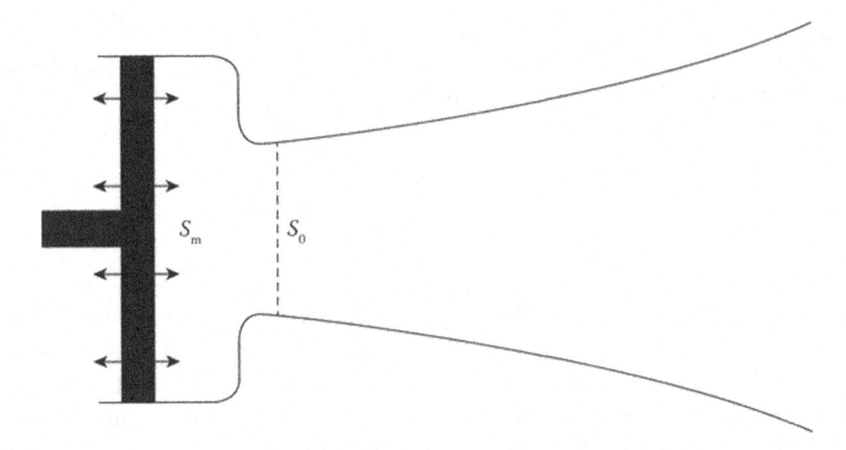


*Figure 11.5* Horn loudspeaker with pressure chamber driven by a moving piston (longitudinal section).

A well-tried method to achieve directivity of sound radiation is combining several loudspeakers, which are fed by the same electrical signal, in a regular array. The sound pressure obtained in some point $R$ is the sum of the contributions. Depending on the direction of $R$ and of the sound frequency, phase differences between these contributions may lead to partial or total mutual cancellation.

As an example, we consider a linear array consisting of $N$ point sources each of them producing a spherical wave according to Equation 1.27. They are arranged along a straight line at equal distances $d$, and they all are fed with the same electrical signal. Suppose its frequency is $\omega = kc$, then the directional factor of this array is given by

$$\Gamma_a(\vartheta) = \frac{\sin\left(\frac{1}{2}Nkd\sin\vartheta\right)}{N\sin\left(\frac{1}{2}kd\sin\vartheta\right)} \tag{11.4}$$

where $\vartheta$ is the angle in which the considered direction includes with the normal of the array. This is illustrated in Figure 11.6, which plots $|\Gamma_a(\vartheta)|$ as a polar diagram for kd = $\pi/2$ and $N = 8$. The subscript $a$ is to indicate that the directional pattern of Equation 11.4 is a property just of the regular arrangement of the elements. In real loudspeaker arrays, each element has a directivity on its own with the directivity factor $\Gamma_0$. Then, the total directivity factor is simply obtained by multiplying $\Gamma_a$ with $\Gamma_0$:

$$\Gamma(\vartheta) = \Gamma_0(\vartheta) \cdot \Gamma_a(\vartheta)$$

The three-dimensional directivity pattern is obtained by rotating the diagram in Figure 11.6 around the vertical axis of the array. It is noteworthy that the radiated sound is concentrated into the plane perpendicular to the array axis. As in the case of the circular piston, the directional pattern $\Gamma_a(\vartheta)$ contains a main lobe which becomes narrower with increasing frequency. Furthermore, for $f > c/Nd$, it shows smaller satellite lobes, the number of which grows with increasing number of elements and with the frequency.

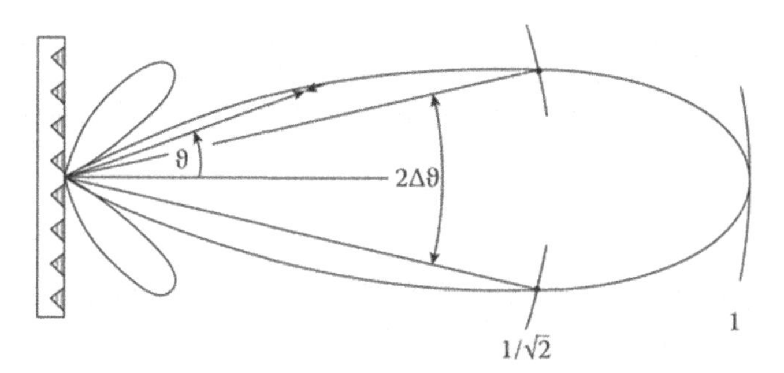


*Figure 11.6* Linear loudspeaker array with $N = 8$ and its directional factor (magnitude $|\Gamma_a(\vartheta)|$) for $kd = \pi/2$.

For $N > 3$, the largest of these side lobes is at least 10 dB lower than the maximum of the main lobe. The angular half-power width of the main lobe is

$$2\Delta\vartheta \approx \frac{\lambda}{Nd} \cdot 50^\circ \tag{11.5}$$

This relationship, however, holds only if the resulting half-width $2\Delta\vartheta$ is less than 30°.

It should be noted that by introducing phase shifts into the signals feeding the elements of the array, the direction of the main lobe can be changed. This is of practical interest because, in this way, a fine adjustment of the direction of maximum radiation can be affected. Digital loudspeaker systems as 'line arrays' became very efficient in terms of an optimized directivity which can be steered by using phase and amplitude weighting for the array elements. They require A/D conversion if the input signals are not in a digital format already. The signal processing in the digital signal domain can be performed in a very flexible way by applying all kinds of filters, delays, and limiters. Sometimes, it may be advantageous to modify the concept of an array with elements arranged along a straight line. In fact, it is easy to imagine that a loudspeaker array with a slightly convex curvature permits a more uniform irradiance of the audience area. The effect of this measure is twofold: firstly, there is always one loudspeaker directed toward listeners, and secondly, by the curvature, the idea of several contributions arriving at a listener with equal amplitudes and phases is given up, and this the more the larger the angle between the axes of neighbouring loudspeakers. Obviously, it is the component $\Gamma_a(\vartheta)$ of the directivity which is most affected by the modified shape of the array.

## 11.2 REQUIRED POWER AND REACH OF A LOUDSPEAKER

A question of crucial importance for the practical design of a sound amplification system is that of which sound power must be provided by the sound source, because this determines the choice of the loudspeakers and their arrangement.

The sound pressure level which must be achieved by the loudspeakers of a sound reproduction system depends on the type of sound signal (speech or music) and on the noise in the auditorium where the sound is to be presented. This noise may be due to a restless audience or to technical installations, such as the air-conditioning system, or to insufficient insulation against exterior noise sources. In any case, the level of the amplified signal should exceed the noise level by 10 dB at least. Under normal conditions, a level of 70–75 dB should be adequate for speech transmission. For music, the required sound level is considerably higher, namely, 95–105 dB, depending on the sort of music.

In the free field, the relation between the energy density $w_d$ and the acoustic power $P$ of a sound source with the directivity factor $\gamma$ reads

$$w_\mathrm{d} = \frac{\gamma P}{4\pi c r^2} \tag{11.6}$$

To calculate the energy density in an enclosure, one might be tempted to apply Equation 5.5,

$$w = \frac{4P}{cA} = \frac{P}{2\delta V} \quad \text{with } A = \sum_i S_i \alpha_i \tag{11.7}$$

denoting the total absorbing area in the room. However, then one would disregard the fact that the impulse response between two points of a room consists of 'early reflections', which are highly welcome because of the support they give to the direct sound (among other beneficial effects), and of 'late reflections', which make up the reverberant tail of the response. This latter part is an indispensable ingredient of the acoustics of fine concert halls, but for the intelligibility of speech, it is rather disturbing in that it forms a kind of background noise which does not carry any useful information. Obviously, this 'noise' cannot be overcome just by increasing the gain of the amplifier, since this would increase not only the useful components (direct sound and early reflections) but also the reverberant part of the impulse response. The only way to reduce the 'reverberant noise' is to reduce the reverberation time of the room and/or to improve the design of the whole system, especially the position and the directivity of the loudspeaker(s), with the goal to achieve a more favourable ratio of the early and the late energy transported in the impulse response.

To estimate the energy contained in the 'reverberant tail' of the impulse response, we model the latter as an exponential decay of sound energy with a decay constant $\delta = cA/8V$

$$E(t) = E_0 \exp(-2\delta t)$$

Now, suppose a sound source supplies the constant sound power $P$ to the room, resulting in the steady-state energy density $w$ according to Equation 11.7. We regard as detrimental those contributions to $w$ which are conveyed by all reflections with delays exceeding 100 ms with respect to the direct sound, in some rough accordance with the definition (7.10) of the clarity index. Thus, the energy they carry is

$$w_r' = \frac{P}{V}\int_{0.1\mathrm{s}}^{\infty} \exp(-2\delta t)\,dt = \frac{\mathrm{P}}{2\delta \mathrm{V}} \exp(-0.2\delta) \tag{11.8}$$

This expression will now be compared with the energy density $w_d$ contained in the direct sound as given by Equation 11.6. We postulate that satisfactory speech intelligibility can only be achieved if $w_\mathrm{d}$ is at least equal to $w_1'$ or, more explicitly,

$$\frac{\gamma P}{4\pi c r^2} \geq \frac{P}{2V\delta} \exp(-0.2\delta) \tag{11.9}$$

In the following, we denote with $r_{max}$ the largest distance $r$ for which this condition can be fulfilled; it indicates the range in which we can expect good intelligibility, provided the loudness of the sound signal is high enough. Introducing the reverberation time by $T = 3 \cdot \ln 10/\delta$ and observing that $\exp(0.3 \cdot \ln 10) \approx 2$, we obtain from Equation 11.9

$$r_\mathrm{max} \approx 0.06 \left(\frac{\gamma V}{T}\right)^{1/2} \cdot 2^{1/T} \tag{11.10}$$

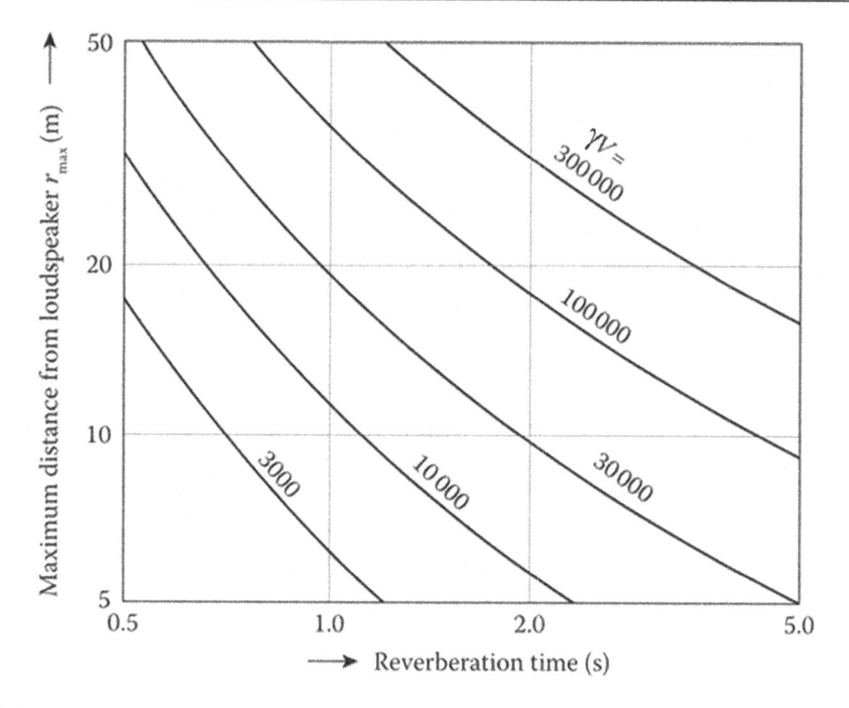


*Figure 11.7* Maximum distance of listeners from a loudspeaker ('reach of a loudspeaker') as a function of the reverberation time for various values of $\gamma V$ ($V$ in m³).

where $r_{max}$ is in metres, and $V$ in cubic metres. This quantity differs from the critical distance $r_c$ in Equation 5.48 by the factor $2^{1/T}$.

In Figure 11.7, $r_{\max}$ is plotted as a function of the reverberation time; the product $\gamma V$ of the loudspeaker gain and the room volume are the parameters of the curves.

It should be noted that Equation 11.10 represents rather a rough estimate of the maximum distance $r_{max}$ than an exact limit. This is a consequence of the somewhat-crude assumptions in the derivation before. In particular, it underestimates the distance $r_{max}$ for two reasons:

1. In most cases, the loudspeaker will also produce early wall reflections which contribute to the left-hand side of Equation 11.9, but there is no general way to account for them.
2. It is safe to assume that the main lobe of the loudspeaker's directional pattern will be directed towards the audience, which is highly absorbing at mid and high frequencies. This reduces the power available for the excitation of the reverberant field roughly by a factor

$$\frac{1}{\gamma'} = \frac{1-\alpha_a}{1-\overline{\alpha}} \qquad 11.11$$

with $\alpha_a$ denoting the absorption coefficient of the audience, while $\overline{\alpha} = A/S$ (see Equation 11.7) is the mean absorption coefficient of the boundary. Consequently, the gain $\gamma$ in Equation 11.10 and Figure 11.6 can be replaced with $\gamma\gamma'$, which is larger than $\gamma$.

The most important result of this consideration is that the reach of a sound source in a room is limited by the reverberant sound field and cannot be extended just by increasing the sound power. To illustrate this, let us consider a hall with a volume of 15,000 m³ and a reverberation time of 2 s. If the product $gg$ can be made as high as 16, the maximum distance which can be acoustically bridged will be $r_{max} \approx 28$ m.

At first glance, it seems that the condition $r < r_{max}$ with $r_{max}$ from Equation 11.10 is not too stringent. This holds, indeed, for medium and high frequencies. At low frequencies, however, $g$ as well as $g'$ are close to unity, and often, the reverberation time of the room is higher than that at mid-frequencies. As a consequence, most of the low-frequency energy supplied by the sound source will feed the reverberant part of the energy density where it is not of any use. This is the reason that so many halls equipped with a sound reinforcement system suffer from a low-frequency background which is unrelated to the transmitted signal and is perceived as a kind of noise. The simplest remedy against this evil is to mute the low-frequency components of the signal which do not contribute to speech intelligibility in any way by a suitable electrical filter.

Of course, Equation 11.10 is not the only condition which must be met if the system is to perform satisfactorily. Other ones are a well-controlled directivity of the loudspeaker and the acoustical power it can produce. So we are coming back to the question already raised at the beginning of this section. We require that the sound pressure level of the direct loudspeaker signal in all points closer to the source than $r_{max}$ is at least $L_d$. Using Equations 11.6 and 1.69 (with SPL = $L_d$ and $cw_d = I_d$), one obtains for the necessary sound power

$$P = \frac{4\pi c r_{max}^2}{\gamma} \cdot w_d \approx \frac{4\pi}{\gamma} r_{max}^2 \cdot 10^{(L_d/10)-12} \text{ watts} \tag{11.12}$$

This is the same formula as would be adequate for outdoor sound amplification — apart from the maximum distance $r_{max}$ based on Equation 11.10.

## 11.3 REMARKS ON LOUDSPEAKER POSITIONS

In this section, we consider again a sound reinforcement system which, in the simplest case, consists of a microphone, an amplifier, and a loudspeaker, as presented in Figure 11.8. Additional electrical components, such as equalizers, delay units, and limiters, are not shown. The amplified microphone signal is supplied to the room, and hence to the audience, either by one central loudspeaker or by several or many loudspeakers distributed throughout the room. (The term 'central loudspeaker' includes, of course, the possibility of combining several loudspeakers closely together, for instance, in a cluster or a linear array as described in

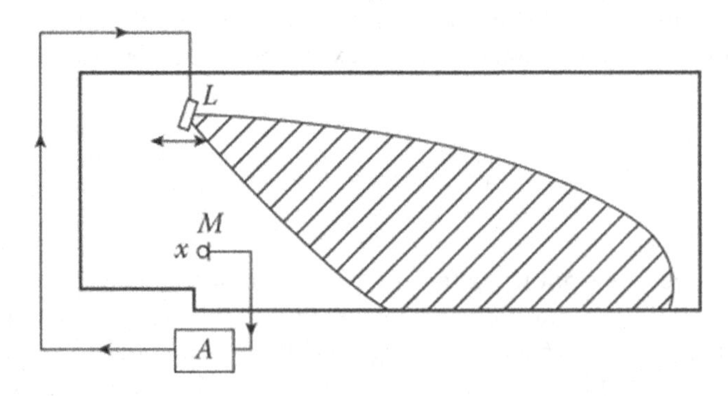


*Figure 11.8* Central loudspeaker system: *L* = loudspeaker, *M* = microphone, *A* = amplifier, and *x* = source.

Section 11.1.) This section deals with several factors which should be considered when loudspeaker locations are selected in a room.

In any case, the loudspeakers must ensure that all listeners are supplied with sufficient sound energy and that a satisfactory speech intelligibility is attained. Furthermore, a sound reinforcement system should yield a natural hearing impression. In the ideal case (possibly with the exception of the presentation of electronic music), the listener would be unable to notice the electroacoustical support at all. To achieve this, it is necessary, apart from using high-quality microphones, loudspeakers, and amplifiers, that the sounds produced by the loudspeakers reach the listener from about the same direction in which he is seeing and hearing the actual speaker or the natural sound source.

Since the microphone and the loudspeaker are operated in the same room, it is inevitable that the microphone will pick up not only sound produced by the natural source, for instance, by a speaker's voice, but also sound arriving from the loudspeaker. This phenomenon, known as 'acoustical feedback', can result in instability of the whole system and lead to the well-known howling or whistling sounds. We shall discuss acoustical feedback in a more detailed manner in the next section.

In most cases, the loudspeaker will be mounted more or less above the natural sound source. This arrangement has the advantage that the direct sound, coming from the loudspeaker, will always arrive from roughly the same direction (with regard to a horizontal plane) as the sound arriving directly from the original sound source. The vertical deviation of directions is not very critical, since our ability to discriminate sound directions is not as sensitive in a vertical plane as in a horizontal one. The subjective impression is even more natural if care is taken that the loudspeaker sound reaches the listener simultaneously with the unamplified sound or even 10–30 ms later. In the latter case, the listener benefits from the law of the first wave front (precedence effect), which raises the illusion that all the sound he hears is produced by the natural sound source, that is, no electroacoustical system is in operation. This illusion can be maintained even if the level of the loudspeaker signal at the listener's position surpasses the level due to the natural source by 5–10 dB, provided the latter precedes the loudspeaker signal by about 10–15 ms (Haas effect; see Section 7.4). The delayed arrival of the loudspeaker's signals at the listener's seat can be achieved by increasing the distance between the loudspeaker and the audience. However, the application of this simple measure is limited by the increasing risk of acoustical feedback, which was already mentioned before. A more elegant and flexible way is to employ an electrical delay device inserted in the electrical signal path.

Very good results in sound amplification, even in large halls, have occasionally been obtained by using a speaker's desk which has loudspeakers built into the front-facing panel. These loudspeakers are arranged in properly inclined vertical columns with suitable directionality. With this arrangement, the sound from the loudspeakers will travel in almost the same direction as the sound from the speaker himself. This reduces problems due to feedback provided that the propagation of structure-born sound is prevented by resiliently mounting the desk loudspeakers and the microphone.

In very large or long halls, or in halls consisting of several sections, the supply of sound energy by one single loudspeaker only will become increasingly difficult because condition (11.9) cannot be met without unreasonable expenditure. The use of several loudspeakers at different positions has the consequence that each loudspeaker supplies a smaller area, which makes it easier to satisfy Equation 11.9. Two simple examples are shown in Figure 11.9. If all loudspeakers are fed with identical electrical signals, however, confusion zones may be caused in which listeners are irritated by hearing sound from more than one source. In these areas, not only is the natural localization of the sound source impaired but also the intelligibility is significantly diminished. Again, this undesirable effect is avoided by electrically

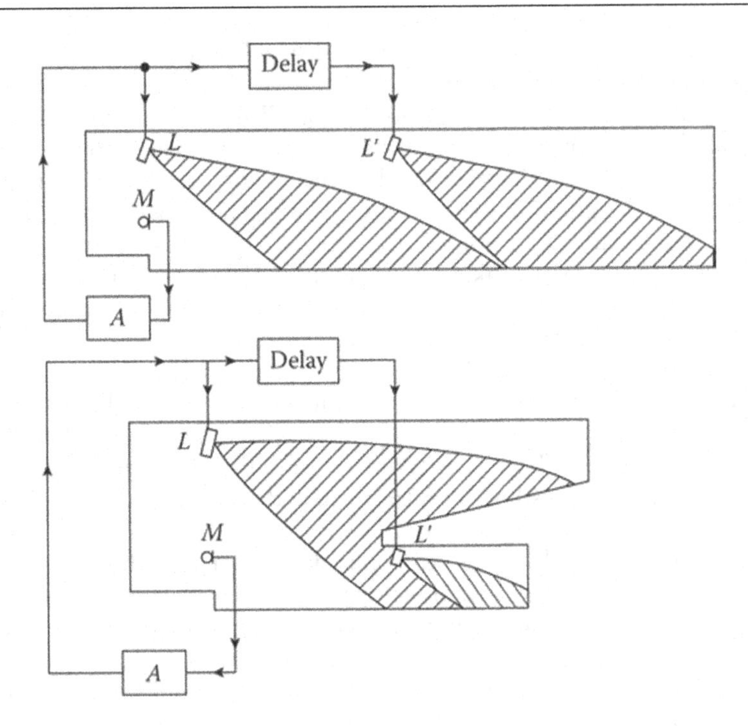


*Figure 11.9* Public address system with more than one loudspeaker ($L$, $L$ = loudspeakers, $M$ = microphone, $A$ = amplifier).

delaying the signals applied to the subsidiary loudspeakers. The delay time should at least compensate for the distances between the auxiliary loudspeaker(s) and the main loudspeaker. Furthermore, the power of the subsidiary loudspeakers must not be too high, since this again would make the listener aware of it and, hence, destroy the illusion that all the sound he receives is arriving from the stage. The delay times applied in sound reinforcement systems are typically in the range of 10–100 ms, sometimes even more.

In halls where a high noise level must be expected but where announcements or other information must be clearly understood by those present, the ideal of a natural-sounding sound transmission, which preserves or simulates the original direction of sound propagation, must be sacrificed. Accordingly, the amplified signals are reproduced by many loudspeakers which are distributed fairly uniformly and are fed with identical electrical signals. In this case, it is important to ensure that all the loudspeakers which can be mounted on the ceiling or suspended from it are supplied with equally phased signals. The listeners are then, so to speak, in the near field of a vibrating piston. Sound signals of opposite phases would be noticed in the region of superposition in a very peculiar and unpleasant manner.

If the sound irradiation is effected by directional loudspeakers from the stage towards the back of the room, the main lobe of one loudspeaker will inevitably project sound towards the rear wall of the room, since the listeners seated in the most remote parts of the hall are also meant to benefit from the sound system. Thus, a substantial fraction of the sound energy will be reflected from the rear wall and can cause echoes in other parts of the room, disturbing listeners as well as speakers. For this reason, it is recommended that such walls or wall portions are rendered highly absorbent. In principle, an echo could also be avoided by a diffusely reflecting wall treatment which scatters the sound in all possible directions. But then the scattered sound would excite the reverberation of the room, which, as was explained earlier, is not favourable for speech transmission.

The performance of an electroacoustical system can be tested by applying suitable test signals to the amplifier input and measuring the response of the system at many points distributed over the area where the audience is seated. In the simplest case, stationary random noise is used as a test 'signal', and the received signal is analyzed using an octave or a third-octave filter. The result can be used for adjusting an equalizing filter which is inserted into the electrical signal path in order to compensate for linear distortions as caused by the frequency dependence of the reverberation time or of the loudspeaker(s)' directivity. More detailed information can be gained with impulses or impulse-equivalent signals (maximum length sequences, sine sweeps; see Section 8.2). Such measurements yield the impulse response of the transmission path and permit us to discriminate the 'useful' signal components from the reverberant ones. In particular, the indicators of speech intelligibility, as discussed in Section 7.4, can be evaluated from the impulse response either of the reinforcement system itself or of its combination with a loudspeaker, which simulates a natural speaker. Furthermore, the correct function of the delay devices can be checked.

The preceding discussions concern mainly the transmission of speech. The electroacoustical amplification of music — apart from entertainment or dance music — is rejected by many musicians and music lovers for reasons which are partly irrational. Obviously, many of these people have the suspicion that the music could be manipulated in an undue way which is outside the artists' influence. On the other hand, almost everybody has experienced the poor performance of technically imperfect reinforcement systems. If, in spite of objections, electroacoustical reinforcement is mandatory in a performance hall, the installation must be carefully designed, and the system must be equipped with first-class components. Furthermore, it must preserve under all circumstance the natural direction of sound incidence and the natural timbre. Care must be taken to avoid linear as well as non-linear distortions, and the amplification should be kept at a moderate level. A particular problem is to comply with the large dynamical range of symphonic music. For entertainment music, the requirements are not as stringent; in this case, people have long been accustomed to the fact that a singer has a microphone in his or her hand, and the audience will more readily accept that it will be conscious of the sound amplification.

These remarks have no significance for the presentation of electronic music; here, the acoustician can safely leave the arrangement of loudspeakers and the operation of the whole equipment to the performers and their technical staff.

## 11.4 ACOUSTICAL FEEDBACK AND ITS SUPPRESSION

Acoustical feedback in sound reinforcement systems has already been mentioned in the preceding section. In principle, feedback will occur whenever the loudspeaker of a public address system is located in the same room as the microphone, which inevitably will pick up a portion of the loudspeaker signal. Only if this portion is sufficiently small are the effects of feedback negligible; higher amplification of the feedback signal may cause substantial linear distortions, such as ringing effects or colouration. At still higher amplification, the whole system will perform self-sustained oscillations at some frequency, which makes the system useless.

Before discussing measures for the reduction or suppression of feedback effects, we shall deal with its mechanism in a somewhat more detailed manner.

We assume that the original sound source, for instance, a speaker, produces a sound signal with spectrum $S(\omega)$ at the microphone (see Figure 11.10). The output voltage of the microphone is amplified with a frequency-independent gain $q$ and is fed to the loudspeaker. The loudspeaker signal will reach the listener via a transmission path in the room with a complex transfer function $\overline{G}(\omega)$ at the same time it will reach the microphone by a different path with

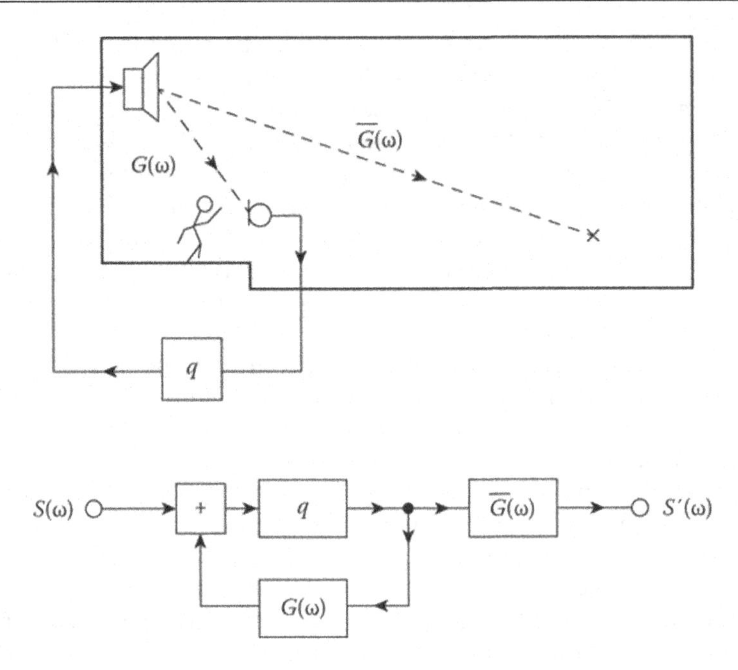


*Figure 11.10* Acoustical feedback in a room. The lower part of the figure represents the block diagram of the system.

the transfer function $G(\omega)$. The latter one, together with the microphone, the amplifier, and the loudspeaker, constitutes a closed loop which the signal passes through repeatedly.

The lower part of Figure 11.10 shows the mechanism of acoustical feedback in a more schematic form. The complex amplitude spectrum of the output signal (i.e. of the signal at the listener's seat) is given by

$$S'(\omega) = \bar{G}(\omega) \cdot \left(qS + q^2GS + q^3G^2S + \cdots\right) = \frac{q\bar{G}}{1 - qG} \cdot S(\omega) \tag{11.13}$$

The first of these expressions clearly shows that acoustical feedback is brought about by the signal repeatedly passing through the closed loop. The second one shows us that the acoustical feedback has changed the original transfer function $q\bar{G}(\omega)$ into

$$G'(\omega) = \frac{S'(\omega)}{S(\omega)} = \frac{q}{1 - qG(\omega)} \cdot \bar{G}(\omega) \tag{11.14}$$

The factor $qG(\omega)$, which is characteristic for the amount of feedback, is called the 'open loop gain' of the system. Depending on its magnitude, the spectrum $S'(\omega)$ of the received signal, and hence the signal itself, may be quite different from the original signal with the spectrum $S(\omega)$.

The properties of the 'effective transfer function' $G'$ can be illustrated by means of the Nyquist diagram, in which the locus of the open loop gain is represented in the complex plane (see Figure 11.11). Each point of this curve corresponds to a particular frequency; abscissa and ordinate are proportional to the real part and the imaginary part of $qG$, respectively. The arrows point in the direction of increasing frequency. The whole system will remain stable as long as this curve does not include the point +1. This condition is certainly fulfilled if $|qG| < 1$

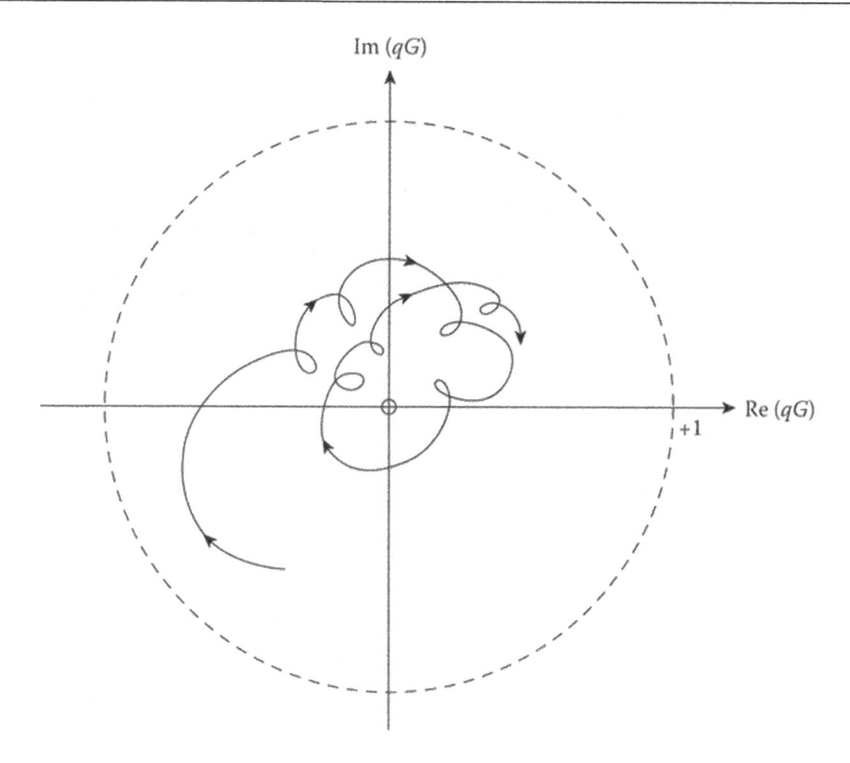


*Figure 11.11* Nyquist diagram illustrating stability of acoustical feedback.

for all frequencies. Mathematically, this condition guarantees the convergence of the series in Equation 11.13.

Suppose that we start with a very small amplifier gain, that is, with $|qG|$ being small compared to unity. If we increase the gain gradually, the curve in Figure 11.11 will be inflated, keeping its shape. In the course of this process, the distance between the curve and the point +1, that is, the quantity $|1 - qG|$, could become very small for certain frequencies. Consequently, the magnitude of the transfer function $G'$ will become very large at these frequencies. Then, the signal received by the listener will sound 'coloured', or if the system is excited by an impulsive signal, ringing effects are heard. With a further increase of $q$, $|qG|$ will exceed unity, and this will happen at a frequency close to that of the absolute maximum of $|G(\omega)|$. Then, the system becomes unstable and performs self-excited oscillations at that frequency.

The effect of feedback on the performance of a public address system can also be demonstrated by plotting $|G'(\omega)|$ on a logarithmic scale as a function of the frequency. This leads to 'frequency curves' similar to that shown in Figure 3.8b. Figure 11.12 presents several such curves for various values of the open loop gain $|qG|$, obtained by simulation with a digital computer (Kuttruff and Hesselmann 1976). With increasing gain, one particular maximum starts growing more rapidly than the other maxima and becomes more and more dominating. This is the condition of audible colouration. When a critical value $q_0$ of the amplifier gain is reached, this leading maximum will become infinitely high, which means that the system will start to perform self-sustained oscillations. (In real systems, the amplitude of these oscillations remains finite because of inevitable non-linearities of its components.)

A question of great practical importance concerns the amplifier gain $q$, which must not be exceeded if colouration is to be avoided or to be kept within tolerable limits. According to

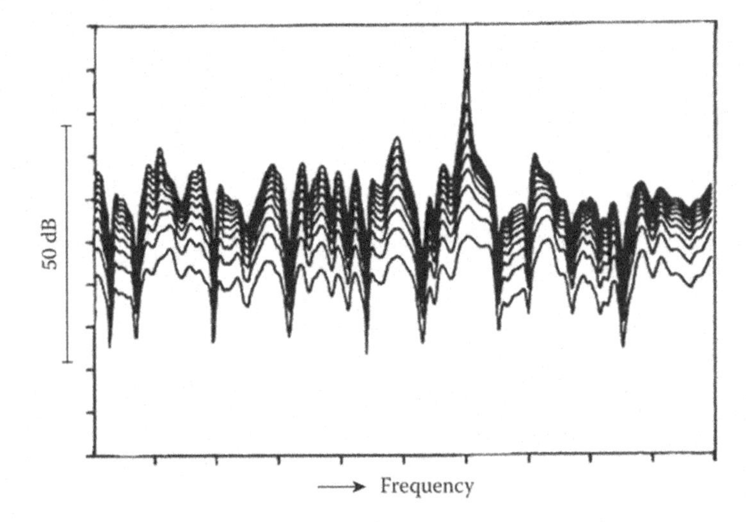


*Figure 11.12* Frequency curves of a room equipped with an electroacoustical sound reinforcement system, simulated for different amplifier gains. The latter varies in steps of 2 dB from −20 dB to 0 dB relative to the critical gain $q_0$. The total frequency range is 90/*T* hertz (*T* = reverberation time).

listening tests as well as to theoretical considerations, colouration remains imperceptible as long as

$$20\log_{10}(q/q_0) \leq -12\text{dB} \tag{11.15}$$

For speech transmission, it is sufficient to keep the relative amplification 5 dB below the instability threshold to avoid audible colouration.

Another effect of acoustical feedback is the increase of reverberance, which, however, is restricted to those frequencies for which $G'(\omega)$ is particularly high. To show this, we first simplify Equation 11.14 by setting $\overline{G} = G$:

$$G'(\omega) = \frac{qG(\omega)}{1-qG(\omega)} = \frac{q}{\dfrac{1}{G(\omega)} - q} \tag{11.16}$$

To demonstrate the increase of reverberation time by acoustical feedback, we consider a single component of the natural sound decay in a room (see Equation 3.60):

$$g(t) = \exp(i\omega_0 t) \cdot \exp(-\delta t) \quad \text{for} \quad t \geq 0$$

The inverse Fourier transform of this 'impulse response' reads

$$G(\omega) = \frac{1}{\delta + i(\omega - \omega_0)} \tag{11.17}$$

Inserting this expression into Equation 11.16 yields the impulse response of the complete system:

$$G'(\omega) = \frac{q}{\delta - q + i(\omega - \omega_0)} \tag{11.18}$$

The comparison of Equations 11.17 and 11.18 tells us that feedback reduces the original decay constant from $\delta$ to $\delta' = \delta - q$ and thus increases the reverberation time of the system by a factor

$$\frac{T'}{T} = \frac{\delta}{\delta'} = \frac{1}{1 - q/\delta} \tag{11.19}$$

In principle, acoustical feedback can be avoided by keeping the open loop gain $qG$ in Equation 11.14 small compared to unity. However, if we would try to achieve this by reducing the amplifier gain $q$, this would reduce the loudness of the loudspeaker signal and the system would become useless. Thus, the only way to improve the stability of the system is to make the magnitude of $G(C)$ as small as possible in the interesting frequency range. For this purpose, the loudspeaker must be given a suitable and well-adjusted directivity; the main lobe of the radiation pattern should point towards the listeners without feeding much energy into the reverberant sound field. Likewise, a microphone with a suitable directivity may be used, for instance, a cardioid microphone oriented in such a way that it gives the original sound source much more weight than the signal arriving from the loudspeaker, which is more or less suppressed. If there are more than one microphones, the 'idle' ones should be switched off. And also, the number of loudspeakers should be kept as small as possible. In any case, it is advantageous to arrange the active microphone close to the original source. With these rather-simple measures, acoustical feedback cannot be completely eliminated, to be sure, but often, the point of instability can be raised high enough so that it will never be reached during normal operation.

In view of the irregular shape of room transfer functions (see Figure 3.8), further increase of feedback stability should be possible by smoothing the function $|G(\omega)|$. A first step in this direction is to suppress the frequency at which feedback instability is expected by means of a narrow bandwidth band-pass filter. The possible increase of stable amplifier gain is not very high because after this step there will emerge another feedback frequency with the same property. To achieve noticeable increase in stability, the combination of many adjustable band-pass filters would be needed. Nowadays, the adjustment of such a filter bank can be carried out automatically, which is important because the transfer properties of a room may show temporal variations. The total increase in feedback stability achieved in this way may be as high as 5–8 dB, provided the linear distortions due to the loudspeaker or the frequency-dependent reverberation time have been removed beforehand by an equalizing network.

Theoretically, by smoothing or averaging the room transfer function, the region of stable operation can typically be augmented by about 10 dB. The reason for this limit is Equation 3.53. It tells us that the level difference $\Delta L_{max}$ between the absolute maximum of a frequency curve and its energetic average amounts typically to about 10 dB.

In an early attempt to average the transfer function of the room, the microphone was moved on a circular path during its operation (Franssen 1968). Likewise, the use of a rotating gradient microphone has also been proposed, with the axis of rotation being perpendicular to the direction of its maximum sensitivity. However, the mechanical movement makes it difficult and uncontrollable to pick up a speaker's voice with such a microphone; furthermore, it produces amplitude modulation of the signal.

A more practical method of virtually flattening the frequency characteristics of the open loop gain has been proposed and demonstrated by Schroeder (1959). It is effected by modifying the signal during its repeated round trips in the feedback loop instead of varying the transfer function. As we saw, acoustical feedback is brought about by particular spectral components which always experience the same 'favourable' amplitude and phase conditions when circulating in the closed loop in Figure 11.10. If, however, the feedback loop contains a device which shifts the frequencies of all spectral components by a small amount $\Delta\omega$, then a particular component will experience favourable as well as unfavourable conditions, which, in effect, is tantamount to frequency averaging the transfer function. After $N$ trips around the feedback loop, the signal power is increased or decreased by a factor

$$K(\omega) = |qG(\omega+\Delta\omega)|^2 \cdot |qG(\omega+2\Delta\omega)|^2 \cdot \ldots \cdot |qG(\omega+N\Delta\omega)|^2$$

Setting $L(\omega') = 10 \cdot \log_{10} |qG(\omega')|^2$, we obtain the change in level:

$$10 \cdot \log_{10}\left[K(\omega)\right] = L(\omega+\Delta\omega) + L(\omega+2\Delta\omega) + \ldots + L(\omega+N\Delta\omega) \approx N \cdot \langle L \rangle \tag{11.20}$$

where $\langle L \rangle$ is the average of the logarithmic frequency curve in the range from $\omega$ to $\omega + N\Delta\omega$. The system will remain stable if $N\langle L \rangle \to -\infty$ as $N$ approaches infinity, that is, if $\langle L \rangle$ is negative.

Now it is no longer the absolute maximum of the frequency curve which determines the onset of instability but a certain average value. As mentioned, the difference $\Delta L_{max}$ between the absolute maximum and the average of a frequency curve is about 10 dB for most large rooms; it is this level difference by which the amplifier gain may be increased theoretically without the risk of instability. Note that the frequency shift $\Delta f = \Delta\omega/2\pi$ must be small enough to remain inaudible. On the other hand, it must be high enough to yield effective averaging after a few round trips. The best compromise is to choose for $\Delta f$ the mean spacing of frequency curve maxima according to Equations 3.48 and 3.49:

$$\Delta f = \left(\Delta f_{max}\right) = \frac{4}{T} \tag{11.21}$$

This method works quite well with speech; with music, however, even very small frequency shifts are noticed, since they change the musical intervals. In practice, the theoretical increase in amplification of about 10 dB cannot be reached without unacceptable reduction of sound quality; as soon as the increase exceeds 5–6 dB, beating effects of the repeatedly modified signals become audible. Practically, the frequency shift is achieved by single side-band modulation of the signal.

Another method of reducing the risk of acoustical feedback by employing time-variable signals was proposed by Guelke and Broadhurst (1971), who replaced the frequency-shifting device by a phase modulator. The effect of phase modulation is to add symmetrical side lines to each spectral line. By suitably choosing the width of phase variations, the centre line (i.e. the carrier) can be removed altogether. In this case, the authors were able to obtain an additional gain of 4 dB. They stated that the modulation is not noticeable even in music if the modulation frequency is as low as 1 Hz.

## 11.5 REVERBERATION ENHANCEMENT WITH EXTERNAL REVERBERATORS

As discussed in Section 9.6, many large halls have to accommodate quite different events, such as meetings, lectures, performance of concerts, theatre and opera pieces, and sometimes even sports events, banquets, and balls. It is obvious that the acoustical design of such a multipurpose hall cannot create optimum conditions for each type of presentation. At best, some compromise can be reached which necessarily will not satisfy all expectations.

A much better solution is to provide variable reverberation time that could be adapted to the different requirements. One way to achieve this is by changing the absorption within the hall by mechanical devices, as described in Section 9.6. Such devices, however, are costly and subject to mechanical wear. A more versatile and less-expensive solution to this problem is offered by electroacoustical systems designed for the control of reverberation.

Principally, electroacoustical enhancement of reverberation time can be achieved in two different ways. The first method employs the regenerative reverberation within the room brought about by acoustical feedback, as mentioned in the preceding section. With the second method, artificial reverberation is created by some external reverberator and is imposed on the sound signal, the reverberance of which is to be augmented. At first, we shall describe the second method in more detail, while the discussion of the first one is postponed to the next section.

The principle of reverberation enhancement with separate reverberators is depicted in Figure 11.13. The sounds produced by the orchestra are picked up by microphones which are close to the performers. The output signals of the microphones are fed into a reverberator. This is a linear system with an impulse response fairly similar to that of an enclosure and hence provides the signals with its inherent reverberation. The signals modified in this manner are re-radiated in the original room by loudspeakers distributed in a suitable way within the hall. In addition, delaying devices must usually be inserted into the electrical circuit in order to ensure that the reverberated loudspeaker signals will not reach any listener's place earlier than the direct sound signal from the natural sound source.

It should be noted that the selection of loudspeaker locations has a great influence on the effectiveness of the system and on the quality of the reverberated sound. Another important point is that not all the loudspeakers should be fed with identical signals; instead, these signals must be mutually incoherent, since this is the condition for creating the impression of spaciousness described in Section 7.7. Therefore, the reverberator must have several output

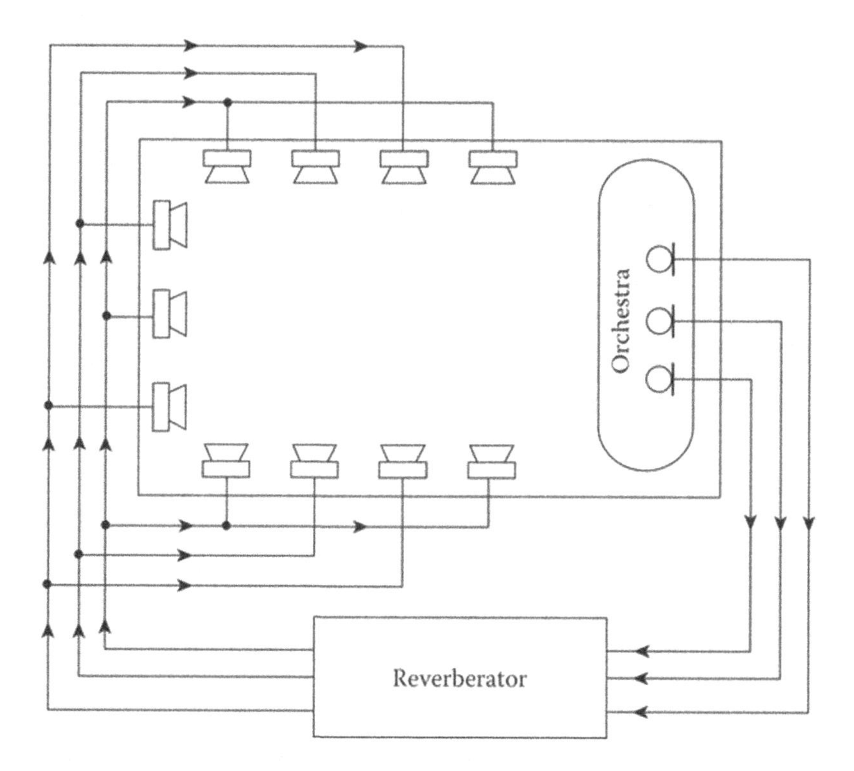


*Figure 11.13* Electroacoustic reverberation enhancement with an external reverberator.

terminals yielding incoherent signals which are all derived from the same input signal. In order to provide each listener with sound incident from several substantially different, mainly lateral, directions, it may be necessary to use more loudspeakers than incoherent signals. It is quite obvious that all the loudspeakers must be sufficiently distant from all the listeners in order to prevent one particular loudspeaker being heard much louder than the others. Finally, care must be taken to avoid noticeable acoustical feedback.

Let us now discuss the various methods to reverberate the electrical signals provided by the stage microphone(s). The most natural way is to apply them to one or several loudspeakers in a separate reverberation chamber which has the desired reverberation time, including its frequency dependence. The sound signal in the chamber is again picked up by microphones which are far apart from each other to guarantee the incoherence of the output signals. The reverberation chamber should be free of flutter echoes, and its volume should not be less than 150–200 m$^3$; otherwise, the density of eigenfrequencies would be too small at low frequencies.

A system of this type was installed in 1963 for permanent use with musical performances in the 'Jahrhunderthalle' of the Farbwerke Hoechst AG (today Aventis S. A.) at Hoechst near Frankfurt am Main (Meyer and Kuttruff 1964). This hall, the volume of which amounts to about 75,000 m$^3$, has a cylindrical side wall with a diameter of 76 m; its roof is a spherical dome. In order to avoid the acoustical risks of this shape, the dome as well as the side wall were treated with highly absorbing materials. In this state, the auditorium has a natural reverberation time of about 1 s. To increase the reverberation time, the sound signals are picked up by several microphones on the stage, passed through a reverberation chamber, and finally, fed to a total of 90 loudspeakers, which are distributed in a suspended ceiling and along the cylindrical side and rear walls. This system, which underwent several modifications in the course of time, raised the reverberation time to about 2 s.

Other reverberators that have found wide application in the past employed bending waves propagating in metal plates, or torsional waves travelling along helical springs, excited and picked up with suitable electroacoustical transducers. The reverberation was brought about by repeated reflections of the waves from the boundary or the terminations of these waveguides.

The essential property of these devices is the finite travelling time of a sound ray or particle between successive reflections. Therefore, in order to produce some kind of reverberation, we only need, in principle, a delaying device and a suitable feedback path by which the delayed signal is transferred again and again from the output to the input of the delay unit (see Figure 11.14a). As before, we denote the open loop gain in the feedback loop with $q$, which must be smaller than unity for stable conditions, and $t_0$ denotes the delay time. Then, the impulse

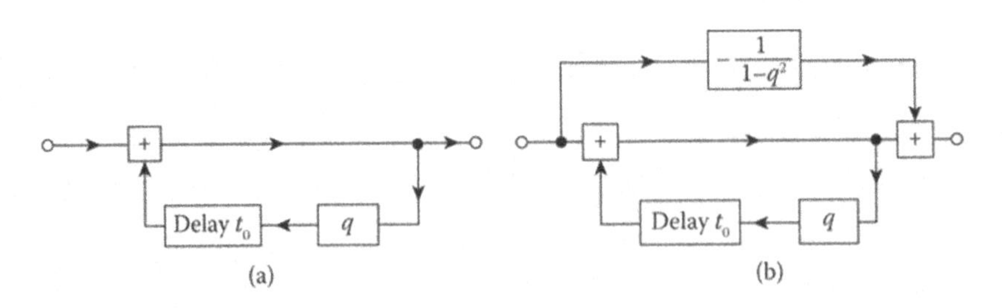


*Figure 11.14* Electronic reverberators with one delay unit: (a) comb filter type and (b) all-pass type.

response of the circuit is given by Equation 7.7. With each round trip, the signal is attenuated by $-20\log_{10} q$ decibels, and hence after $-60/(20 \log_{10} q)$ passages, the level has fallen by 60 dB. The time in which this happens is, by definition, the reverberation time of the reverberator:

$$T = -\frac{3t_0}{\log_{10} q} \tag{11.22}$$

In order to obtain a sufficiently long reverberation time, either $q$ must be fairly close to unity, which makes the adjustment of the open loop gain very critical, or $t_0$ must be relatively long. Suppose we aim at a reverberation time of 2 s. With a delay time of 10 ms, the gain $q$ must be set at 0.966, while with $t_0$ = 100 ms, the required gain is still 0.708. In both cases, the reverberation has an undesirable tonal quality. In the first case, the reverberator produces 'coloured' sounds due to the regularly spaced maxima and minima of its transfer function, as shown in Figure 7.9. In the second case, the regular succession of 'reflections' is perceived as flutter.

The performance of such a reverberator can be improved to a certain degree, according to Schroeder and Logan (1961), by modifying it such that it has all-pass characteristics. For this purpose, the fraction $1/(1 - q^2)$ of the input signal is subtracted from its output (see Figure 11.14b). The impulse response of the modified reverberator is

$$g(t) = -\frac{1}{1-q^2}\delta(t) + \sum_{n=0}^{\infty} q^n \delta(t - nt_0) \tag{11.23}$$

Its Fourier transform, that is, the transfer function of the reverberator, is given by

$$G(\omega t) = -\frac{1}{1-q^2} + \frac{1}{1-q\,\exp(-i\omega t_0)} \tag{11.24}$$

It is easily verified that this transfer function $G(\omega)$ has no maxima and minima — the system is an all-pass. Subjectively, however, the undesirable properties of the reverberation produced in this way have not completely disappeared, since our ear does not perform a Fourier analysis in the mathematical sense but, rather, a 'short-time frequency analysis', thus also being sensitive to the temporal structure of a signal.

A substantial improvement can be achieved by combining several reverberation units with and without all-pass characteristics and with different delay times. These units are connected partly in parallel, partly in series. An example is shown in Figure 11.15. Of course, simple ratios between the various delay times must be avoided; moreover, the impulse response of the reverberator should be free of long repetition periods. This can be checked by performing an autocorrelation analysis of its impulse response (see Section 8.3).

Still better results are achieved with reverberators employing units with time-variable delays. They are the basis of a reverberation enhancement system named LARES (Lexicon Acoustic Reverberance and Enhancement System), invented by Griesinger (1991). For picking up the sounds, only two microphones are needed, which are positioned outside the critical distance of the original sound source and hence are in the diffuse sound field of the hall. Each microphone is connected to many loudspeakers in such a way that neighbouring loudspeakers are fed with different signals. Suppose there are two microphones and eight loudspeakers. Then, there are 16 connections between microphones and loudspeakers. The essential point is that in each of these connections, a time-variant reverberator is inserted. These reverberators randomize the microphone signals; accordingly, the processed signals are mutually incoherent. This has about the same effect as the frequency shifter described in the preceding section, namely, to smooth the effective room transmission function. It is

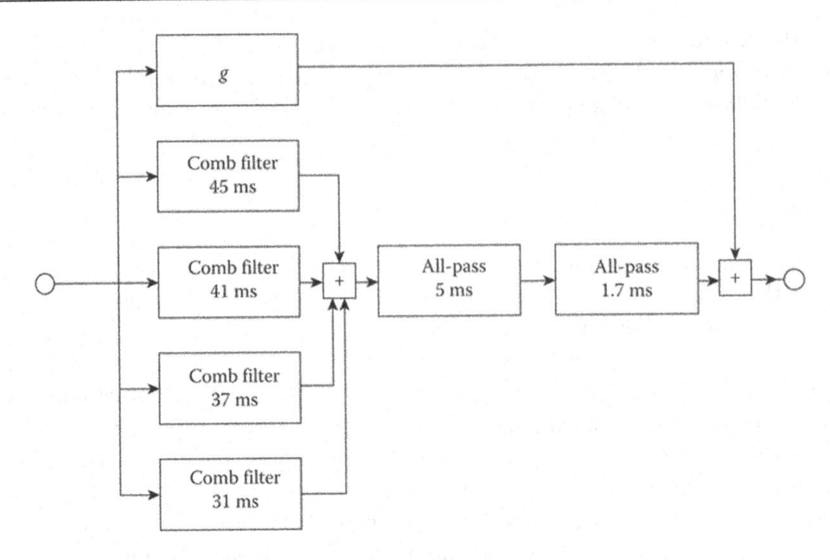


*Figure 11.15* Electronic reverberator consisting of four comb filter units and two all-pass units. The numbers indicate the delay of each unit. Optionally, the unreverberated signal attenuated by some factor g can be added to the output.

Source: Schroeder and Logan (1961).

important that the time variations are so fast that the autocorrelation function of the reverberated signals vanishes after about 1 s.

A quite sophisticated system has been developed by Berkhout (1988) and Berkhout et al. (1993) which attempts to modify the original signals in such a way that they contain, and hence transplant, not only the reverberation but also the complete wave field of the hall or — what may be even more promising — the wave field from a fictive hall (preferably one with excellent acoustics) into the actual environment. This system, called the acoustic control system (ACS), is based on Huygens's principle, according to which each point hit by a wave may be considered as the origin of a secondary wave which effects the propagation of the wave to the next points. The ACS is intended to simulate this process by hardware components, that is, by microphones, amplifiers, filters, and loudspeakers. In the following explanation of 'wave front synthesis', we describe all signals in the frequency domain, that is, as functions of the angular frequency $\omega$.

Let us consider, as shown in Figure 11.16, an auditorium in which a plane and regular array of $N$ loudspeakers $LS$ is installed. The purpose of these loudspeakers is to synthesize the wave fronts originating from the real sound sources. For this purpose, the original sounds are picked up by $M$ microphones which are regularly arranged next to the stage (e.g. in the ceiling above the stage). These microphones have some directional characteristics, each of them covering a particular subarea of the stage, with one 'notional sound source' in its centre which is at a point $r_m$. If we denote the signals produced by these sound sources with $S(\mathbf{r}_m, \omega)$, the signal received by the $m$th microphone at the location $\mathrm{r}'_\mathrm{m}$ is

$$\mathrm{M}(\mathrm{r}'_\mathrm{m}, \omega) = W\left(\mathrm{r}_m, \mathrm{r}'_\mathrm{m}\right) \cdot S\left(\mathrm{r}_m, \omega\right) \quad (m = 1, 2, \ldots, M) \tag{11.25}$$

where $W$ is a 'propagator' describing the propagation of a spherical wave from a source point $r_m$ to a microphone position $\mathrm{r}'_\mathrm{m}$. Each of these propagators involves an amplitude change by a factor $A$ and a delay $\tau$:

$$\tau = \left|\mathrm{r}_m - \mathrm{r}'_\mathrm{m}\right| / \mathrm{c} \tag{11.26}$$

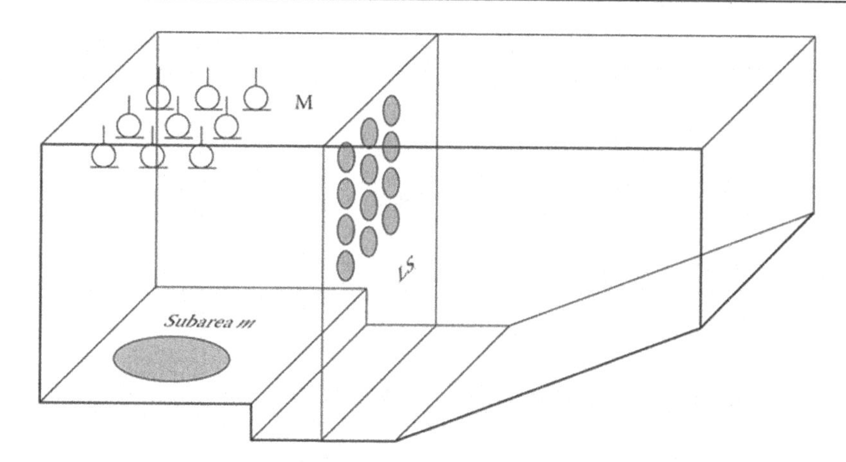


*Figure 11.16* Principle of wave front synthesis (*M* = microphones, *LS* = loudspeakers).

The microphone signals $M(\mathbf{r}'_m,\omega)$ are fed to the loudspeakers after processing them as if the source signals $S(\mathbf{r}_m,\omega)$ would have reached the loudspeaker locations $r_K$ directly, that is, as sound waves. Hence, we have to undo the effect of the propagator $W(\mathbf{r}_m,\mathbf{r}'_m)$ in Equation 11.25 and to replace it with

$$W(\mathbf{r}_m,\mathbf{r}_n)=\frac{\exp(-ik|\mathbf{r}_m-\mathbf{r}_n|)}{|\mathbf{r}_m-\mathbf{r}_n|} \tag{11.27}$$

describing the direct propagation from the *m*th notional source to the *n*th loudspeaker. Finally, the input signal of this loudspeaker is obtained by adding the contributions of all notional sources:

$$L(\mathbf{r}_n)=\sum_{m=1}^{M}W^{-1}(\mathbf{r}_m,\mathbf{r}'_m)\cdot W(\mathbf{r}_m,\mathbf{r}_n)\cdot M(\mathbf{r}'_m,\omega) \tag{11.28}$$

The loudspeakers will correctly synthesize the original wave fronts if their mutual distances are small enough and if they have dipole characteristics. (The latter follows from Kirchhoff's formula, which is the mathematical expression of Huygens's principle but will not be discussed here.) For the practical application of this principle, it is sufficient to substitute the planar loudspeaker array by a linear one in horizontal orientation, since our ability to localize sound sources in vertical directions is rather limited.

This relatively simple version of ACS can be used not only for enhancing the sounds produced onstage but also for improving the balance between different sources, for instance, between singers and an orchestra. It has the advantage that it preserves the natural localization of the sound sources. Systems of this kind have been installed in many halls and theatres. Although the derivation presented earlier neglects all reflections from the boundary of the auditorium, the system works well if the reverberation time of the hall is not too long.

Sound reflections from the boundaries can be accounted for by constructing the mirror images of the notional sources at $\mathbf{r}_m$ and including their contributions into the loudspeaker input signals. However, it may be more interesting to construct image sources not with respect to the actual auditorium but to a virtual hall with desired acoustical conditions, and hence to transplant these conditions into the actual hall. This process is illustrated in Figure 11.17. It presents the ground plan of the actual auditorium (assumed as fan-shaped) drawn in the system of mirror images of a virtual rectangular hall. (For the sake of simplicity, the images of only one notional source are shown.) Suppose the positions and the relative strengths of these image sources are numbered in some way:

$$\mathrm{r}_m^{(1)}, \mathrm{r}_m^{(2)}, \mathrm{r}_m^{(3)} \ldots \quad \text{and} \quad B_m^{(1)}, B_m^{(2)}, B_m^{(3)} \ldots$$

Then, the propagator $\mathrm{W}(\mathrm{r}_m, \mathrm{r}_n)$ in Equation 11.28 must be replaced with

$$W = \left(\mathrm{r}_m, r_n\right) + B_m^{(1)} \mathrm{W}\left(r_m^{(1)}, r_n\right) + B_m^{(2)} \mathrm{W}\left(\mathrm{r}_m^{(2)}, \mathrm{r}_n\right) + \ldots \; (m = 1, 2, \ldots, M)$$

which leads to the following loudspeaker input signal:

$$L(\mathrm{r}_n) = \sum_{m=1}^{M} W^{-1}\left(\mathrm{r}_m, \mathrm{r}_m'\right) \cdot \left[\mathrm{W}\left(\mathrm{r}_m, \mathrm{r}_n\right) + \sum_k B_m^{(k)} \mathrm{W}\left(\mathrm{r}_m^{(k)}, \mathrm{r}_n\right)\right] \cdot S\left(\mathrm{r}_{m,} \omega\right) \tag{11.29}$$

In practical applications, it is useful to arrange loudspeaker arrays along the side walls of the actual auditorium and to allocate to each of them the right-hand and the left-hand image sources, respectively, as indicated in Figure 11.17.

The coefficients $B_m^{(k)}$ contain in a cumulative way the absorption coefficients of all walls involved in the formation of a particular image source (see Section 4.1). Therefore, the reverberation time and the reverberation level, including their frequency dependence, are easily controlled by varying these coefficients. Similarly, the shape and the volume of the virtual hall can be changed. Thus, an ACS permits the simulation of a great variety of different environments.

However, since the number of image sources increases rapidly with the order of reflection, a vast number of propagators $W\left(\mathrm{r}_m^{(k)}, \mathrm{r}_n\right)$ would be required to synthesize the whole impulse response of the virtual room. Therefore, this treatment must be restricted to the early part

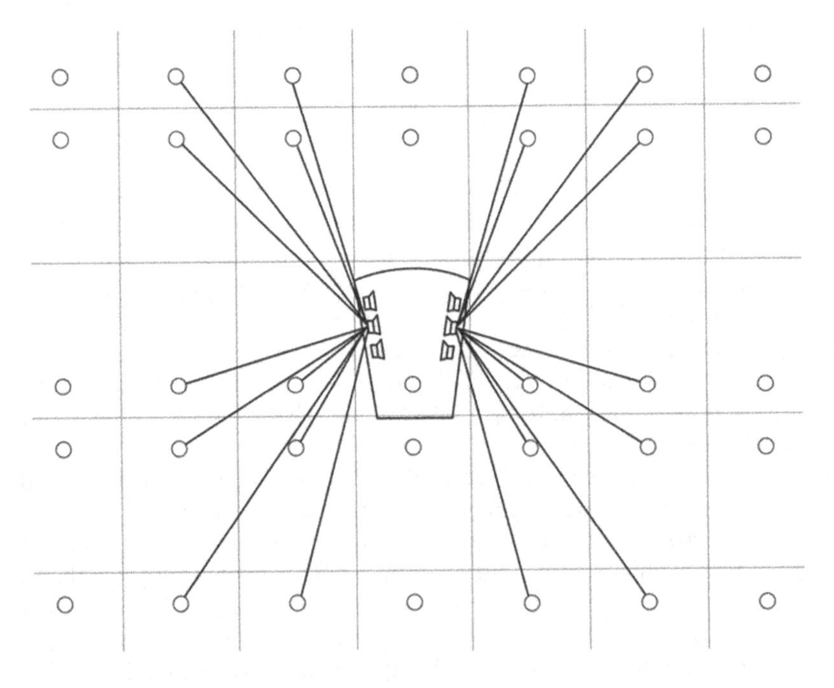

*Figure 11.17* Actual hall and image sources of a virtual rectangular hall.

Source: Berkhout et al. (1995).

of the impulse response. The later parts, that is, those corresponding to reverberation, can be synthesized in a more statistical way because the auditive impression conveyed by them does not depend on individual reflections. More can be found on this matter in the work by Berkhout et al. (1993).

## 11.6 REVERBERATION ENHANCEMENT BY CONTROLLED FEEDBACK

In Section 6.4, we learned that each sound reinforcement system is subject to positive feedback, provided the microphone of the system is located in the same room as the loudspeaker. Moreover, it has been shown that acoustical feedback in a room is necessarily accompanied by an increase in reverberation time, which depends on the open loop gain of the system (see Equation 11.19). However, in a usual sound reinforcement system, this effect is restricted to one frequency (see Figure 11.12), and so it cannot be used for wide-band enhancement of reverberation with satisfactory tonal quality. One way to overcome this problem is to provide for numerous independent acoustical transmission channels which are simultaneously operated in the auditorium.

In Figure 11.18, the multi-channel system invented by Franssen (1968) is depicted. It consists of $N$ (>>1) microphones, amplifiers, and loudspeakers, each of the latter being connected with one microphone. All microphone–loudspeaker distances are significantly larger than

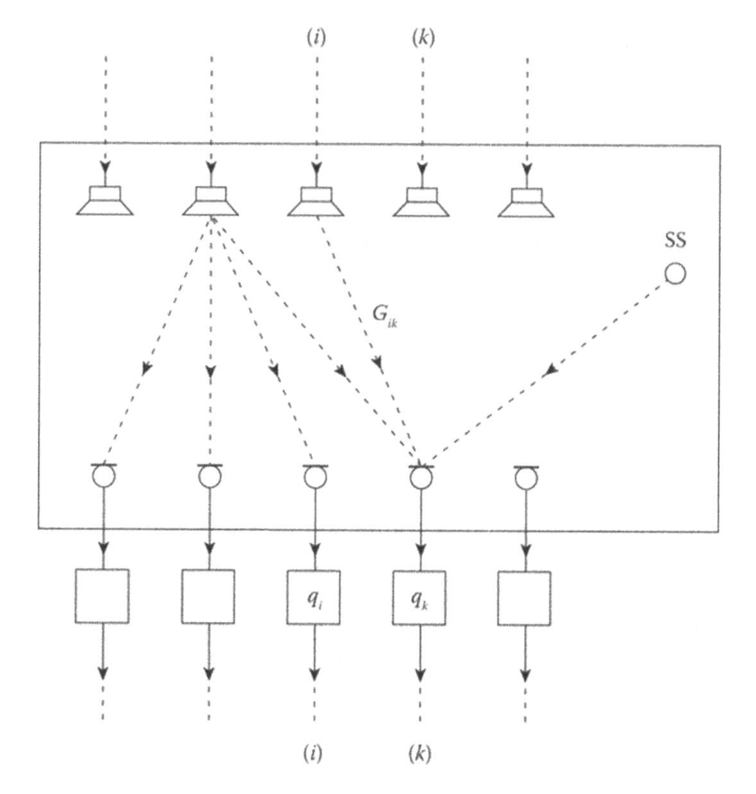


*Figure 11.18* Multi-channel system.

Source: Franssen (1968).

the critical distance (see Equation 5.48). Thus, there are $N^2$ transmission paths which are interconnected by the sound field. Accordingly, the output signal $S_k(\omega)$ of the $k$th microphone contains the contribution $S_0(\omega)$ made by the original sound source SS as well as the contributions of all loudspeakers. Therefore, its amplitude spectrum is given by

$$S_k(\omega) = S_0(\omega) + \sum_{i=1}^{N} q_i G_{ik}(\omega) S_i(\omega) \text{ (for } k = 1,2,\ldots,N) \tag{11.30}$$

where $q_i$ denotes the gain of the $i$th amplifier and $G_{ik}(\omega)$ is the transfer function of the acoustic transmission path from the $i$th loudspeaker to the $k$th microphone, including the properties of both transducers.

Equation 11.30 represents a system of $N$ linear equations from which the unknown signal spectra $S_k(\omega)$ can be determined, at least in principle. To get an idea of what the solution of this system is like, we neglect all phase relations and hence replace all complex quantities by their squared magnitudes averaged over a small frequency range, for example,

$$s_0 = \langle |S_0|^2 \rangle, \quad s_i = \langle |S_i|^2 \rangle, \quad \text{and } \gamma_{ik} = \langle |G_{ik}|^2 \rangle$$

By these operations, we obtain from Equation 11.30:

$$s_k = s_0 + \sum_{i=1}^{N} q_i^2 \gamma_{ik} s_i \quad (k = 1,2,\ldots,N) \tag{11.31}$$

These equations mean that we are superimposing energies instead of complex amplitudes. This may be justified if the number $N$ of channels is sufficiently high. In the second step, we assume as a further simplification that all amplifier gains and energetic transfer functions are equal, $q_i \approx q$, and $\gamma_{ik} \approx \gamma$ for all subscripts $i$ and k; furthermore, we suppose that $s_i \approx s$ (except $s_0$). Then, we obtain immediately from Equation 11.31

$$s = \frac{s_0}{1 - Nq^2\gamma} \tag{11.32}$$

The ratio $s/s_0$ characterizes the increase of the energy density caused by the electroacoustical system. On the other hand, the simple diffuse-field theory presented in Section 5.1 tells us that the steady-state energy density in a reverberant space is inversely proportional to the equivalent absorption area and, hence, proportional to the reverberation time (see Equations 5.5 and 5.9). Therefore, the ratio of reverberation times with and without the system is

$$\frac{T'}{T} = \frac{1}{1 - Nq^2\gamma} \tag{11.33}$$

This formula is similar to Equation 11.19. However, in the present case, one can afford to keep all (energetic) open loop gains low enough to exclude the risk of sound colouration by feedback, due to the large number $N$ of channels. Franssen recommended making $q^2\gamma$ as low as 0.01; then, 50 independent channels would be needed to double the reverberation time.

However, investigations by Behler (1989) and Ohsmann (1990) into the properties of such multi-channel systems have shown that Equation 11.33 is too optimistic. According to the latter author, a system consisting of 100 amplifier channels will increase the reverberation time by slightly more than 50% if all channels are operated with gains 3 dB below the instability limit.

For the performance of a multi-channel system of this type, it is of crucial importance that all open loop gains are virtually frequency-independent within a wide frequency range. To a certain degree, this can be achieved by carefully adjusted equalizers which are inserted into the electrical paths. In any case, there remains the problem that such a system comprises $N^2$ feedback channels, but only $N$ amplifier gains and equalizers to control them.

Nevertheless, systems of this kind have been successfully installed and operated at several places, for instance, in the Concert House at Stockholm (Dahlstedt 1974). This hall has a volume of 16,000 m$^3$ and seats 2,000 listeners. The electroacoustical system consists of 54 dynamic microphones and 104 loudspeakers. That means some microphones are connected to more than one loudspeaker. It increases the reverberation time from 2.1 s (without audience) to about 2.9 s. The tonal quality is reportedly so good that unbiased listeners are not aware that an electroacoustical system is in operation.

An electroacoustical multi-channel system of quite a different kind, but to be used for the same purpose, has been developed by Parkin and Morgan (1970) and has become known as 'assisted resonance system'. Unlike Franssen's system, each channel has to handle only a very narrow frequency band. Since the amplification and the phase shift in each channel can be adjusted independently (or almost independently), all unpleasant colouration effects can be avoided. Furthermore, electroacoustical components, that is, the microphones and loudspeakers, need not meet very high-fidelity standards.

The 'assisted resonance system' was originally developed for the Royal Festival Hall in London. This hall, which was designed and built mainly as a concert hall, has a volume of 22,000 m$^3$ and a seating capacity of 3,000 persons. It has been felt, since its opening in 1951, that the reverberation time is not as long as it should be for optimum conditions, especially at low frequencies. For this reason, an electroacoustical system for increasing the reverberation time was installed in 1964; at first, this was on an experimental basis, but in the following years, several aspects of the installation have been improved, and so it has been made a permanent fixture.

In the final state of the system, each channel consists of a condenser microphone, tuned by an acoustical resonator to a certain narrow frequency band; an electrical phase shifter; a very stable 20 W amplifier; a broadband frequency filter; and a 10- or 12-inch loudspeaker, which is tuned by a quarter wavelength tube to its particular operating frequency at frequencies lower than 100 Hz. (For higher frequencies, each loudspeaker must be used for two different frequency bands in order to save space and therefore has to be left untuned.) The feedback loop is completed by the acoustical path between the loudspeaker and the microphone. For tuning the microphone, Helmholtz resonators with a $Q$-factor of 30 are used for frequencies up to 300 Hz; at higher frequencies, they are replaced by quarter wave tubes. The loudspeaker and the microphone of each channel are positioned in the ceiling in such a way that they are situated at the antinodes of a particular room resonance.

There are 172 channels altogether, covering a frequency range of 58–700 Hz. The spacing of operating frequencies is 2 Hz from 58 to 150 Hz, 3 Hz for the range 150–180 Hz, 4 Hz up to 300 Hz, and 5 Hz for all higher frequencies. In Figure 11.19, the reverberation time of the occupied hall is plotted as a function of frequency with both the system on and the system off. These results were obtained by evaluating recordings of suitable pieces of music which were taken in the hall. The difference in reverberation time below 700 Hz is quite obvious. Apart from this, the system has the very desirable effect of increasing the overall loudness of the sounds perceived by the listeners and of increasing the variety of directions from which sound reaches the listeners' ears. In fact, from a subjective point of view, the acoustics of the hall seem to be greatly improved by the system, and well-known performers have commented enthusiastically on the achievements, particularly on a more resonant and warmer sound (Barron 1993).

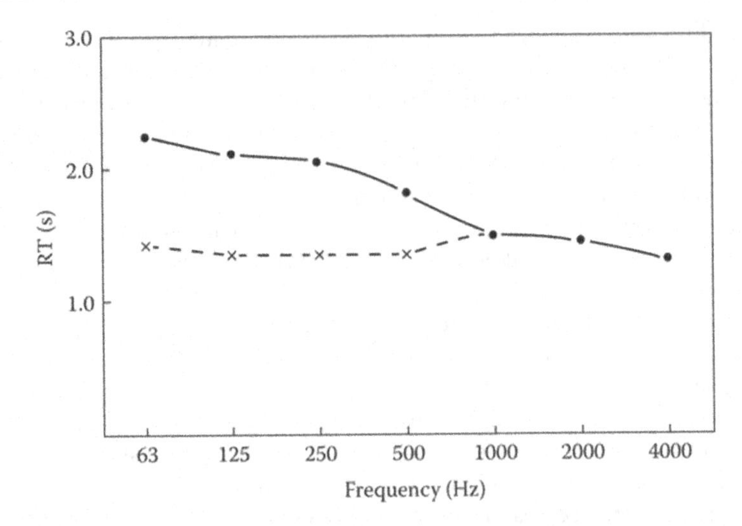


*Figure 11.19* Reverberation time of occupied Royal Festival Hall, London, both with (·———·) and without (×———×) 'assisted resonance system'.

During the past years, assisted resonance systems have been installed successfully in several other places. These more recent experiences seem to indicate that the number of independent channels need not be as high as was chosen for the Royal Festival Hall.

New sound engineering technologies have led to novel concepts that can combine the support of direct and reverberant sounds independently of each other. Such systems are subject to the complex international standardization of digital audio data in groups dealing with digital telecommunication and broadcasting formats. As a result, some new approaches of object-based audio and immersive systems have become the next evolutionary stage of electroacoustic sound reinforcement systems, as explained in the next section.

## 11.7 IMMERSIVE AUDIO SYSTEMS

The introduction of immersive audio systems has been a major step forward, particularly for multipurpose halls. These audio technologies have redefined the possibilities for live performances and offer more flexibility in customising the sound field to the desired auditory perception (Roginska and Geluso 2017). In so-called 'object-based audio systems' (Smith 2021), immersive audio systems typically utilise technology, where individual sound sources are treated as objects in a 3D sound environment.

Such a representation method transfers the task of reconstructing the sound events to an audio engineer who uses a so-called software-based 'renderer', which processes the metadata (positions, size, etc.) and the object-related audio streams and assembles them into a representation of the sound event. The rendering process is very similar to the computer simulation described in Chapter 10, in which the sources are separated from the sound propagation phenomena and the propagation effects are introduced by individually filtering the source signals. The choice and quality of the 3D playback system in suitable audio formats is also decisive for the quality of the immersive audio system. Besides wave field synthesis (see Section 11.5), VBAP (Pulkki 1997) or higher-order Ambisonics technologies (Daniel and Moreau 2004) are examples of well-suited 3D audio systems. In all these cases, in order to reproduce the desired sound field, algorithms solving the inverse problem for 3D sound

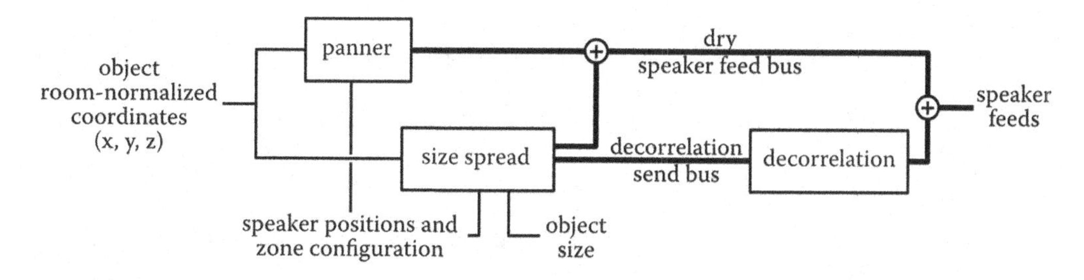


*Figure 11.20* Object-based audio processing.

synthesis are applied to calculate the required gains and filters for the speakers arranged in arrays surrounding the audience.

An example of object-based signal processing is shown in Figure 11.20 (after Tsingos 2017). 'Panning' in this context means choosing the position of the object on the digital data stream. Typically, these 3D audio renderers can also include control of the perceptual object to create spatially extended sound sources. This is achieved using decorrelation techniques by distributing the object to several neighbouring speakers.

Such systems are designed in an integrative process, including the room-acoustic design and the electroacoustic design. They have been very successful in creating immersive sound illusions in multipurpose halls (Ellison and Schwenke 2010). Several halls around the world have successfully integrated immersive audio systems. Case studies illustrate the benefits and challenges of implementing immersive audio technology, such as the Royal Albert Hall in London, and countless examples of performance venues where a high variety in the music programme is of high importance. With immersive audio technology, these venues can adapt to a wide range of performances, from symphony orchestras to Broadway shows and theatre plays.

Further advances in immersive audio technology can be expected in terms of stability and artefact-free playback as well as audio quality in general. This includes the avoidance of colouration and distortion and the production of an even distribution of sound quality across the entire listening area. Advances in this highly dynamic area of audio technology are being developed jointly between music production (or content production in general) on the one hand and room acoustics on the other, with 3D audio system technology forming the interface.

The foregoing discussions should have made clear that there is a great potential in electroacoustical systems for creating acoustical environments which can be adapted to nearly any type of performance. Their widespread and successful application depends, of course, on the technical perfection of their components and on further technical progresses, but equally on the skill and experience of the persons who operate them. In the future, however, the 'human factor' will certainly be reduced by more sophisticated systems, allowing application also in places where no specially trained personnel are available.

## REFERENCES

Ahnert W., Steffen F. *Sound Reinforcement Engineering*. London: E & FN Spon, 1999.

Barron M. *Auditorium Acoustics and Architectural Design*. London: E & FN Spon, 1993.

Behler G. Investigation of multichannel loudspeaker installations for increasing the reverberation of enclosures (in German). *Acustica* 1989; *69*: 95.

Berkhout A.J. A holographic approach to acoustic control. *J Audio Eng Soc* 1988; *36*: 977.

Berkhout A.J., de Vries D., Boone M.M. Application of wave field synthesis in enclosed spaces: nNew developments. Proceedings of 15th International Congress on Acoustics, Vol. II, Trondheim, 1995, p. 377.

Berkhout A.J., de Vries D., Vogel P. Acoustical control by wave field synthesis. *J Acoust Soc Am* 1993; *93*: 2764.

Dahlstedt S. Electronic reverberation equipment in the Stockholm Concert Hall. *J Audio Eng Soc* 1974; *22*: 626.

Daniel J., Moreau S. Further study of sound field coding with higher order ambisonics. Proceedings of 118th AES Convention, Berlin, 2004.

Ellison S., Schwenke R. The case for widely variable acoustics. Proceedings of International Symposium on Room Acoustics, Melbourne, 2010, pp. 29–31.

Franssen N.V. On the amplification of sound fields (in French). *Acustica* 1968; *20*: 315.

Griesinger D. Improving room acoustics through time-variant synthetic reverberation. 90th AES Convention, Paris, 1991.

Guelke R.W., Broadhurst A.D. Reverberation time control by direct feedback. *Acustica* 1971; *24*: 33.

Kuttruff H., Hesselmann N. On tone colouration by acoustical feedback in public address systems (in German). *Acustica* 1976; *36*: 105.

Meyer E., Kuttruff H. On the room acoustics of a large festival hall (in German). *Acustica* 1964; *14*: 138.

Ohsmann M. Analytic treatment of multiple-channel reverberation systems (in German). *Acustica* 1990; *70*: 233.

Parkin P.H., Morgan K.J. 'Assisted Resonance' in the Royal Festival Hall, London: 1965–1969. *J Acoust Soc Am* 1970; *48*: 1025.

Pulkki V. Virtual sound source positioning using vector base amplitude panning. *J Audio Eng Soc* 1997; *45*: 456.

Roginska A., Geluso R. (eds.). *Immersive Sound — The Art and Science of Binaural and Multi-Channel Audio*. New York: eBook Routledge, 2017.

Schroeder M.R. Improvement of acoustic feedback stability in public address systems. Proceedings of the Third International Congresses on Acoustics, Elsevier, Stuttgart, Amsterdam, 1961/1959, p. 771.

Schroeder M.R., Logan B.F. Colorless artificial reverberation. *J Audio Eng Soc* 1961; 9: 192.

Smith J. Advancements in object-based audio technology. Proceedings of the International Audio Engineering Conference, Audio Engineering Society, 2021, pp. 45–56.

Tsingos N. Object-based audio. Chapter 8. In *Immersive Sound — The Art and Science of Binaural and Multi-Channel Audio*. New York: eBook Routledge, 2017.

# Index

Note: Page numbers in *italics* indicate a figure and page numbers in **bold** indicate a table.

## A



## B



## C

## S

## T



## U



## V



## W

Printed in the United States
by Baker & Taylor Publisher Services